Collins

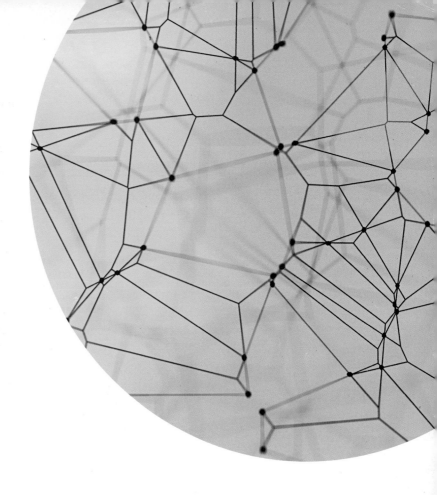

Student Book

EDEXCEL INTERNATIONAL GCSE (9-1) BIOLOGY

Jackie Clegg, Sue Kearsey,
Price and Mike Smith

William Collins' dream of knowledge for all began with the publication of his first book in 1819. A self-educated mill worker, he not only enriched millions of lives, but also founded a flourishing publishing house. Today, staying true to this spirit, Collins books are packed with inspiration, innovation and practical expertise. They place you at the centre of a world of possibility and give you exactly what you need to explore it.

Collins. Freedom to teach.

Published by Collins
An imprint of HarperCollins*Publishers*
The News Building
1 London Bridge Street
London
SE1 9GF

Browse the complete Collins catalogue at
www.collins.co.uk

© HarperCollins*Publishers* 2017

10 9 8 7 6 5 4 3 2 1

ISBN 978-0-00-823619-9

Jackie Clegg, Sue Kearsey, Gareth Price and Mike Smith assert their moral rights to be identified as the authors of this work.

British Library Cataloguing in Publication Data
A catalogue record for this publication is available from the British Library.

Authors: Jackie Clegg, Sue Kearsey, Gareth Price and Mike Smith
Original material by Jackie Clegg, Sue Kearsey, Gareth Price and Mike Smith
New material by Mike Smith
Commissioning Editor: Joanna Ramsay
Development Editor: Rebecca Ramsden
Project Manager: Maheswari PonSaravanan
Project editor: Vicki Litherland
Copy editor: Diana Anyakwo
Proofreader: Jan Schubert
Indexer: Jane Henley
Answer checker: Nick Mason
Typesetting: Jouve India
Artwork: Jouve India
Cover design: ink-tank
Production: Rachel Weaver
Printed by: Grafica Veneta

MIX
Paper from
responsible sources
FSC® C007454

This book is produced from independently certified FSC paper to ensure responsible forest management.

For more information visit: **www.harpercollins.co.uk/green**

Contents

Getting the best from the book

Welcome to *Edexcel International GCSE Biology*.

This textbook has been designed to help you understand all of the requirements needed to succeed in the Edexcel International GCSE Biology course. Just as there are five sections in the Edexcel specification, so there are five sections in the textbook: The nature and variety of living organisms, Structure and functions in living organisms, Reproduction and inheritance, Ecology and the environment and Use of biological resources.

Each section is split into topics. Each topic in the textbook covers the essential knowledge and skills you need. The textbook also has some very useful features which have been designed to really help you understand all the aspects of Biology which you will need to know for this specification.

SAFETY IN THE SCIENCE LESSON

This book is a textbook, not a laboratory or practical manual. As such, you should not interpret any information in this book related to practical work as including comprehensive safety instructions. Your teachers will provide full guidance for practical work and cover rules that are specific to your school.

A brief introduction to the section to give context to the science covered in the section.

Starting points will help you to revise previous learning and see what you already know about the ideas to be covered in the section.

The section contents shows the separate topics to be studied matching the specification order.

Knowledge check shows the ideas you should have already encountered in previous work before starting the topic.

Learning objectives cover what you need to learn in this topic.

The blue side panels and background shading indicate content for Biology International GCSE students only.

Examples of investigations are included with questions matched to the investigative skills you will need to learn.

Questions to check your understanding.

Excretion

INTRODUCTION

The shedding of leaves by trees, either all together in the autumn by deciduous trees or a few at a time by evergreens, is a form of excretion. Trees store metabolic waste substances in cells in the leaves, out of the way so that they do not interfere with other life processes. When the leaves are shed, this waste is shed also – we say it has been excreted because it has been removed from the body of the tree.

△ Fig. 2.103 Shedding leaves is a way of getting rid of waste.

KNOWLEDGE CHECK

✓ Plants produce oxygen from photosynthesis, and plants and animals release carbon dioxide from respiration – these are waste substances if they are not used in other processes.
✓ Excess amino acids from digestion are broken down to form urea in the liver.

LEARNING OBJECTIVES

✓ In plants, understand the origin of carbon dioxide and oxygen as waste products of metabolism and their loss from the stomata of a leaf.
✓ In humans, know the excretory products of the lungs, kidneys and skin (organs of excretion).
✓ Understand how the kidney carries out its roles of excretion and osmoregulation.
✓ Describe the structure of the urinary system, including the kidneys, ureters, bladder and urethra.
✓ Describe the structure of a nephron, to include Bowman's capsule and glomerulus, convoluted tubules, loop of Henle and collecting duct.
✓ Describe ultrafiltration in the Bowman's capsule and the composition of the glomerular filtrate.
✓ Understand how water is reabsorbed into the blood from the collecting duct.
✓ Understand why selective reabsorption of glucose occurs at the proximal convoluted tubule.
✓ Describe the role of ADH in regulating the water content of the blood.
✓ Understand that urine contains water, urea and ions.

EXCRETION IN FLOWERING PLANTS

Excretion is defined as the process or processes by which an organism eliminates the waste products of its metabolic activities. (Remember that excretion is different from egestion.) In flowering plants two waste products that need to be excreted are carbon dioxide and oxygen. Carbon dioxide is produced in respiration while oxygen is a product of photosynthesis. Excess amounts of these gases (not needed for other processes) are excreted through the stomata of the leaves.

EXCRETION IN HUMANS

The metabolic activities in human cells produce many waste products that need to be excreted.

Carbon dioxide is the waste product from respiration. If it remained in cells, it would change their pH and affect the activity of enzymes. It diffuses from respiring cells into the plasma of the blood and is carried around the body until it reaches the lungs. There it diffuses through the capillary and alveoli walls and is breathed out.

The skin plays a minor part in excretion. Sweat, which is secreted on to the skin surface from special cells in the skin, contains water and some minerals such as sodium and chloride ions (salt).

Waste products of many cell processes dissolve in the blood and are carried to the **kidneys**, where they are excreted. These products include **urea**, produced from the breakdown of excess amino acids by the liver.

△ Fig. 2.104 Sweat is water and ions that have been secreted from the body via the skin.

THE URINARY SYSTEM

Humans have two kidneys situated just under the rib cage at the back of the body, about halfway down the spine. The kidneys are well supplied with blood, which enters through the renal arteries and leaves through the renal veins. Inside the kidneys, the blood is filtered to remove waste substances no longer needed by the body. These include excess water, urea and mineral ions, which together form **urine**. Urine flows out of the kidneys down the **ureters** and into the **bladder**. The urine is stored in the bladder until a ring of muscle at the base is relaxed (usually when you go to the toilet). The urine then flows out of the bladder, through the **urethra** to the environment.

△ Fig. 2.105 The human urinary system.

QUESTIONS

1. Which is the main organ of excretion in plants? Explain your choice.
2. Which are the main organs of excretion in humans? Explain your choices.
3. Draw up a table to list the main structures of the urinary system and their functions.

Developing investigative skills

You can investigate the effect of light on photosynthesis by shining a light on a water plant and measuring how quickly bubbles are given off, as shown in Fig. 2.45.

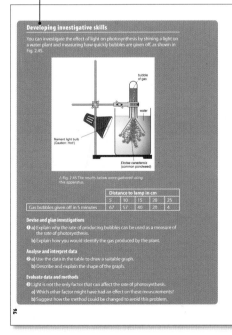

△ Fig. 2.45 The results below were gathered using this apparatus.

	Distance to lamp in cm				
	5	10	15	20	25
Gas bubbles given off in 5 minutes	67	57	40	20	4

Devise and plan investigations

❶ a) Explain why the rate of producing bubbles can be used as a measure of the rate of photosynthesis.
b) Explain how you would identify the gas produced by the plant.

Analyse and interpret data

❷ a) Use the data in the table to draw a suitable graph.
b) Describe and explain the shape of the graph.

Evaluate data and methods

❸ Light is not the only factor that can affect the rate of photosynthesis.
a) Which other factor might have had an effect on these measurements?
b) Suggest how the method could be changed to avoid this problem.

QUESTIONS

1. List three factors that affect the rate of photosynthesis.
2. Explain how each of these factors affects the rate of photosynthesis.
3. Explain as fully as possible why a variegated leaf tested for starch only causes the iodine/potassium iodide solution to turn from brown to blue-black where the leaf was green.

MINERAL IONS IN PLANT GROWTH

Photosynthesis produces carbohydrates, but plants contain many other types of chemical. Carbohydrates contain just the elements carbon, hydrogen and oxygen, but the amino acids that make up proteins also contain nitrogen. So plants need a source of nitrogen, which they take in in the form of nitrate ions.

Other chemicals in plants contain other elements: for example, chlorophyll molecules contain magnesium and nitrogen. Without a source of magnesium and nitrogen, a plant cannot produce chlorophyll and so cannot photosynthesise.

These additional elements are dissolved in water in the soil as **mineral ions**. Plants absorb the mineral ions through their roots, using active transport because the concentration of the ions in the soil is lower than in the plant cells.

Plants that are not absorbing enough mineral ions show symptoms of deficiency. For example, a plant with a nitrogen deficiency has stunted growth, and a plant with magnesium deficiency has leaves that are yellow between the veins, particularly in older leaves as the magnesium ions are transported in the plant to the new leaves.

△ Fig. 2.46 A plant with nitrogen deficiency.

△ Fig. 2.47 A plant with magnesium deficiency.

QUESTIONS

1. Explain why plants need a supply of mineral ions.
2. Describe the deficiency symptoms in a plant for the following mineral ions
a) nitrogen
b) magnesium.
3. Explain why plants show the deficiency symptoms for a) nitrogen and b) magnesium that you described in Question 2.

Getting the best from the book *continued*

Science in context boxes put the ideas you are learning into a historical or modern context.

Extension boxes take your learning even further.

Remember boxes provide tips and guidance to help you during the course and in your exam.

Sample page 344–345

Improved yield can be produced in many ways:

- increasing the size of the part of the plant we eat, such as seeds in wheat, maize and rice; tubers in potatoes and yams; leaves in cabbages
- decreasing the size of the parts of the plant we do not eat, such as stalks in wheat, because less energy is then 'wasted' by the plant growing parts that we do not want and it is easier to harvest
- improving pest and disease resistance, as less damage to the plant means it will grow faster
- improved growth in adverse conditions, such as drought or cold
- improving the taste or colour of the crop.

Other factors can also help, such as reducing stalk length so that rice and wheat plants aren't blown over as easily in strong winds and so are easier to harvest.

TULIP MANIA

Plants are also bred in horticulture, for gardens, for houseplants and cut flowers, to improve the colour, shape and form of the flowers and leaves. This is because people like new things.

For example, tulips were introduced to Europe in the 1500s from Turkey. They were so exotic that they became a luxury item that all wealthy people had to have. Plant breeders rapidly developed new varieties through selective breeding, such as flowers with different-coloured lines or specks on the petals.

At the peak of 'tulip mania' in the Netherlands in the 1630s, single tulip bulbs were being sold for more than 10 times the annual income of a skilled craftsman. Prices suddenly collapsed in 1637.

△ Fig. 5.19 A completely black flower is almost impossible to breed, but that does not stop people trying to produce it because many people would pay a lot of money for something so rare.

Selective plant breeding is not all a success story. For example, rice plants from around the world were crossed in breeding experiments to produce so-called 'miracle rices'. But the plants required extra fertiliser and plenty of water to produce the high yields. The modern seeds were also very expensive. If conditions were not perfect, the new varieties could sometimes do worse than the traditional varieties, and in some countries productivity actually went down. Scientists began to appreciate how important the environment was to the way the genes worked. The old-fashioned varieties had evolved over thousands of years to cope with local environmental conditions.

QUESTIONS

1. a) Explain why some characteristics can be bred for in selective breeding programmes.

   b) Explain why some characteristics cannot be bred for in selective breeding programmes.

2. Give three characteristics that have been selectively bred for in crop plants to improve crop yield.

3. For each of the characteristics you have given in Question 2, explain how these improve crop yield.

4. Explain why plants are selectively bred in horticulture.

EXTENSION

One of the problems with selective breeding is that, when you breed from only a small number of individuals, you reduce not only the variation in the characteristics you are selecting for, but also the variation in other alleles. This means that you can lose other characteristics which might be useful in the future.

To protect against this, many wild varieties of rice, wheat, potatoes and other plants are collected and grown in case we need their characteristics in the future.

△ Fig. 5.20 There are many wild varieties of rice, but we eat only a few varieties selectively bred for particular characteristics such as larger grain size.

1. Using what you know about sexual reproduction, suggest why the amount of variation between selected individuals is smaller than in wild populations of a plant.

2. Why is it useful that selectively bred varieties have only limited genetic variation? Explain your answer as fully as you can.

3. Why could it be a problem in the future that selectively bred varieties have only limited genetic variation? Explain your answer as fully as you can.

4. Growing wild varieties of crop plants to keep them for the future takes a lot of space and time to look after them. This space and time could be used to grow varieties that produce more food. Do you think it is worth keeping wild varieties like this? Explain your answer as fully as you can.

Sample page 136–137

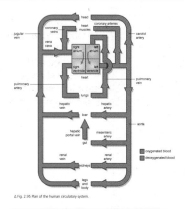

△ Fig. 2.95 Plan of the human circulatory system.

REMEMBER

The heart muscles have their own blood supply: the coronary arteries that branch from the aorta and link to the coronary veins that drain into the right atrium.

The name of a major blood vessel is often related to the organ it supplies: *coronary* for heart (from the Latin *corona* for 'crown' because the blood vessels surround the top of the heart like a crown), *hepatic* for liver (from the Greek *hepatos* meaning 'liver'), *renal* for kidneys (from the Latin *renes* meaning 'kidneys'), *pulmonary* for lungs (from the Latin *pulmonis*, 'lungs'). Learn the names of these blood vessels that are associated with the heart, the lungs, liver and kidneys.

EXTENSION

The circulatory system in mammals such as humans is a *double circulatory system*. This means that the blood flows twice through the heart for every one time it flows through the body tissues. The advantage of this is that the blood pressure in the circulation through the body can be kept higher than the blood pressure in the circulation through the lungs. A lot of force is needed to pump the blood down to the legs and back, but this force could damage the tiny capillaries in the lungs, which are much closer to the heart.

BLOOD VESSELS

The blood vessels are grouped into three different types: arteries, capillaries and veins.

REMEMBER

Remember: **a** for **a**rteries that carry blood **a**way from the heart. **Ve**ins carry blood **in**to the heart and contain **v**alves.

Arteries

Arteries are large blood vessels that carry blood flowing away from the heart. Blood in the arteries is at higher pressure than in the other vessels. The highest pressure is in the aorta, the blood vessel that leaves the left ventricle.

Arteries have thick muscular and elastic walls, with a narrow lumen (centre) through which the blood flows. The thick walls protect the arteries from bursting when the pressure increases as the pulse of blood enters them. The recoil of the elastic wall after the pulse of blood has passed through the artery helps to maintain the blood pressure and even out the pulses.

By the time the blood enters the fine capillaries, the change in pressure during and after a pulse has been greatly reduced.

Capillaries

Capillaries are the tiny blood vessels that flow through every tissue and connect arteries to veins. Capillaries have very thin walls, which helps to increase the rate of diffusion of substances. All exchange of substances between the blood and tissues happens in the capillaries.

artery:
thick-walled, carrying blood at high pressure

△ Fig. 2.96 Arteries vary in diameter from about 10 to 25 mm.

vein:
thin-walled, carrying blood at low pressure

capillary:
very small; the walls may be just one cell thick

△ Fig. 2.97 Veins vary in diameter from about 5 to 15 mm. Capillaries are very small, with a diameter of around 0.01 mm.

The first question is a student sample with examiner's comments to show best practice.

Each section includes exam-style questions to help you prepare for your exam in a focussed way and get the best results.

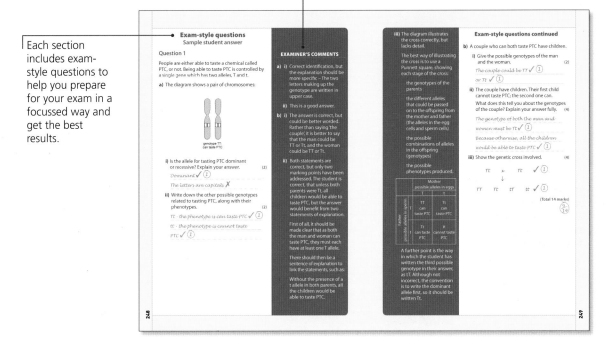

A full checklist of all the information you need to cover the complete specification requirements for each topic.

End of topic questions allow you to apply the knowledge and understanding you have learned in the topic to answer the questions.

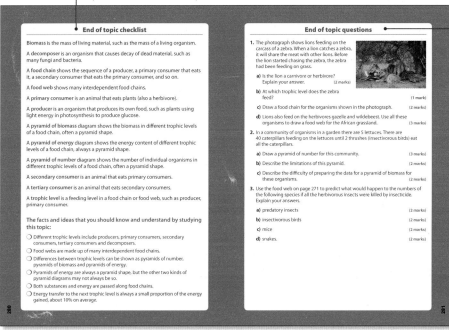

Around 1.9 million living species have been described and named on Earth. Around 350 000 of these species are classified as plants and around 1.37 million species are classified as animals. Over 66 000 of the animal species are vertebrates (they have bony skeletons), and the rest are invertebrates of which the majority (around 1 million species) are insects.

It is difficult to know how many species are still to be discovered, although scientists reckon they have discovered most living mammals, birds and coniferous trees. The smaller the organism, the greater the chance that there are species we don't yet know about. So although over 4000 species of bacteria have been identified, there could be more species of bacteria than of all the other kinds of organisms put together.

STARTING POINTS

1. What are the characteristics shared by living organisms?
2. Crystals can grow in size, but does that mean they are alive?
3. We talk of 'feeding' a fire when we add fuel, but does that mean fire is a living thing?
4. Why is it useful to group organisms?
5. What features are the most useful for grouping organisms?

CONTENTS

a) Characteristics of living organisms
b) Variety of living organisms
c) Exam-style questions

1

The nature and variety of living organisms

Δ Many species of different kinds of organisms live on a coral reef.

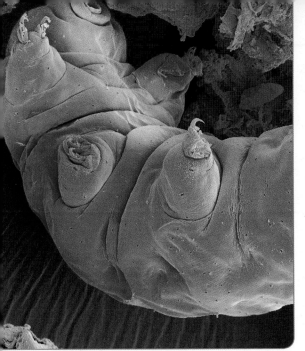

△ Fig. 1.1 Tiny tardigrades (about 1 mm long) are one of the toughest organisms known. They can survive temperatures below −200 °C, 10 days in the vacuum of space and over 10 years without water!

Characteristics of living organisms

INTRODUCTION

Sometimes it is easy to tell when something dies: an animal stops moving around; a plant wilts and all the green parts collapse. But does a tree die in winter, when its leaves have dropped off? Are animals 'dead' when they hibernate underground for months? As technology gets increasingly sophisticated, and we can create machines with 'brains' and grow new organs in a laboratory, distinguishing between living and non-living could get even more difficult. We need a set of 'rules' that work for most organisms, most of the time.

KNOWLEDGE CHECK

✓ Living organisms show a range of characteristics that distinguish them from dead or non-living material.
✓ The life processes are supported by the cells, tissues, organs and systems of the body.

LEARNING OBJECTIVES

✓ Name the eight characteristics shown by living organisms.
✓ Describe each of the characteristics of living organisms.
✓ Explain that not all living organisms show every characteristic all of the time.

THE EIGHT CHARACTERISTICS OF LIFE

There are eight life characteristics that most living **organisms** will show at some time during their lives.

- **Movement:** In all living cells, structures in the **cytoplasm** move. In more complex organisms, the whole structure may move. Animals may move their entire bodies; plants may move parts of their body in response to external stimuli such as light.

△ Fig. 1.2 Sunflowers follow the Sun as it moves across the sky through the day.

- **Respiration:** This is a series of reactions that take place in living cells to release energy from nutrients. This energy is used for all the chemical reactions that keep the body alive.
- **Sensitivity:** Living organisms are able to detect and respond to changes in their external and internal conditions.
- **Homeostasis:** This is the control of internal conditions, to provide the best conditions inside cells for all the reactions needed for life to exist. For example, when we eat and drink we take in water – our body controls how much water is absorbed and removed from the blood, so that cell processes can continue to work efficiently.
- **Growth:** This is the permanent increase in the size and/or dry mass (mass without water content) of cells or the whole body of an organism. Your mass changes throughout the day, depending on how much you eat and drink, but your growth is the amount by which your body increases in size when you take nutrients into cells to increase their number and size. As organisms grow, they may also change or develop.
- **Reproduction:** This includes all the processes that result in making more individuals of that kind of organism, such as making gametes and the fertilisation of those gametes.
- **Excretion:** Living cells produce many products from the reactions that take place inside them. Some of these are waste products – materials that the body does not use. For example, animals cannot use the carbon dioxide produced during respiration. Waste products may also be toxic, so they must be removed from the body by excretion.
- **Nutrition:** The taking of nutrients, such as organic substances and mineral ions, into the body. Nutrients are the raw materials that cells need to release energy and to make more cells.

QUESTIONS

1. For each of the eight characteristics, give one example for:

   a) a human

   b) an animal of your choice

   c) a plant.

2. For each of the eight characteristics, explain why they are essential to a living organism.

REMEMBER

An easy way to remember all eight characteristics is to take the first letter from each process. This spells MRS H GREN. Instead, you may make up a sentence in which each word begins with same letter as one of the processes: for example, My Revision System Here Gets Really Entertaining Now.

Viruses are very simple structures, consisting of an outer protein coat that protects the genetic material inside. They have no cell structures or cytoplasm, so they do not respire or sense their surroundings. They also do not take in substances to build more cells, or excrete anything. In many ways they behave like simple crystalline chemicals. However, when they infect a cell, such as a bacterial, plant or animal cell, they cause that cell to produce many copies of the virus. So they do reproduce.

Not everyone agrees on whether viruses can be called *living* organisms.

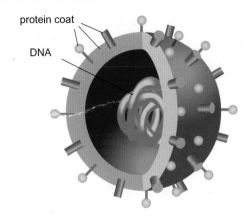

△ Fig. 1.3 The structure of a virus.

1. Which characteristic of living organisms do viruses have?

2. List the other characteristics of living organisms, and for each one describe what viruses can and cannot do.

3. Using what you know about viruses, prepare an argument for classifying them as living organisms.

4. Using what you know about viruses, prepare an argument for *not* classifying them as living organisms.

REMEMBER

Be prepared to make a decision and use your knowledge to argue your point of view about difficult examples such as viruses.

End of topic checklist

Excretion is the removal of waste (often toxic) substances that have been produced from chemical reactions inside the body, such as carbon dioxide and urea in animals.

Growth is the permanent increase in body size and dry mass of an organism, usually from an increase in cell number or cell size (or both).

Homeostasis is the maintenance of a constant internal environment, such as body water content and body temperature.

Movement happens in all living cells: both plants and animals.

Nutrition is the taking in of substances for use in the body as food or to make food.

Reproduction is the production of new organisms.

Respiration is the chemical process in which glucose is broken down inside cells, releasing energy and producing carbon dioxide and water.

Sensitivity refers to the detection of changes (stimuli) in the surroundings by a living organism, and its responses to those changes.

The facts and ideas that you should know and understand by studying this topic:

⬭ All living organisms show the eight characteristics of life at some point in their lives.

⬭ The characteristics of life are: movement, respiration, sensitivity, homeostasis, growth, reproduction, excretion and nutrition.

End of topic questions

1. Name the eight processes of life. Try making up your own sentence to help you remember them all. **(9 marks)**

2. Name two life processes necessary for an organism to release energy. **(2 marks)**

3. Explain why dry mass is used to measure growth. **(2 marks)**

4. When you place a crystal of copper(II) sulfate in a saturated solution of the same compound, the crystal will increase in size. Does this mean that the crystal is alive? Explain your answer. **(1 mark)**

5. Plants cannot move about, as animals can. Does that mean animals are more alive than plants? Explain your answer. **(2 marks)**

6. During winter, an oak tree will lose its leaves and not grow. Is the tree still living during this time? Explain your answer using all the characteristics of life.

   (4 marks)

Variety of living organisms

△ Fig. 1.4 The tallest plants are giant redwood trees, capable of growing to over 90 m high.

INTRODUCTION

The many different kinds of living organisms come in a confusing variety of forms. Classifying (grouping) organisms using their similar characteristics helps us to make sense of all the variation. This information can help us understand which organisms are most closely related to each other, which groups have evolved from other groups, and which groups play the most important roles in an ecosystem.

KNOWLEDGE CHECK

✓ Living organisms show great variety.
✓ Organisms can be classified according to their characteristics.

LEARNING OBJECTIVES

✓ Describe the common features shown by eukaryotic organisms: plants, animals, fungi and protoctists.
✓ Describe the common features shown by prokaryotic organisms such as bacteria.
✓ Understand the term *pathogen* and know that pathogens may include fungi, bacteria, protoctists or viruses.

EUKARYOTES AND PROKARYOTES

Living organisms are either **eukaryotic** or **prokaryotic**. Eukaryotic organisms have cells that contain a **nucleus** and other organelles such as **mitochondria** and **chloroplasts**. Plants, animals, fungi and protoctists are all eukaryotic organisms. Prokaryotic organisms have cells that are much smaller and simpler than eukaryotic cells. Prokaryotic cells do not contain a nucleus or organelles such as mitochondria or chloroplasts. Bacteria are prokaryotic organisms.

PLANTS

Plants are **multicellular** organisms, which means they are made up of more than one (usually thousands or millions) of **cells**.

Plant cells have a **cell wall** as well as a **cell membrane**. The cell wall is made of **cellulose** and gives the cell shape and support.

Many plant cells have a large central **vacuole** that contains cell sap, which is water with various substances dissolved in it. The vacuole may also be a storage space for some substances. In a healthy plant the vacuole is large and helps support the cell when it is full of sap.

Plant cells may contain **chloroplasts**, which are able to carry out **photosynthesis** – a process in which they use the Sun's energy to produce **carbohydrates**. Carbohydrates, stored as **starch** or **sucrose**, store energy for the plant but can also be used as food by animals that eat plants.

Plants vary greatly in size and shape, from tall rainforest trees to tiny flowers like violets. We use many plants as food, including cereals such as rice and maize, and herbaceous legumes such as lentils, peas and beans.

QUESTIONS

1. Explain what is meant by the term *multicellular*, and give one example.

2. Plants do not have skeletons, as animals do, but are still able to stand upright. Explain why.

3. One kind of structure found in some plant cells makes plants able to produce their own food. What is this structure called?

ANIMALS

Animals are multicellular organisms. Unlike plants, their cells do not contain chloroplasts and so cannot carry out photosynthesis. This means that they have to eat other organisms (plants or other animals) to get their food. Animal cells have a cell membrane but no cell wall. Many animals are able to coordinate their movement using **nerves** and are able to move from one place to another. For energy, animals also store carbohydrates, often in the form of **glycogen**. They also store **lipids**, often as a layer of fat below the skin or around body organs, as a store of energy.

As with plants, the variety of animals is huge, from enormous whales and elephants to tiny ants.

▷ Fig. 1.5 Ants belong to the insects, as do the housefly and mosquito. Whales and elephants are classified as mammals, as are humans.

QUESTIONS

1. If you compared a plant cell and an animal cell under the microscope, which features would you see:

   a) in both cells

   b) only in the plant cell?

2. Describe two differences between plants and animals in terms of the structure of their bodies.

3. One difference between plants and animals is that many animals can move from place to place, but plants cannot. Explain this difference.

FUNGI

Some fungi (such as yeast) are single-celled but most have a structure consisting of fine threads known as **hyphae**. Each hypha may contain many nuclei. Several hyphae together form a **mycelium**. Many fungi can be seen without a microscope. Their cell walls are made of **chitin**, a fibrous carbohydrate which is different to the cellulose used in plants. Their cells do not contain chlorophyll so they cannot carry out photosynthesis. To obtain energy they secrete digestive enzymes outside the cells (**extracellular secretion**), onto living or dead animal or plant material, and absorb the digested nutrients. This is called **saprotrophic nutrition**. Like animals, fungi may store carbohydrate in the form of glycogen.

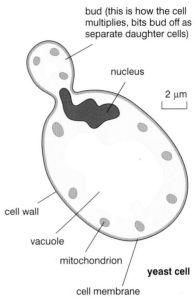

△ Fig. 1.6 The internal structure of a yeast cell that is reproducing by budding.

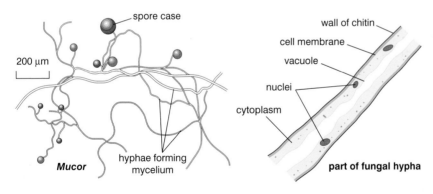

△ Fig. 1.7 Left: the mycelium and spore cases of *Mucor*, a mould. Right: detail of a hypha of *Mucor*.

Examples of fungi include yeast, a single-celled fungus used by humans in baking, and *Mucor*, a fungus with the typical hyphal structure. *Mucor* is often seen as a mould growing on spoiled foods.

Some species of fungi are **pathogens**, which means they cause disease in other organisms. For example, ringworm is caused by a fungus that produces rings of itchy skin in humans. Also many plants are damaged by rusts and moulds which are different kinds of fungi.

MUSHROOMS AND TOADSTOOLS

We normally think of a mushroom or toadstool as the whole of a fungus, because this is usually all we can see. However, these are only the reproductive organs, where spores are produced. The mycelium of the fungus is usually hidden below ground or within rotting materials, where it is moist and where the hyphae can digest the surrounding tissue and absorb the nutrients that are released. The reproductive structures have to be large enough so that the wind can carry the spores away to other places, and tough enough to survive the drying conditions of the air until the spores have been dispersed.

▷ Fig. 1.8 A mushroom or toadstool is only the visible part of this fungus. The rest of the structure is hidden from view.

QUESTIONS

1. Which characteristics do fungi share with **a)** plants and **b)** animals?

2. Describe what is meant by saprotrophic nutrition and how it differs from the way animals get their nutrition.

3. Most people think of toadstools and mushrooms as the main part of fungi. Explain why this is incorrect.

PROTOCTISTS

Protoctists are also single-celled microscopic organisms, usually much larger than bacteria. Their cells contain a nucleus and many have features of animal cells or plant cells. One example is *Amoeba,* which looks like an animal cell and is found in ponds and feeds on other microscopic organisms. Other protoctists, such as *Chlorella,* look more like plant cells because they contain chloroplasts and so can photosynthesise. A few protoctists are pathogens, such as *Plasmodium,* the organism that causes the disease malaria in humans.

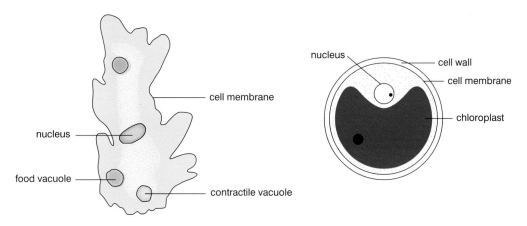

△ Fig. 1.9 *Amoeba* and *Chlorella*.

MALARIA

SCIENCE
IN
CONTEXT

Malaria is one of the greatest causes of death from infectious illness in the world today. Around 750 000 people die of the disease each year, mostly young children and mostly in sub-Saharan Africa. The disease is caused by the protoctist *Plasmodium*, which has a clever way of getting from one person to the next: it hitches a lift in the alimentary canal of an *Anopheles* mosquito. The female mosquitoes suck blood from humans to provide the nutrients they need to lay eggs. As a mosquito pierces into a blood vessel, it inserts a little liquid to prevent the blood from clotting. If the mosquito has fed recently on a person infected with *Plasmodium*, this liquid will contain some of the parasites and so infect the new person. This protects the protoctist from the harsh conditions of the environment and allows it to be passed on to a new host.

QUESTIONS

1. Explain why some protoctists were once classified as plants and others as animals.

2. What features do all protoctists have in common?

3. Is it correct to describe mosquitoes as the cause of malaria? Explain your answer.

BACTERIA

Bacteria are single-celled, microscopic organisms that are smaller than plant and animal cells and come in many different shapes. Their cells have no **nucleus**, so the single circular **chromosome** of DNA lies free in the cytoplasm inside the cell. Many bacteria have additional circles of genetic material, called **plasmids**. Bacterial cells are surrounded by a cell membrane and cell wall, although in different groups of bacteria the cell wall is made of different chemicals. Some bacteria can carry out photosynthesis but most feed off other living or dead organisms.

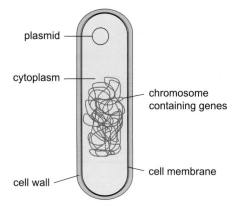
△ Fig. 1.10 General structure of a bacterial cell.

Some bacteria are useful to humans. For example, *Lactobacillus bulgaricus*, a rod-shaped bacterium, is used to make yoghurt from milk. Other bacteria are pathogens, causing diseases in plants and animals. An example of a pathogen is *Pneumococcus*, a spherical bacterium that can cause pneumonia in humans.

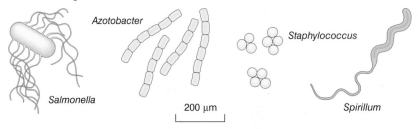

△ Fig. 1.11 Different bacteria can be recognised from their shape and structure.

BACTERIAL PLASMIDS

Bacterial plasmids have become very useful to us in genetic engineering, where they are used as vectors (see Topic 5c: Genetic modification (genetic engineering)). Not all bacteria have them, but those that do transfer these small circles of genetic material to other bacteria quite easily. Plasmids may even be transferred between bacteria of different species. This is not true reproduction as the transfer is not of the main chromosome and may not lead to production of new individuals. However, this kind of transfer may be important in the spread of antibiotic resistance between bacterial species, because some of the genes for antibiotic resistance are found in the plasmids.

QUESTIONS

1. Describe three differences between a plant cell and a bacterial cell.

2. Compare the structure of bacteria and protoctists.

3. How are bacterial chromosomes different from the chromosomes of eukaryotes?

VIRUSES

Viruses are small particles rather than cells. They have a wide variety of shapes and sizes, but they all consist of a protein coat containing one type of **nucleic acid**, either **DNA** or **RNA**. They are even smaller than bacteria: they may only be seen with an electron microscope. Viruses are **parasites** and can only reproduce inside the living cells of an organism they have infected. They are not living organisms themselves: although they can reproduce, they do not show other characteristics of life.

There are viruses that can infect every type of living thing. Many viruses are pathogenic and cause disease in the organisms they infect. The tobacco mosaic virus is a plant virus. It prevents the formation of chloroplasts in the tobacco plant cells, which causes discolouring of the leaves.

Influenza viruses are a group of many closely related viruses that can cause 'flu' in many different animals including humans. The human immunodeficiency virus (HIV) causes the disease called AIDS (acquired immune deficiency syndrome) in humans.

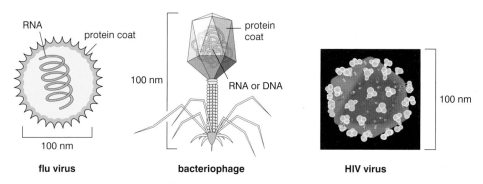

△ Fig. 1.12 Some viruses. 1 m = 1000 mm (millimetres). 1 mm = 1000 μm (micrometres). 1 μm = 1000 nm (nanometres).

SCIENCE IN CONTEXT

HIV AND AIDS

The HIV virus is one of a group of viruses that attack and destroy cells in the immune system. This leaves the body open to infection by other pathogens – in the case of HIV, this causes the disease called AIDS. Many AIDS patients do not die from the HIV virus, but from other diseases such as tuberculosis (caused by a bacterium) that take advantage of the weak immune system.

The HIV virus does not survive well outside the body and is mainly transmitted from one person to another through body fluids. The most common transmission is during sexual intercourse. However, transmission in blood is also possible, such as through blood transfusion, or sharing of injection needles between drug users. An infected mother can pass the HIV virus to a fetus in her uterus through the placenta, or through breast milk after birth.

QUESTIONS

1. Describe the basic structure of a virus.

2. Why are viruses described as particles rather than cells?

3. Compare the size of a virus with the size of a bacterium.

EXTENSION

The five groups of living organisms described here are often defined as the five kingdoms of organisms, where a kingdom is the largest group in biological classification. Kingdoms are subdivided into increasingly smaller groups:
- kingdom
- phylum
- class
- order
- family
- genus
- species

by the characteristics of the organisms within the groups until, at the lowest level, there is only one species. A species is a group of individuals that share most of their characteristics and are capable of breeding with each other to produce fertile offspring.

End of topic checklist

A **chloroplast** is an organelle found in plant cells and some protoctist cells that can capture energy from light for use in photosynthesis.

A **hypha** is a single thread of fungal mycelium (plural hyphae).

A **mycelium** is a mass of hyphae that form the body of a fungus.

A **pathogen** is an organism that causes disease in another living organism. Pathogens occur in fungi, bacteria, protoctists and viruses.

A **plasmid** is a small circle of genetic material found in some bacteria in addition to the circular chromosome.

Saprotrophic nutrition is the digestion of dead food material outside the body, as in fungi.

The facts and ideas that you should know and understand by studying this topic:

○ Organisms are grouped by their common features into the major groups: eukaryotes, consisting of plants, animals, fungi and protoctists; prokaryotes, consisting of bacteria; and viruses.

○ Plants are multicellular organisms that have cells with cell walls, vacuoles and may have chloroplasts. They photosynthesise to make their food and store carbohydrate as starch or sucrose.

○ Animals are multicellular organisms that do not have chloroplasts so they cannot photosynthesise; they usually can move and have a nervous system; and store carbohydrates, often as glycogen.

○ Most fungi are made from multicellular hyphae, though a few, like yeast, are single-celled. The cells have no chloroplasts, so fungi feed by saprotrophic nutrition; their cells have cell walls and may store carbohydrate as glycogen.

○ Protoctists are microscopic, single-celled organisms; some have no chloroplasts and feed off other organisms, others have chloroplasts and can photosynthesise.

○ Bacteria are microscopic, single-celled organisms that have no nucleus; they have a circular chromosome and some may have additional genetic material in plasmids; they have cell walls and most feed off other living or dead organisms.

○ Viruses are infective particles made of a protein coat surrounding nucleic acid; they have no true cell structure and can only reproduce when inside a cell of another organism.

○ The groups of fungi, bacteria, protoctists and viruses all contain pathogens, which cause disease in other organisms.

End of topic questions

1. Put the following organisms into size order, starting with the smallest:
   bacteria   protoctists   viruses   (2 marks)

2. Copy and complete the table below to compare the different groups of organisms.
                                                                                        (20 marks)

	Multicellular or single-celled	Key cell structures	Food store	Other distinguishing features
plants				
animals				
fungi				
bacteria				
protoctists				
viruses				

3. Explain why a large tree and a crop such as rice or maize are both classified
   as plants.   (2 marks)

4. A new organism is discovered. It is formed from cells that have no cell wall and
   no chloroplasts. Which group should it be classified in and why?   (2 marks)

5. **a)** State what is meant by the term *pathogen*.   (1 mark)

   b) Give one example of a pathogen from each of the following groups:
       fungi, bacteria, protoctists, viruses.   (4 marks)

   c) Name the disease each example causes.   (4 marks)

6. Explain why all viruses are parasitic (live off other living organisms).   (2 marks)

7. Before fungi were classified in their own separate group, they were sometimes
   grouped with plants and sometimes with animals.

   a) Explain why they could be grouped either way.   (2 marks)

   b) Suggest why they are no longer classified with either plants or animals.   (2 marks)

a) i) It is important to know the features of different groups of organisms and be able to label these.

A: Correct

B: Correct

C: Incorrect – the outer layer of the hypha is the wall

D: Incorrect – the membrane is the next layer within the wall, pushed up against the wall.

E: Correct

ii) Correct – the hyphae of moulds such as *Mucor* have a large central vacuole.

The answer is correct in that the hyphae have a wall, but this cannot be described as a 'cell wall' as the hyphae are not divided into cells. The student simply had to repeat this information from part a) i) to be absolutely correct.

iii) Correct. The student could have chosen from a selection of features but was only asked to give one.

Exam-style questions
Sample student answer

Question 1

The diagram shows the structure of a part of an organism called *Mucor*.

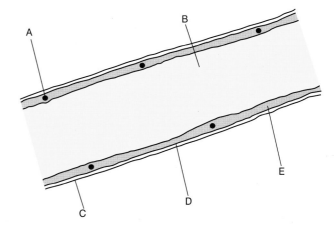

a) i) Name each of the parts, A–E. Use the words in the box below. **(5)**

cytoplasm	membrane	nucleus
starch grain	**vacuole**	**wall**

A	*nucleus*	✓
B	*vacuole*	✓
C	*membrane*	✗
D	*wall*	✗
E	*cytoplasm*	✓

ii) Give **two** features the organism has in common with plants. **(2)**

Mucor has a large central vacuole ✓

Mucor has a cell wall ✓

iii) Give one feature that tells you that *Mucor* is a fungus. **(1)**

It has many nuclei lying in the cytoplasm. ✓

Exam-style questions continued

b) Describe how moulds such as *Mucor* feed. (4)

Mucor lives on its food, e.g. bread, and secretes enzymes into it. ✓

The food is absorbed over the surface of the fungus. ✓

c) Yeast is another type of fungus. State one major difference between *Mucor* and yeast. (1)

Yeast is single-celled. ✓

(Total 13 marks)

9/13

b) The answer is correct, but the student could have added that food is digested outside the mould, and that this process is called saprotrophic nutrition.

c) Correct – the mycelium of Mucor has many nuclei distributed through the cytoplasm, with no cell boundaries; yeast is made up of single cells.

Question 2

This question is about the variety of living organisms.

a) Five types of living organism are listed A–D below.

A: *Amoeba*	B: *Chlorella*	C: *Lactobacillus*	D: *Pneumococcus*

   i) Which **two** organisms are protoctists? (2)

  ii) Which organism is a pathogen? (1)

 iii) Which organism has chloroplasts? (1)

 iv) Which **two** organisms are bacteria? (2)

   v) Which organism is used to produce yoghurt? (1)

b) The diagram below shows the structure of one of the stages in the life cycle of *Plasmodium*.

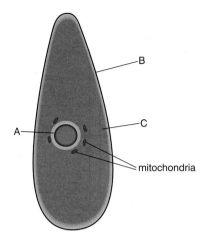

    i) Name each of the parts, A–C. (3)

    ii) Explain one way that you can tell that *Plasmodium* is not a type of bacteria. (1)

(Total 11 marks)

Question 3

Not everyone agrees that viruses should be called living things. Use your knowledge of viruses, and the characteristics of living things, to discuss whether or not viruses should be classed as living. (4)

(Total 4 marks)

Question 4

a) Name two examples of a flowering plant, to include one cereal crop and one legume. (2)

b) Name two examples of animals, to include one mammal and one insect. (2)

c) Plants and animals respond to their surroundings.

    i) Name a body system that animals use to respond to their surroundings. (1)

    ii) Describe one way that plants can respond to their surroundings. (1)

(Total 6 marks)

When we talk about the 'heart' of something, in a general way, we mean the centre of something – not only where it is, but also the role it plays. And for good reason: the human heart is not only positioned in the middle of the body; it also plays a central role in maintaining life.

The heart and circulatory system function to circulate blood around the body and so deliver oxygen from the lungs, and nutrients from the digestive system, to all cells so that they can respire and carry out all the processes needed for life. The blood then removes waste products from these processes, for example delivering carbon dioxide to the lungs and urea to the kidneys for excretion and removal from the body. The heart plays a central role in staying alive and staying healthy.

STARTING POINTS

1. How is the body organised so that it can carry out the life processes effectively?

2. What do all cells have in common, and how are some cells different from others?

3. What are the basic molecules of life?

4. How do cell membranes control what can get into and out of the cell?

5. How do plants and humans get the food they need for growth?

6. What is cellular respiration and how do the body systems support it?

7. How are gas exchange surfaces adapted for rapid exchange of gases into and out of the body?

8. How are materials transported around the bodies of plants and humans?

9. What are the waste materials of metabolism and how are they removed from the body?

10. How do plants and humans respond to changes in the environment around them?

SECTION CONTENTS

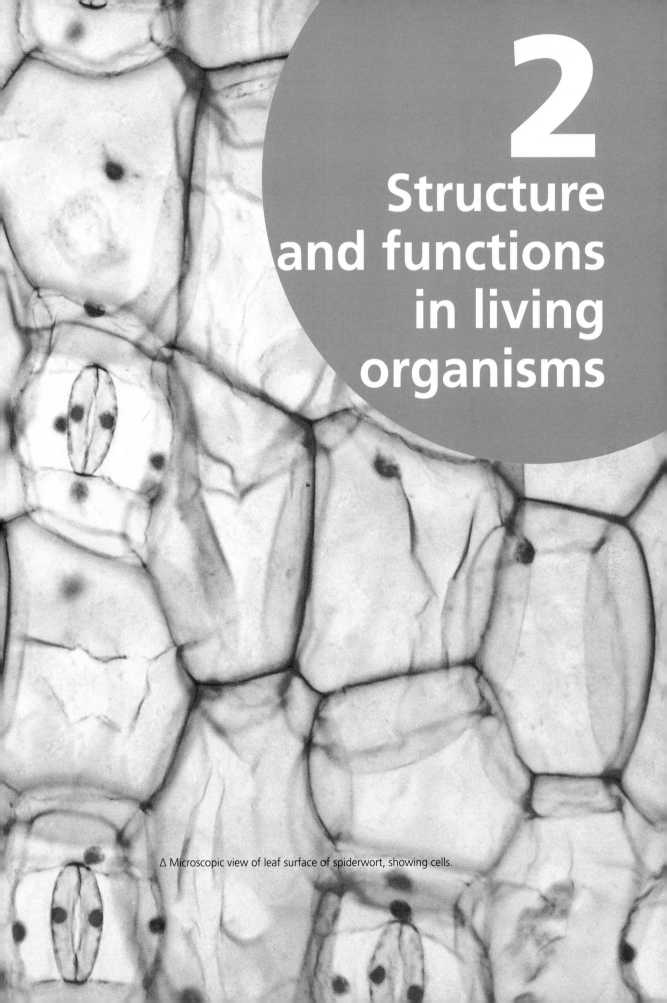

2
Structure and functions in living organisms

△ Microscopic view of leaf surface of spiderwort, showing cells.

Level of organisation

INTRODUCTION

Bringing together similar activities that have the same purpose can make things much more efficient. For example, bringing teachers and students together in a school helps more students to learn more quickly than if each teacher travelled to each student's home for lessons. The same is true in the body. Having groups of similar cells in the same place – as a tissue, and grouping tissues into organs, helps the body carry out all the life processes much more efficiently and so stay alive.

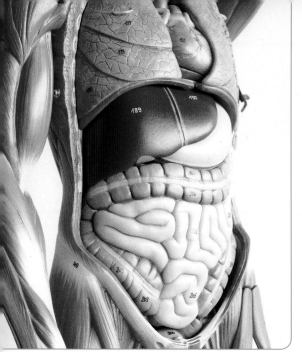

△ Fig. 2.1 The human body is made up of several systems of grouped organs, including the digestive system, the nervous system, the muscle/skeletal system and the respiratory system.

CELLS AND ORGANELLES

The cells that make up organisms all have certain things in common. Each cell is surrounded by a cell membrane. Inside the cell membrane is a jelly-like substance called the cytoplasm. Cells may be specialised to carry out particular roles, such as secreting enzymes or carrying electrical impulses. Within the cytoplasm are structures called **organelles**. The most obvious organelle is usually the nucleus, which contains the cell's genetic material. Other organelles, such as chloroplasts and vacuoles, have specific roles within the cell.

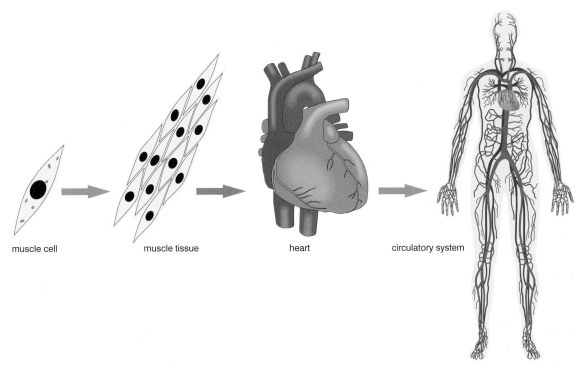

muscle cell   muscle tissue   heart   circulatory system

Δ Fig. 2.2 The human body is organised at cell level. Muscle cells are found in muscle tissue, which may be found in the heart, which is part of the circulatory system.

TISSUES

In multicellular organisms, most cells are specialised and are organised into **tissues**. A tissue is a group of similar cells with the same function. Muscle cells, for example, are specially adapted to produce movement and are arranged in large groups to make muscle tissue; nerve cells transmit impulses from one nerve cell to another and are organised into nervous tissue such as the brain, spinal cord and nerves. In plants, the cells are also organised into tissues, such as **xylem** tissue, which forms long tubes that transport water through a plant; **epidermal** tissue that covers surfaces; and **mesophyll** tissue, which packs the spaces between other tissues.

QUESTIONS

1. Give two examples of organelles.

2. a) Give two examples of tissues in the human body.

   b) Explain how they are adapted for their function.

3. a) Give two examples of tissues in a plant.

   b) Explain how they are adapted for their function.

ORGANS

Organs are structures within larger organisms that are adapted to do a specific function. They are formed from different tissues that work together to carry out that function. For example, the function of the stomach is to digest food. To do this, secretory tissue lining the stomach produces **enzymes** to break down the food. In the wall of the stomach there is muscle tissue that contracts and relaxes. This helps to churn up the food in the stomach, to mix it with the enzymes and moves the food through to the intestines. Other organs in the human body include the heart, liver, lungs and kidneys. In plants, leaves are the organs adapted for photosynthesis, and flowers are the organs that are adapted for reproduction.

SYSTEMS

Organs like the stomach form part of larger structures called **systems**. The stomach is part of the digestive system, which has many parts including the teeth, the oesophagus and the intestines. Other systems in the human body include the nervous system, the circulatory system and the reproductive system.

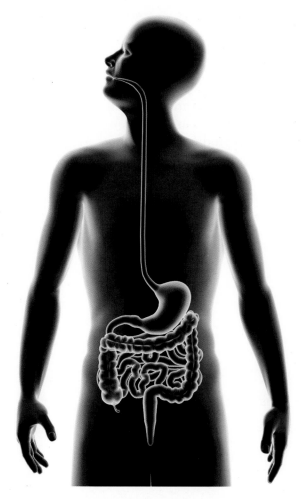

Δ Fig. 2.3 The human digestive system.

QUESTIONS

1. Give two examples of organs in the human body, and describe their function.

2. Give two examples of organs in a plant, and describe their function.

3. Explain what we mean by a body *system*.

REMEMBER

As you study body systems in more detail through your course, remember to identify the organs, tissues and cell types involved in each system, so that you build a range of examples that you can use to answer questions with.

End of topic checklist

An **organelle** is a structure within a **cell** that carries out a particular function, such as a nucleus or vacuole.

A **tissue** is a group of similar cells that have a similar function, such as muscle tissue.

An **organ** is a group of tissues that work together to carry out a particular function, such as the stomach or the heart.

A **system** is a group of organs that work together to carry out a particular function, such as the mouth, stomach and intestines in the digestive system.

The facts and ideas that you should know and understand by studying this topic:

○ Within the body, organelles are found in cells, cells are grouped into tissues, tissues are grouped into organs and organs are grouped into systems.

○ Give examples of organelles, cells, tissues, organs and systems.

○ These levels of organisation help the body to function efficiently.

End of topic questions

1. Put the following in order of size, starting with the largest:

cell   system   organ   tissue   **(3 marks)**

2. Write definitions for each of these words:

  a) tissue   **(1 mark)**

  b) organ   **(1 mark)**

  c) system   **(1 mark)**

3. Give one example of each of the following in a) a plant and b) a human:

cell   tissue   organ   **(6 marks)**

4. Draw a table with the following headings.

System	Function	Organs in this system	Tissues in these organs	Cells in these tissues

Complete your table as fully as you can, using up to three examples of systems in the human body.   **(max. 15 marks)**

5. The diagram shows part of a leaf, which is a plant organ. The diagram is labelled to show some of the tissues. Describe the functions of the tissues in this organ.

(4 marks)

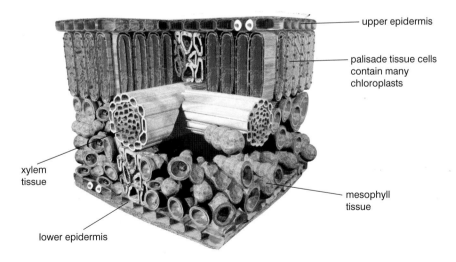

upper epidermis

palisade tissue cells contain many chloroplasts

xylem tissue

mesophyll tissue

lower epidermis

Cell structure

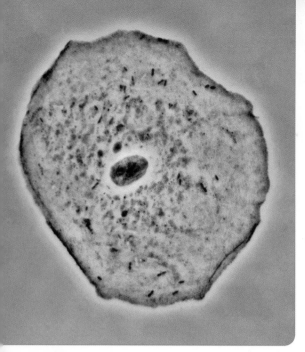

△ Fig. 2.4 All plant, protoctist and animal cells (apart from some very specialised cells) have a cell nucleus like this human cheek cell.

INTRODUCTION

It's an amazing fact that the basic structure of a 'complex cell' (one that contains a nucleus) is the same in all animals, plants and protoctists on Earth. Scientists say that this is because they have all evolved from a single complex cell. This cell evolved from a simple bacteria-like cell (without a nucleus) around 2 billion years ago. This is the origin of all the millions of different species of plants, animals and protoctists that live on Earth today.

KNOWLEDGE CHECK

✓ Most organisms are formed from many cells.
✓ Cells may be specialised in different ways to carry out different functions.

LEARNING OBJECTIVES

✓ Describe cell structures, including the nucleus, cytoplasm, cell membrane, cell wall, mitochondria, chloroplasts, ribosomes and vacuole.
✓ Describe the functions of the nucleus, cytoplasm, cell membrane, cell wall, mitochondria, chloroplasts, ribosomes and vacuole.
✓ Know the similarities and differences in the structure of plant and animal cells.
✓ Explain the importance of cell differentiation in the development of specialised cells.
✓ Understand the advantages and disadvantages of using stem cells in medicine.

PLANT AND ANIMAL CELLS

The diagrams below show a typical animal cell and typical plant cells. These cells all have a **nucleus**, **cytoplasm**, **cell membrane**, **mitochondria** and **ribosomes**.

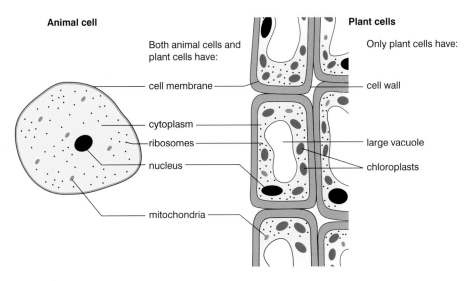

Animal cell

Plant cells

Both animal cells and plant cells have:

Only plant cells have:

cell membrane

cell wall

cytoplasm

ribosomes

large vacuole

nucleus

chloroplasts

mitochondria

△ Fig. 2.5 The basic structures of an animal cell and plant cells.

- The cell membrane holds the cell together and controls substances entering and leaving the cell.
- The cytoplasm is more complicated than it looks. It contains many small organelles and is where many different chemical processes happen.
- The nucleus contains genetic material in the chromosomes. These control how a cell grows and works. The nucleus also controls cell division.
- The mitochondria are organelles within the cytoplasm. They are the site of respiration.
- Ribosomes are very small organelles. They are where proteins are made (**protein synthesis**).

QUESTIONS

1. a) Using the photograph in the Introduction, make a careful drawing of the cell using a sharpened pencil to make clear lines.

b) Use Fig. 2.5 to help you label your drawing to show the nucleus, cytoplasm and cell membrane.

Plant cell structure

Plant cells also have features that are not found in animal cells, such as a **cell wall** made of **cellulose** that supports the cell and defines its shape. Many plant cells have a large central **vacuole** in the cytoplasm that contains **cell sap**. The vacuole is used for storage of some materials, and to support the shape of the cell. If there is not enough cell sap in the vacuole, the cell may collapse and the whole plant may wilt.

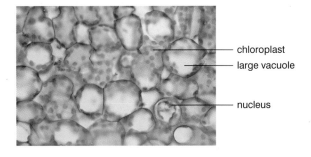

chloroplast
large vacuole

nucleus

△ Fig. 2.6 Plant mesophyll cells as seen under a light microscope.

Many plant cells also contain organelles called **chloroplasts**. These contain the green pigment **chlorophyll**, which absorbs the light energy that plants need to make food in the process known as **photosynthesis**.

QUESTIONS

1. Name the part of a plant cell that does the following:

    a) carries out photosynthesis

    b) contains cell sap

    c) stops the cell swelling if it takes in a lot of water.

Developing investigative skills

The photograph shows the view of some cells seen through a light microscope.

Demonstrate and describe techniques

❶ a) Describe how to set up a slide on a microscope so that the image is clearly focused.

    b) Describe and explain what precautions should be taken when viewing a slide at high magnification.

    c) Describe what precaution should be taken if using natural light to illuminate the slide, and explain why this is important.

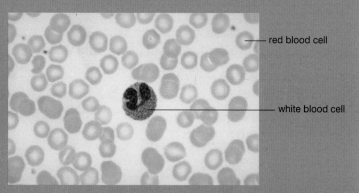

△ Fig. 2.7 Real light micrograph of red and white blood cells.

Make observations and measurements

❷ a) Draw and label a diagram of the white blood cell shown in the light micrograph above.

    b) If the ×4 eyepiece was used, and the ×20 objective, calculate the magnification of the image compared with the specimen on the slide.

Analyse and interpret data

❸ Are the cells shown plant cells or animal cells? Explain your answer.

CELL DIFERENTIATION AND SPECIALISED CELLS

Single-celled organisms obviously do not have tissues, organs or systems. Their single cell carries out all the life processes. In multicellular organisms, because the cells can support each other, they can develop differently. This is called **cell differentiation**. This means different cells can become **specialised** for particular purposes, such as contraction in muscle cells or the ability to carry oxygen in red blood cells.

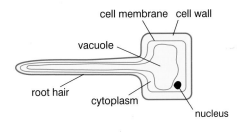

◁ Fig. 2.8 Plant root hair cells have a large surface area to absorb water and mineral ions from the soil.

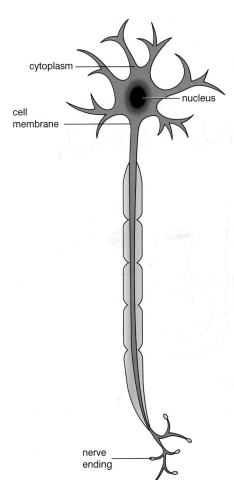

◁ Fig. 2.9 Nerve cells can be very long to carry electrical impulses from one part of the body to another.

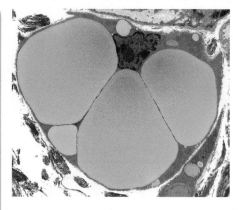
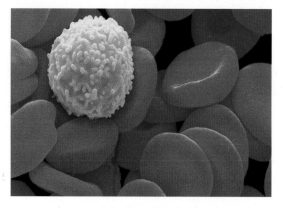

△ Fig. 2.10 A range of different cells from the human body. Left: A fat cell, with the nucleus in purple. Right: red blood cells and a white blood cell.

STEM CELLS

Cells that have not yet differentiated are called **stem cells**. As an embryo develops, to begin with the cells are all very similar, but as these stem cells continue to divide, they differentiate and develop into all the different types of specialised cells found in the adult. In adults, there are also still some stem cells but they are not able to differentiate into as many types of cells as embryo stem cells. For example, adult human stem cells found in the bone marrow can differentiate into red blood cells, white blood cells and platelets.

Stem cells have the potential to treat a range of medical conditions. For example, bone marrow can be taken from a donor and given to a patient to help treat leukaemia (blood cancer). Stem cells in the bone marrow differentiate into new healthy blood cells. Stem cells may also have the potential to treat other conditions such as diabetes, arthritis, Parkinson's and Alzheimer's disease.

There are also some disadvantages to using stem cells, for example a patient's own immune system may attack the stem cells. This is called rejection. There may also be other as yet unknown long-term side effects.

QUESTIONS

1. Look at Fig. 2.9, the diagram of a nerve cell. Describe how it is specialised for its function.

2. Where possible, bone marrow transplants to treat leukaemia are taken from close relatives of the patients. Suggest why this is.

End of topic checklist

The **cell membrane** is the structure surrounding cells that controls what enters and leaves the cell.

A **cell wall** of cellulose surrounds plant cells, giving them support and shape.

Chloroplasts are organelles found only in plant (and some protoctist) cells and are where photosynthesis takes place in the cell.

Chromosomes are long DNA molecules found in the nucleus of a cell.

Cytoplasm is the jelly-like liquid inside the cell which contains the organelles and where many chemical reactions take place.

Mitochondria (singular **mitochondrion**) are organelles in plant and animal cells, and are where respiration takes place.

The **nucleus** is the organelle in plant and animal cells that contains the genetic material.

Photosynthesis is the chemical process by which plants use light, water and carbon dioxide, to create glucose and oxygen.

Ribosomes are small organelles found in cells, and are where protein synthesis takes place.

Specialised cells develop by the process of **cell differentiation**.

Stem cells are cells that have not yet differentiated. They have many potential medical uses.

A large **vacuole** is found in the middle of many plant cells, and contains **cell sap**.

The facts and ideas that you should know and understand by studying this topic:

○ Structures inside cells include the nucleus, cytoplasm, cell membrane, mitochondria, ribosomes, cell wall, chloroplasts and vacuole.

○ The nucleus, cytoplasm, cell membrane, mitochondria, ribosomes, cell wall, chloroplasts and vacuole, have specific roles in cells.

○ Plant and animal cells have many structures in common, but plant cells also have cell walls, and may have chloroplasts and large central vacuoles that animal cells do not have.

○ Stem cells can develop into specialised cells by the process of differentiation.

○ There are advantages and disadvantages of using stem cells in medicine.

End of topic questions

1. Describe the role of the following cell structures:

   a) nucleus (1 mark)

   b) cell membrane (1 mark)

   c) cytoplasm (1 mark)

   d) mitochondria (1 mark)

   e) ribosomes. (1 mark)

2. Draw a table to compare the structures found in plant and animal cells.

   (16 marks)

3. Here are some examples of statements written by students. Each statement contains an error. Identify the error and rewrite the statement so that it is correct.

   a) Animal cells are surrounded by a cell wall that controls what enters and leaves the cell. (1 mark)

   b) All plant cells contain chloroplasts. (1 mark)

   c) Both animal cells and plant cells contain a large central vacuole in the middle of the cell. (1 mark)

4. Red blood cells are unusual because they contain no nucleus. When they are damaged, they have to be replaced with new cells from the bone marrow. Explain how this is different from other cells. (2 marks)

5. State what is meant by the term cell differentiation. (1 mark)

6. Describe one advantage and one disadvantage of using stem cells in medicine. (2 marks)

Biological molecules

INTRODUCTION

Around 65% of your body mass is oxygen, another 18% is carbon and 10% is hydrogen. The remainder of your mass is made up of a large range of other elements, including nitrogen, sulfur, calcium and iron. These elements are combined in different ways to form all the compounds in your body. For example, sugar contains carbon, hydrogen and oxygen. Protein contains carbon, oxygen, hydrogen and nitrogen.

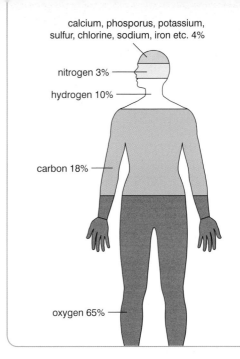

calcium, phosporus, potassium, sulfur, chlorine, sodium, iron etc. 4%

nitrogen 3%

hydrogen 10%

carbon 18%

oxygen 65%

△ Fig. 2.11 What makes up the mass of the human body?

CARBOHYDRATES, PROTEINS AND LIPIDS

Most of the molecules found in living organisms fall into three groups: **carbohydrates**, **proteins** and **lipids** (fats and oils). All of these molecules contain carbon, hydrogen and oxygen. In addition, all proteins contain nitrogen and some also contain sulfur.

Carbohydrate molecules are made up of small basic units called **simple sugars**. These are formed from carbon, hydrogen and oxygen atoms, sometimes arranged in a ring-shaped molecule. One example of a simple sugar is **glucose**. Simple sugar molecules can link together to form larger molecules. They can join in pairs, such as sucrose (the 'sugar' you are familiar with). They can also form much larger molecules called polysaccharides such as **starch** and **glycogen**, which are both long chains of glucose molecules.

Protein molecules are made up of long chains of **amino acids** linked together. There are 20 different kinds of amino acids in plant and animal cells, and they can join in any order to make all the different proteins within the plant or animal body. Examples include the structural proteins in muscle, as well as enzymes that help control cell reactions.

A lipid is what we commonly call a fat or oil. Fats are solid at room temperature whereas oils are liquid, but they have a similar structure. Both are made from basic units called **fatty acids** and **glycerol**. Commonly, there are three fatty acids in each lipid, and the fatty acids vary in different lipids. Lipids are important in forming cell membranes and many other molecules in the body such as fats in storage cells.

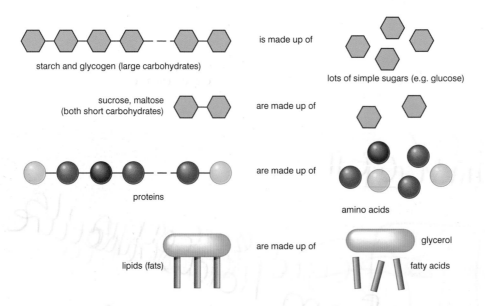

starch and glycogen (large carbohydrates)

is made up of

lots of simple sugars (e.g. glucose)

sucrose, maltose
(both short carbohydrates)

are made up of

proteins

are made up of

amino acids

lipids (fats)

are made up of

glycerol

fatty acids

△ Fig. 2.12 Large biological molecules are formed from small sub-units.

TESTS FOR GLUCOSE, STARCH, PROTEIN AND FAT

The following simple tests can be used to test foods for the presence of starch, protein, fat or for the simple sugar glucose.

Glucose and other reducing sugars

Glucose is a 'reducing sugar' that is important in respiration and photosynthesis. So it is commonly found in plant and animal tissues, and therefore in our food. Its presence can be detected using Benedict's reagent. The pale blue Benedict's solution is added to a prepared sample that contains glucose and heated to 95 °C. If it changes colour or forms a precipitate; this indicates the presence of reducing sugars. A green colour means there is only a small amount of glucose in the solution. A medium amount of glucose will produce a yellow colour. A significant amount of glucose produces a precipitate that is an orange-red colour.

The test using Benedict's reagent will produce an orange-red precipitate for any 'reducing sugar', such as the simple sugars fructose and galactose, and the disaccharides (made from two basic units; *di-* means two or double) lactose and maltose. So it is not exclusively a test for glucose. But as glucose is the usually the most common sugar, this is the test most commonly used for it. Some sugars, such as sucrose, are 'non-reducing sugars' and do not produce an orange-red colour with Benedict's solution.

◁ Fig. 2.13 Benedict's reagent with a range of concentration of sugars (very low in the tube on the left, getting more concentrated towards the right).

Developing investigative skills

Benedict's solution is used to test for the presence of glucose and other 'reducing sugars', but it will only show a positive result if there is at least a certain minimum concentration of sugar present. You can investigate the sensitivity of the Benedict's test by testing a range of glucose concentrations.

The table shows the results of one investigation where Benedict's solution was added to different concentrations of glucose solution and then heated in a water bath.

Glucose concentration %	Colour before heating	Colour after heating
10	pale blue	orange-red
1	pale blue	yellow
0.1	pale blue	green-blue
0.01	pale blue	pale blue
0.001	pale blue	pale blue

Devise and plan investigations

❶ Starting with a glucose solution of concentration 10%, explain how could you make up the other concentration solutions.

❷ What variables would you keep constant in this experiment?

Analyse and interpret data

❸ What was the minimum concentration of glucose detected in this experiment? Explain your answer.

Evaluate data and methods

❹ Suggest how you could improve this experiment to find more accurately the minimum concentration of glucose that can be detected. What safety precautions should you take?

Starch

Starch is the storage molecule of plants, and is found in many foods that are made from plant tissue. When iodine/potassium iodide solution is mixed with a solution of food containing starch, or dropped onto food containing starch, it changes from brown to blue-black. This happens when even small amounts of starch are present and can be used as a simple test for the presence of starch. This change is easiest to see if the test is examined against a white background, such as on a white spotting tile.

Δ Fig. 2.14 The blue-black colour shows there is starch in the biscuit.

Protein

Protein is a part of all cells so is found in both animal and plant tissues. Its presence can be shown by using Biuret reagent (copper sulfate and sodium hydroxide). When Biuret reagent is added to a sample of food that contains protein, it changes from a pale blue to a purple colour.

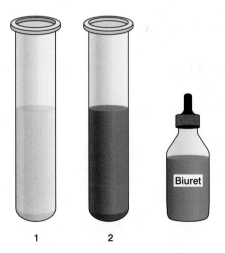

Δ Fig 2.15 The purple colour shows there is protein in the test tube on the right.

Fat

Lipids (fats and oils) are found in many foods. You can test for the presence of lipids by placing a small amount of food into a test tube. Add a few drops of ethanol, shake the test tube and leave for one minute. Pour the solution into a test tube of water and you will see a cloudy white layer if lipid is present. The cloudy layer forms because, although lipids will dissolve in ethanol, they will not dissolve in water and so form an emulsion (milky-white liquid).

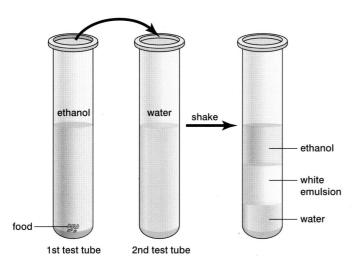

Δ Fig 2.16 The cloudy white emulsion in water shows the presence of a lipid.

QUESTIONS

1. What are the basic units of **a)** lipids, **b)** carbohydrates and **c)** proteins?

2. Using the diagram of large biological molecules in Fig. 2.12 on page 44, give two differences between the structure of a protein and a carbohydrate.

3. Describe what you would see if you tested a sample of the following with **i)** Benedict's solution, **ii)** iodine solution: **a)** glucose syrup, **b)** cake made with wheat flour, table sugar (sucrose), fat and eggs. Explain your answers.

ENZYMES

A **catalyst** is a substance that helps to speed up a reaction. Catalysts are often used in industrial processes such as making ammonia. Living cells also use catalysts to speed up the reactions that happen inside them – what are known as **metabolic reactions** because they are the reactions of the metabolism (all the processes that keep a living organism alive).

Catalysts that control metabolic reactions are enzymes, and because they work in living cells they are called **biological catalysts**. Enzymes help cells carry out all the life processes quickly. Without them, these metabolic reactions would happen too slowly for life to carry on.

Enzymes are proteins. Enzymes are also **specific**, which means each enzyme only works with one substance (or a similar group of substances), called its **substrate**.

- **Amylase** is a type of carbohydrase enzyme produced in the mouth which starts the digestion of starch in food into simple sugars.
- **Proteases** are digestive enzymes that break down proteins into smaller units.
- **Lipases** are digestive enzymes that break down lipids in foods.

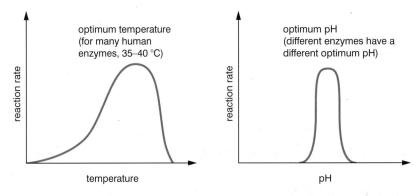

△ Fig. 2.17 Enzymes work best at an optimum temperature and optimum pH.

Enzymes work best at a particular temperature (around 35–40 °C for human digestive enzymes), called their optimum temperature. At lower temperatures they work more slowly than this because the molecules have less energy and move around more slowly. At temperatures that are too high the structure of an enzyme will be changed so that it will not work. This is a permanent change, and when it happens the enzyme is said to be **denatured**.

Enzymes also work best at a particular pH, called their optimum pH. Very high or very low pH can slow down the rate of action of enzymes and even denature them.

Enzyme action and the active site

Like all proteins, enzymes have a three-dimensional shape. This is because an enzyme is made of amino acids, joined together in a chain. The amino acids interact with surrounding amino acids, which causes the chain to fold up into the 3D shape of the enzyme.

Enzymes are unusual proteins in that they have a space in the molecule with a particular 3D shape, called the **active site**. This site matches the shape of the reacting molecule, called the substrate. The substrate fits tightly into the active site of the enzyme, forming an enzyme–substrate complex during the reaction. This makes it easier for the bonds inside the substrate to be rearranged to form the products.

Once the products are formed, they no longer fit the active site, so they are released, leaving the active site free to bind with another substrate molecule. Note that this also happens when two or more substrates are joined by an enzyme to form one product.

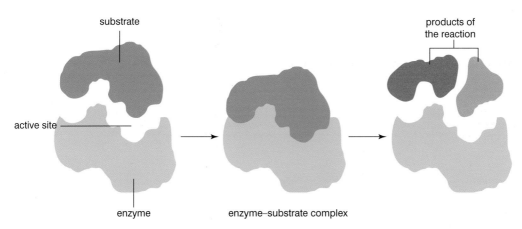

Δ Fig. 2.18 During a reaction, the substrate fits into the active site of the enzyme.

Temperature affects the active site because, beyond the optimum temperature, the atoms in the enzyme molecule are vibrating so much that the bonds between them start to break, changing the shape of the active site permanently and denaturing the enzyme.

The shape of the active site is also affected by pH, because some of the interactions between amino acids in the enzyme molecule depend on the pH of the surrounding solvent. So the shape of the enzyme will depend on the surrounding pH. If the pH changes too much from the optimum pH, the shape of the enzyme, and particularly its active site, will change. So the substrate will not fit as well and the rate of reaction will decrease.

SCIENCE IN CONTEXT

ENZYMES IN WASHING DETERGENTS

Many of the stains on dirty clothes are created by biological molecules. For example, sweat stains include oils from skin, blood stains include the protein haemoglobin, food stains may include grease which is a lipid.

Digestive enzymes are now used in many washing detergents because they help to break down the large biological molecules in stains into smaller ones that wash away easily in water. This speeds up the rate of cleaning, and makes it easier to do at lower temperatures than without enzymes, saving time and energy.

Developing investigative skills

Developed black-and-white negative film consists of a celluloid backing covered with a layer of gelatin. Where the film has been exposed the gelatin layer contains tiny particles of silver which make that area black. Gelatin is a protein and is easily digested by proteases.

Strips of exposed film were soaked in protease solution at different temperatures. When the gelatin had been digested, the silver grains fell away from the celluloid backing, leaving transparent film. The table shows the results.

Tube	Temperature in °C	Time to clear in seconds
1	10	394
2	20	195
3	30	163
4	40	235
5	50	513

Devise and plan investigations

❶ Describe how you would set up this investigation to get results like those shown in the table.

Analyse and interpret data

❷ Draw a graph using the data in the table.

❸ Describe the shape of the graph.

❹ Explain the shape of the graph.

Evaluate data and methods

❺ How could you modify this experiment to get a more accurate estimate of the optimum temperature for this enzyme?

❻ How could you modify this experiment to investigate the effect of pH on enzyme activity?

QUESTIONS

1. Explain what is meant by a biological catalyst.

2. Which group of biological molecules do enzymes belong to?

3. Explain why cells need enzymes.

4. Describe the effect of temperature on the rate of an enzyme-controlled reaction.

End of topic checklist

An **amino acid** is the basic unit of a protein.

Carbohydrates include simple sugars as well as larger molecules, such as starch or glycogen, which are made of many simple sugars joined together.

Enzymes are biological catalysts controlling the rate of metabolic reactions.

A **fatty acid** is one of the basic units of a lipid, along with glycerol.

Glycerol is one of the basic units of a lipid, along with fatty acids.

Lipids are large molecules commonly made of the basic units of three fatty acid molecules and one glycerol molecule and include examples such as olive oil and ear wax.

Proteins are large molecules, such as found in muscle tissue, made of many amino acids.

A **simple sugar** is a basic sugar unit (such as glucose) that can join together with other sugar units to make large carbohydrates such as starch and glycogen.

The facts and ideas that you should know and understand by studying this topic:

○ Carbohydrates, proteins and lipids (fats and oils) are made of carbon, hydrogen and oxygen. Proteins also contain nitrogen and may contain sulfur.

○ Glucose can be tested for using Benedict's reagent because it produces an orange-red precipitate on heating.

○ Starch can be tested for with iodine solution as it turns the solution blue-black.

○ Protein can be tested for using Biuret reagent as it turns the solution purple.

○ Lipids can be tested for using ethanol and then water as they cause a cloudy white layer.

○ Enzymes have an optimum temperature at which the rate of reaction occurs most rapidly: the rate is slower at lower temperatures because the molecules are not moving as quickly; the rate is also slower at higher temperatures when the active site of the enzyme molecule starts to change shape and denature.

○ Enzymes also have an optimum pH at which the rate of reaction happens most rapidly and may be denatured by a pH that is too high or too low.

End of topic questions

1. **a)** Explain why carbon, hydrogen and oxygen are the most common elements found in a human body. **(2 marks)**

   b) Why does the body need other elements, in addition to those in **a)**? **(1 mark)**

2. If enzymes were used to break a lipid down to its basic units, what would those units be? **(2 marks)**

3. Which of the following would test positive with Benedict's reagent? Explain your answer.

   starch   lactose (a disaccharide)   fructose (a simple sugar). **(2 marks)**

4. A sample of bread was ground up. Some of the breadcrumbs were tested with Benedict's reagent and some with iodine/potassium iodide solution. The rest of the crumbs were mixed with Substance A. After 20 minutes, some of the mixture was tested with Benedict's reagent and some with iodine/potassium iodide solution. The results of the tests are shown in the table.

	Test with Benedict's solution	Test with iodine/ potassium iodide solution
Before adding Substance A	no precipitate	change to blue-black colour
After 20 mins with Substance A	orange-red precipitate	no colour change

   a) Describe what the results show. **(2 marks)**

   b) What was Substance A? Explain your answer. **(2 marks)**

5. **a)** Describe how to test a food for the presence of protein. Include what a positive result will show. **(2 marks)**

   b) Describe how to test a food for the presence of lipids. Include what a positive result will show. **(3 marks)**

End of topic questions continued

6. A protease enzyme, like those found in the human digestive system, was tested to find its optimum temperature.

a) What do we mean by *optimum temperature*? **(1 mark)**

b) Sketch a graph of reaction rate and temperature for this enzyme and annotate it to explain its shape. **(4 marks)**

c) Suggest what the optimum temperature is most likely to be for this enzyme, and explain your answer. **(2 marks)**

d) Some protease enzymes work in the stomach, which is very acidic. Would you expect these enzymes to also work in other parts of the digestive system where it is not acidic? Explain your answer. **(2 marks)**

Movement of substances into and out of cells

INTRODUCTION

If you put a red blood cell into pure water, it will absorb water and eventually burst open. If you place the red blood cell into a salty solution instead, it will lose water and shrink. Surrounding every cell is the cell membrane. Imagine the membrane as a leaky layer that is strong enough to hold all the contents in the cell together, but that allows small particles to move through it.

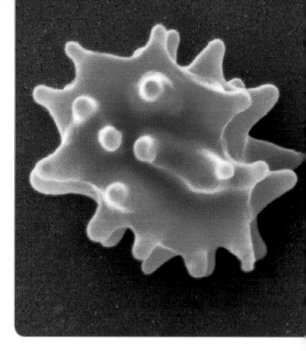

△ Fig. 2.19 A red blood cell that has been placed in a salty solution loses water and shrinks.

The cell membrane also has special 'gates' which allow certain important particles through. Different cells have different kinds of 'gate' in them. So cell membranes play an essential role in controlling what goes in and out of cells, and therefore control the way that the cell functions.

KNOWLEDGE CHECK

✓ Cells need oxygen and glucose for respiration.
✓ Cells need to get rid of waste substances, such as carbon dioxide from respiration.

LEARNING OBJECTIVES

✓ Understand the processes of diffusion, osmosis and active transport by which substances move into and out of cells.
✓ Understand how factors affect the rate of movement of substances into and out of cells, including the effects of surface area to volume ratio, distance, temperature and concentration gradient.
✓ Practical: Investigate diffusion and osmosis using living and non-living systems.

DIFFUSION

Substances like water, oxygen, carbon dioxide and food are made of particles (**molecules**).

In liquids and gases the particles are constantly moving around. This means that they eventually spread out evenly. For example, if you dissolve sugar in a cup of water, even if you do not stir it, the sugar molecules eventually spread throughout the liquid. This is because all the molecules are moving around, colliding with and bouncing off other particles.

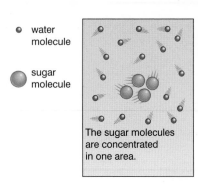 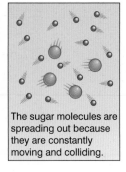 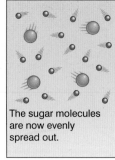

water
molecule

sugar
molecule

The sugar molecules are concentrated in one area.

The sugar molecules are spreading out because they are constantly moving and colliding.

The sugar molecules are now evenly spread out.

△ Fig. 2.20 Diffusion of sugar molecules in a solution.

The sugar molecules have spread out from an area of high concentration, when they were added to the water, to an area of low concentration. Eventually, although all the particles are still moving, the overall picture is that the sugar molecules are evenly spread out and there is no longer a **concentration gradient**.

Only while there is **net movement** (where there are more particles moving in one direction than another) from an area of high concentration to an area of lower concentration is there **diffusion**.

REMEMBER

Diffusion is the net movement of molecules from an area of their higher concentration to an area of their lower concentration.

Diffusion can only occur when there is a difference in concentration between two areas. Particles are said to move *down* their concentration gradient. This happens because of the random movement of particles and needs no energy from the cell. It is a **passive** process.

Diffusion in cells

Cells are surrounded by membranes. These membranes are leaky – they let tiny particles pass through them. Large particles cannot get through, so the cell membranes are said to be **partially permeable**.

Tiny particles can *diffuse* across a cell membrane if there is a difference in concentration on either side of the membrane (a concentration gradient). For example, in the blood vessels entering the lungs there is a *low* oxygen concentration inside the red blood cells (because they have given up their oxygen to cells in other parts of the body) and a *high* oxygen concentration in the alveoli of the lungs. Therefore oxygen diffuses from the alveoli into the red blood cells.

Other examples of diffusion include:

• carbon dioxide entering leaf cells
• digested food substances from the small intestine entering the blood.

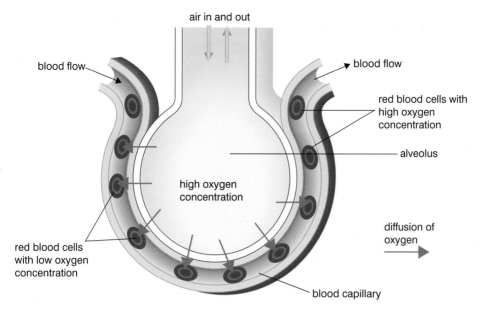

air in and out

blood flow

blood flow

red blood cells with high oxygen concentration

alveolus

high oxygen concentration

red blood cells with low oxygen concentration

diffusion of oxygen

blood capillary

△ Fig. 2.21 In blood vessels in the lungs, oxygen diffuses down its concentration gradient from the air in the alveolus into red blood cells.

SCIENCE IN CONTEXT

KIDNEY FAILURE AND DIALYSIS

The kidneys are organs that depend on filtration and diffusion to produce urine and keep the concentration of many substances in the blood at a fairly constant level. People who suffer from kidney failure are unable to do this, and are very quickly at risk from the build-up of waste products, such as urea, in the body as a result of cell processes. In high concentrations these waste products can damage body cells and lead to death.

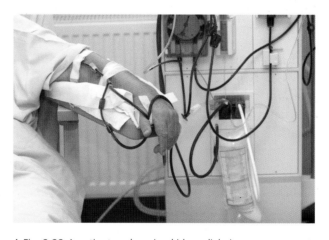

△ Fig. 2.22 A patient undergoing kidney dialysis.

Haemodialysis is an artificial way of cleaning the blood by which substances diffuse out of the blood into dialysis fluid in a machine called a dialyser. The concentration of substances in the dialysis fluid has to be correct, so that all the waste products are removed and other substances are returned to the body at the right concentration.

Investigating diffusion

Diffusion in non-living systems may be investigated in different ways.

It can be studied simply by dropping a coloured crystal of a simple soluble substance such as potassium manganate(VII) into still water, and watching what happens. As the crystal dissolves, the particles of solute diffuse throughout the water until the concentration of solute is the same throughout the water.

△ Fig. 2.23 Potassium manganate(VII) crystal dissolving in still water.

Visking tubing is an artificial partially permeable membrane which can be used to model diffusion across cell membranes. You can create a small bag shape by knotting one end of some tubing. If you place a solution of glucose and starch into the bag, and suspend it in a beaker of water for a few hours, you can then test the water in the beaker for reducing sugar (glucose) and starch. The tests will show that glucose has diffused through the membrane, but the starch has not, because glucose molecules are small enough to pass through the holes but the starch molecules are too large.

Diffusion from cells may be investigated by looking at coloured cells such as red onion or beetroot cells. If slices of tissue from these are left in water, after a while the colour will start to diffuse out of the cells into the water.

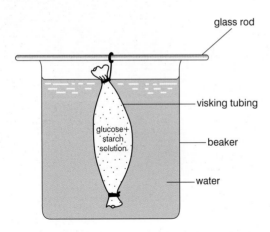

glass rod

visking tubing

glucose + starch solution.

beaker

water

△ Fig. 2.24 Visking tubing can act as a model for the cell membrane.

Developing investigative skills

The effect of temperature on the rate of diffusion can be investigated using the apparatus shown.

Glucose solution inside the visking tubing will diffuse through the tubing into the water surrounding it. Samples of water can be removed and tested for the presence of glucose.

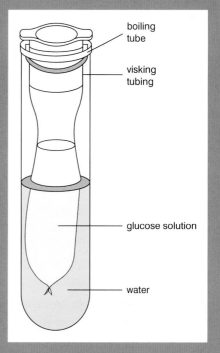

boiling tube

visking tubing

glucose solution

water

Devise and plan investigations

❶ a) Write a plan for an experiment to investigate the effect of temperature on the rate of diffusion, using the apparatus shown. You will need to include: how you will change the temperature and what values you will test; how you will test the water for the presence of glucose and how you will work out the time it has taken for glucose to diffuse into the water; the number of repeats of each temperature you will use to get reliable results.

b) What risks should be considered with this method and how should they be handled?

Make observations and measurements

The following results were obtained in a similar experiment.

Temperature in °C	Time taken to get a positive result for glucose in seconds
20	570
30	480
40	330
50	270
60	120

❷ Use the results to draw a suitable graph.

Analyse and interpret data

❸ Describe any pattern shown in your graph.

❹ Draw a conclusion from the graph.

❺ Explain your conclusion using your scientific knowledge.

1. In your own words, describe the terms *net movement* and *diffusion*.

2. Is diffusion a passive or active process? Explain your answer.

3. a) Which of these molecules can diffuse through a cell membrane: starch, glucose?

   b) Explain why some particles can diffuse through cell membranes but not others.

OSMOSIS

Water molecules are small enough to diffuse through the holes in cell membranes. However, because water molecules are so important to cells, and may be diffusing in a different direction to other molecules, this kind of diffusion has a special name – **osmosis**. Like diffusion, osmosis is a passive process and is a result of the random movement of particles.

Water molecules diffuse from a place where there is a high concentration of water molecules (such as a dilute sucrose sugar solution) to where there is a low concentration of water molecules (such as a concentrated sucrose sugar solution).

Osmosis is the net movement of water molecules from a region of their high concentration to a region of their lower concentration across a partially permeable membrane.

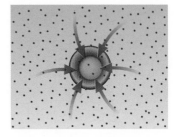

△ Fig. 2.25 A red blood cell in pure water takes in water by osmosis.

Concentrations in solutions

Many people confuse the concentration of the solution with the concentration of the water. Remember, in osmosis it is the *water molecules* that we are considering, so you must think of the concentration of water molecules in the solution instead of the concentration of solutes dissolved in it.

• A low concentration of dissolved solutes means a high concentration of water molecules.
• A high concentration of dissolved solutes means a low concentration of water molecules.

So the water molecules are moving from a *high concentration (of water molecules)* to a *low concentration (of water molecules)*, even though this is often described as water moving from a low-concentration solution to a high-concentration solution.

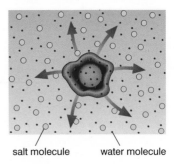

salt molecule   water molecule

△ Fig. 2.26 A red blood cell in a concentrated salt solution will lose water as a result of osmosis.

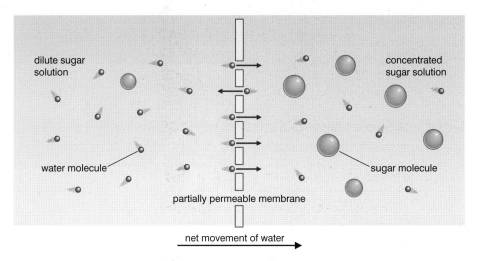

△ Fig. 2.27 Water molecules diffuse from an area of higher concentration (of water molecules) into an area of lower concentration (of water molecules). This sort of diffusion is known as osmosis.

Investigating osmosis

We can investigate osmosis in non-living systems using visking tubing. If we set up a tube containing a sugar solution in the tubing, and suspend it in a beaker of water for a few hours, the level of solution in the tubing will rise.

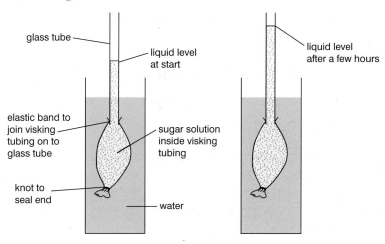

△ Fig. 2.28 Osmosis demonstration.

This is because the concentration of water molecules is greater in the beaker, outside the visking tubing, than inside the tubing. Osmosis results in a net movement of water into the tubing, increasing the volume of the solution.

Osmosis in living systems can be investigated by measuring the change in mass of plant tissue (such as strips of potato) placed in solutions of different concentration. Any change in mass is considered to be the results of loss or gain of water by the cells due to osmosis. An alternative investigation is described in the following box.

Developing investigative skills

Strips of dandelion stem about 5 cm long and 3 mm wide were placed in sodium chloride solutions of different concentration. After 10 minutes, the strips looked as shown in the diagram. (Note that the outer layer of a dandelion stem is 'waterproofed' with a waxy layer to protect it from water loss to the environment.)

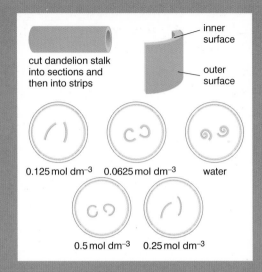

Devise and plan investigations

❶ Write a plan for an experiment to carry out this investigation. Your plan should include:

a) instructions on how to prepare the stem samples

b) instructions on how to keep the stem samples until the experiment starts.

△ Fig. 2.29 Results of putting dandelion stems in different concentrations of sodium chloride solution.

Make observations and measurements

❷ Using the diagram, describe the results of this investigation.

Analyse and interpret data

❸ Explain as fully as you can the results of this investigation.

❹ Use the results to suggest the normal concentration of cell cytoplasm. Explain your answer.

QUESTIONS

1. Explain the term *osmosis* in your own words.

2. Explain how osmosis is a) similar to and b) different from diffusion.

3. Draw a labelled diagram to show what happens to the water molecules when a red blood cell is placed in a solution that has a higher concentration of solute than the cytoplasm of the cell.

REMEMBER

For the highest marks, you will need to explain diffusion and osmosis in terms of particles and their concentration gradients. Be clear that even when diffusion and osmosis stop because there is no concentration gradient, the particles in the solution continue to move – but there is no longer any net movement.

FACTORS THAT AFFECT THE RATE OF MOVEMENT OF SUBSTANCES

The rate of diffusion of a solute, or the rate of osmosis of water molecules, in a solution from one side of a membrane to the other can be affected by different factors.

- Concentration gradient: The greater the difference in concentration on either side of a cell membrane, the more particles will move from the concentrated side to the less concentrated side. So the greater the concentration gradient, the faster the rate of diffusion and osmosis.
- Distance: The greater the distance particles have to move (e.g. if there are several membranes to pass through) the longer it takes. So the greater the distance, the slower the rate of diffusion or osmosis from one side to the other.
- Temperature: At a higher temperature, particles have more energy so they move around faster. This means they are more likely to bump into the cell membrane and pass through to the other side. So increasing the temperature increases the rate of diffusion and osmosis.
- Surface area: The greater the area of membrane across which the particles can move, the more that will cross in a given time. So the larger the surface area, the faster the rate of diffusion and osmosis.

The effect of surface area to volume ratio

Since rate of diffusion increases as surface area increases, it would seem that larger animals should be able to exchange materials with the environment faster than small ones. But there is another problem to consider. If we have a cube with length of side l:

- the surface area of the cube is the sum of the area of each of the 6 sides = $6\,l^2$
- the volume of the cube is l^3.

As surface area increases, volume increases even faster, as shown in the table.

If we then calculate the surface area to volume ratio, we can see that as body size increases so the ratio decreases.

Length of side of cube	1 cm	2 cm	3 cm	4 cm
Surface area of cube	$6\ cm^2$	$24\ cm^2$	$54\ cm^2$	$96\ cm^2$
Volume of cube	$1\ cm^3$	$8\ cm^3$	$27\ cm^3$	$64\ cm^3$
Surface area to volume ratio	6	3	2	1.5

△ Table 2.1 Surface area to volume ratio.

The 'volume' in a body is all the tissue that is making waste products, or needing substances for respiration and other processes. So as body size increases, the need for a higher rate of diffusion of materials into and out of the body increases more rapidly than the increase in surface area. This explains why the areas of exchange in larger animals, such as the lungs, intestines and kidneys, have adaptations to increase the surface area as much as possible. You will learn more about these adaptations later in this course.

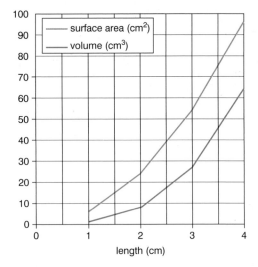

△ Fig. 2.30 Surface area to volume ratio.

QUESTIONS

1. Name four factors that affect the rate of movement of substances across a partially permeable membrane.

2. Explain in your own words why concentration gradient affects the rate of diffusion. (Use drawings to support your answer if you wish.)

3. Explain why surface area to volume ratio decreases as size increases.

ACTIVE TRANSPORT

Sometimes cells need to absorb particles *against a concentration gradient*: from a region of low concentration into a region of high concentration. For example, root hair cells may take in nitrate ions from the soil even though the concentration of these ions is higher in the plant cells than in the soil. Also, glucose is reabsorbed in the kidneys from the kidney tubules, even though the concentration of glucose in the blood and kidney tubules is initially the same.

The nitrate ions in plants and glucose molecules in animals are essential for healthy growth. The cells use energy to absorb these substances, so this is an *active* process and so is called **active transport**.

Active transport occurs when special carrier proteins in the surface of a cell pick up particles from one side of the membrane and transport them to the other side. You can see this happening in Fig. 2.31.

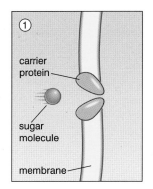 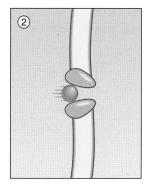 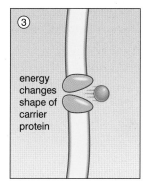

carrier protein

sugar molecule

membrane

energy changes shape of carrier protein

△ Fig. 2.31 Active transport of a sugar molecule through a carrier protein in a membrane.

The energy for active transport comes from cell respiration. A simple test to show whether a substance is being absorbed by an active process or a passive process is to treat the cells with a poison that stops the cells respiring. For example, treating root hair cells with cyanide stops the uptake of nitrate ions but does not affect diffusion or osmosis.

QUESTIONS

1. Explain what is meant by *active transport*.

2. Give one example of active transport in a plant and one in an animal.

End of topic checklist

When molecules move in both directions, but more move from high concentration to low concentration than from low to high, there is a **net movement** from high to low. When the concentrations become equal, molecules still move but now equal numbers are moving in both directions so there is *no* net movement.

Passive is the opposite of *active*: happening without the need for additional energy.

The facts and ideas that you should know and understand by studying this topic:

○ Diffusion is the net movement of particles from a region of their higher concentration to a region of their lower concentration.

○ Osmosis is the net movement of water molecules across a partially permeable membrane from their higher concentration (a dilute solution) to their lower concentration (a more concentrated solution).

○ Active transport is the transfer of molecules across a cell membrane, against their concentration gradient, using energy from cell respiration.

○ Diffusion and osmosis are passive processes, that result from the random movement of particles.

○ The rate of diffusion and osmosis across a membrane increases with a greater concentration gradient, a higher temperature, a greater surface area and a shorter distance.

○ The surface area to volume ratio decreases as size increases.

End of topic questions

1. An old-fashioned way of killing slugs in the garden is to sprinkle salt on them. This kills the slugs by drying them out. Explain why this works. **(2 marks)**

2. Copy and complete to compare diffusion, osmosis and active transport. **(9 marks)**

	Diffusion	Osmosis	Active transport
active or passive?			
which molecules move?			
requires special carrier proteins?			

3. Which of the following are examples of diffusion, osmosis or neither?

   a) Carbon dioxide entering a leaf when it is photosynthesising. **(1 mark)**

   b) Food entering your stomach when you swallow. **(1 mark)**

   c) A dried out piece of fruit swelling up when placed in a bowl of water. **(1 mark)**

4. Explain why a large animal, such as a human, needs special adaptations to organs that exchange materials with the environment (such as the lungs, small intestine and kidney), but a single-celled protoctist does not. **(4 marks)**

5. a) If you measured the rate of respiration of plant root hair cells, would you expect it to be **i)** the same as, **ii)** more than, **iii)** less than the rate of respiration of other plant cells? **(1 mark)**

   b) Explain your answer to part **a)**. **(1 mark)**

6. Students were investigating the effect of a cell membrane on diffusion and osmosis. They suspended a visking tubing bag containing fructose solution and starch solution in a beaker of water for a few hours.

   a) Explain why they used visking tubing as a model for the cell membrane. **(1 mark)**

   b) At the end of the experiment they tested the water in the beaker to see if any starch or fructose had diffused through the bag. What tests would they use to do this? **(2 marks)**

   c) What results do you think they would have had with these tests? **(2 marks)**

   d) Explain your answer to part **c)**. **(2 marks)**

7. There are many membranes in a cell, separating off organelles that produce substances such as hormones and enzymes, or where cell processes such as photosynthesis and respiration occur. Explain fully the importance of these membranes and suggest why it is an advantage to the cell to have them. **(3 marks)**

Nutrition

INTRODUCTION

At a certain time of summer in Canada the rivers are full of salmon returning to breed. Their sudden appearance attracts large numbers of bears, which feed on nuts, grasses, insects and other small animals at other times of the year. The salmon are full of protein and fat and are an essential source of nutrition for the bears.

△ Fig. 2.32 Canadian bears fishing for salmon.

KNOWLEDGE CHECK

✓ Plants make their own food in their leaves using photosynthesis.
✓ Plant structures, such as the leaf and root cells, are adapted for their functions in nutrition.
✓ Animals eat other organisms to get the food they need for their life processes.
✓ The organs, tissues and cells of the digestive system are adapted to digest and absorb nutrients from food.
✓ Food may be chemically or mechanically digested before absorption.
✓ Different groups of people need different diets.

LEARNING OBJECTIVES

✓ Understand the process of photosynthesis and its importance in the conversion of light energy to chemical energy.
✓ Know the word equation and balanced chemical symbol equation for photosynthesis.
✓ Understand how varying carbon dioxide concentration, light intensisty and temperature affect the rate of photosynthesis.
✓ Describe the structure of the leaf and explain how it is adapted for photosynthesis.
✓ Understand that plants require mineral ions for growth, and that magnesium ions are needed for chlorophyll and nitrate ions are needed for amino acids.
✓ Practical: Investigate photosynthesis, showing the evolution of oxygen from a water plant, the production of starch and the requirements of light, carbon dioxide and chlorophyll.
✓ Understand what is meant by a balanced diet.
✓ Identify the sources and describe the functions of nutrients in human nutrition.
✓ Understand how energy requirements vary with activity levels, age and pregnancy.
✓ Describe the structure and function of organs in the human alimentary canal.
✓ Understand how food is moved through the gut by peristalsis.
✓ Understand the role of digestive enzymes.
✓ Understand that bile is produced by the liver and stored in the gall bladder.
✓ Understand the role of bile in neutralising stomach acid and emulsifying lipids.
✓ Understand how the small intestine is adapted for absorption, including the structure of a villus.
✓ Practical: Investigate the energy content in a food sample.

Nutrition in flowering plants

PHOTOSYNTHESIS

Plant tissue contains the same types of chemical molecules (carbohydrates, proteins and lipids) as animal tissue. However, while animals eat other organisms to get the nutrients they need to make these molecules, plants make these molecules from basic building blocks, beginning with the process of photosynthesis.

In photosynthesis, plants combine carbon dioxide from the air with water absorbed from the soil to form glucose, a simple sugar and a carbohydrate. This process transfers light energy (usually from sunlight) into chemical energy in the bonds in the glucose. Photosynthesis is fundamental to most living organisms on Earth, because most organisms other than plants get their energy from the chemical energy in the food that they eat.

Oxygen is also produced in photosynthesis, and although some is used inside the plant for respiration (releasing energy from food), most is not needed and is given out as a waste product.

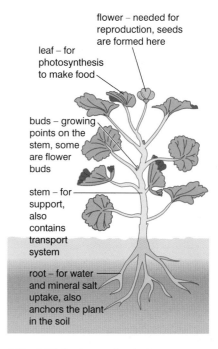

flower – needed for reproduction, seeds are formed here

leaf – for photosynthesis to make food

buds – growing points on the stem, some are flower buds

stem – for support, also contains transport system

root – for water and mineral salt uptake, also anchors the plant in the soil

△ Fig. 2.33 Anatomy of a plant.

The sunlight is absorbed by the green pigment **chlorophyll** in plants.

The process of photosynthesis can be summarised in a word equation:

$$\text{carbon dioxide} \quad + \quad \text{water} \quad \xrightarrow[\text{light energy}]{\text{chlorophyll}} \quad \text{glucose} \quad + \quad \text{oxygen}$$

It can also be summarised as a balanced symbol equation:

$$6CO_2 \quad + \quad 6H_2O \quad \xrightarrow[\text{light energy}]{\text{chlorophyll}} \quad C_6H_{12}O_6 \quad + \quad 6O_2$$

REMEMBER

For higher marks you will need to know and be able to balance the chemical equation for photosynthesis.

Much of the glucose formed by photosynthesis is converted into other substances, including **starch**. Starch molecules are large carbohydrates made of lots of glucose molecules joined together. Starch is insoluble and so can be stored in cells without affecting water movement into and out of the cells by osmosis. Some plants, such as potato and rice, store large amounts of starch in particular parts (tubers or seeds). We use these parts as sources of starch in our food.

Some glucose is converted to **sucrose** (a type of sugar formed from two simple sugar molecules, glucose and fructose, joined together). This is still soluble, but not as reactive as glucose, so can easily be carried around the plant in solution.

The energy needed to join simple sugars to make larger carbohydrates comes from respiration.

The sugars produced by photosynthesis can be converted into cellulose, which is another large carbohydrate used for making cell walls. They can also be converted into the other types of chemical molecule (such as proteins and lipids) in reactions inside the cells.

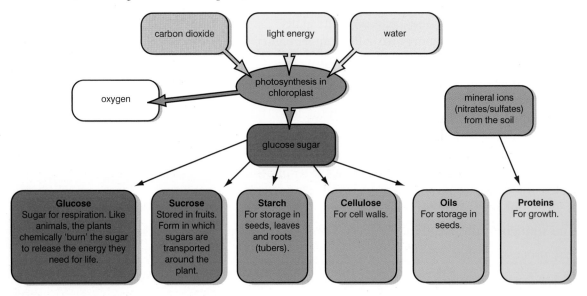

△ Fig. 2.34 The raw materials and products in a plant.

QUESTIONS

1. Write the balanced symbol equation for photosynthesis.

2. Annotate your equation to show where each of the reactants comes from, and where each of the products goes to.

3. Give four examples of how glucose is used in a plant.

ADAPTATIONS FOR PHOTOSYNTHESIS

Photosynthesis takes place mainly in the leaves, although it can occur in any cells that contain green chlorophyll. Leaves are adapted to make them very efficient as sites for photosynthesis.

Fig. 2.35 shows the external adaptations of the leaf that help to maximise the rate of photosynthesis.

The inside structure of the leaf also helps to maximise the rate of capture of light energy and so maximise the rate of photosynthesis.

• The transparent **epidermis** allows as much light as possible to reach the photosynthesising cells within the leaf.

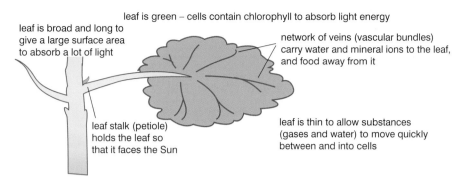

leaf is green – cells contain chlorophyll to absorb light energy

leaf is broad and long to give a large surface area to absorb a lot of light

network of veins (vascular bundles) carry water and mineral ions to the leaf, and food away from it

leaf stalk (petiole) holds the leaf so that it faces the Sun

leaf is thin to allow substances (gases and water) to move quickly between and into cells

△ Fig. 2.35 Adaptations of a leaf.

- The **palisade cells**, where most photosynthesis takes place, are tightly packed together in the uppermost half of the leaf so that as many as possible can receive sunlight.
- **Chloroplasts** containing chlorophyll are concentrated in cells in the uppermost half of the leaf to absorb as much sunlight as possible.
- Air spaces in the **spongy mesophyll** layer allow the movement of gases (carbon dioxide and oxygen) through the leaf to and from cells.
- A leaf has a large internal surface area to volume ratio to allow the efficient absorption of carbon dioxide and removal of oxygen by the photosynthesising cells.
- Many pores or **stomata** (singular: stoma) allow the movement of gases into and out of the leaf.
- **Phloem** tissue transports sucrose, formed from glucose in photosynthesising cells, away from the leaf. **Xylem** tissue transports water and mineral ions to the leaf from the roots.

△ Fig. 2.36 The leaves of trees are arranged so that as far as possible they do not overlap each other, which makes it possible for the tree to capture as much light energy as possible.

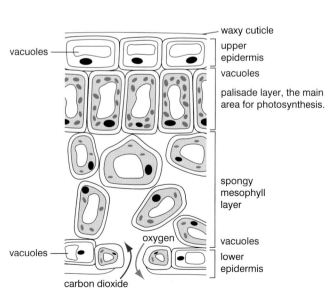

vacuoles

waxy cuticle

upper epidermis

vacuoles

palisade layer, the main area for photosynthesis.

spongy mesophyll layer

oxygen

vacuoles

vacuoles

lower epidermis

carbon dioxide

△ Fig. 2.37 Cells in the section of a leaf.

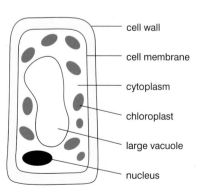

palisade cell

cell wall

cell membrane

cytoplasm

chloroplast

large vacuole

nucleus

1. List as many adaptations of a plant leaf for photosynthesis as you can.

2. Explain why a large surface area inside the leaf is essential for photosynthesis.

3. Explain why a transparent epidermis is an adaptation for photosynthesis.

Factors affecting the rate of photosynthesis

The rate at which a process can occur depends on how quickly the required materials can be supplied. Photosynthesis needs light energy. As light levels fall as night approaches, or on a very cloudy day, the rate at which energy is absorbed by chlorophyll decreases and the rate at which photosynthesis proceeds slows down. Light has become a **limiting factor** for the process.

If it is a sunny day with plenty of light, then light is not a limiting factor. Instead, the rate of photosynthesis may be controlled by the rate of diffusion of carbon dioxide into the photosynthesising cells in the leaf from the air. In this case, carbon dioxide has become the limiting factor.

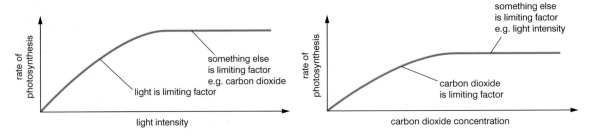

Δ Fig. 2.38 The rate of photosynthesis is affected by (left) light intensity and (right) carbon dioxide concentration, up to a point when something else becaomes a limiting factor.

Increasing the levels of light and carbon dioxide, two of the factors needed for photosynthesis, will increase the rate of photosynthesis until the rate is stopped by some other limiting factor.

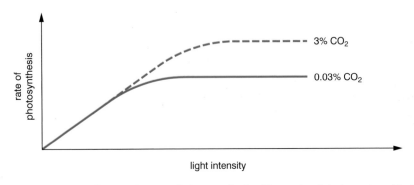

Δ Fig. 2.39 The effect on the rate of photosynthesis of increasing light intensity with different concentrations of carbon dioxide.

We can test whether this is true by adding more carbon dioxide. Farmers sometimes do this in glasshouses to increase the growth rate of crops. If the rate of photosynthesis increases, then carbon dioxide was the limiting factor.

Photosynthesis involves several chemical reactions. Like all chemical reactions, their rate of reaction is affected by temperature, which affects the energy of the reacting particles and how quickly they bump into each other. So on a cool day, or in the early morning when the air has not yet heated up, temperature may be the factor that limits the rate of photosynthesis. If temperature rises too high, however, the enzymes that control the rate of reactions start to become denatured and so the reactions go more slowly.

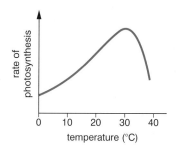

△ Fig. 2.40 The effect of temperature on the rate of photosynthesis.

SCIENCE IN CONTEXT **MAXIMISING PLANT GROWTH**

Farmers and plant growers want their crops to grow well, but in open fields it is not usually possible to control the amount of carbon dioxide or light the plants receive, or the temperature at which they are growing. However, if the plants are grown in sheltered conditions, such as in glasshouses or plastic polytunnels, then it can be possible to change conditions, such as by:

△ Fig. 2.41 Plants in a glasshouse.

- using artificial lighting so that the plants can continue growing at night
- adding carbon dioxide to the atmosphere around the plants by burning coal or oil stoves
- using heating to increase the temperature.

Remember that enzymes have an optimum temperature at which they work, so glasshouses and polytunnels may also need to be ventilated to release hot air if the temperature rises too high, otherwise the rate of photosynthesis will decrease.

INVESTIGATING PHOTOSYNTHESIS

We can use the iodine test to show that photosynthesising parts of a plant produce starch. Before carrying out this test, though, you must start by leaving the plant in a dark place for at least 24 hours. This will make sure that the plant uses up its stores of starch (this is known as destarching). This means that any starch identified by the test is the result of photosynthesis during the investigation.

- The production of starch after photosynthesis can be shown simply by placing a destarched plant in light for an hour. Remove one leaf and place it in boiling water for a few minutes to remove the waterproof waxy cuticle. Then place the leaf in hot ethanol heated using an electrically heated water bath, **not over a Bunsen burner** because ethanol fumes are flammable. This removes the chlorophyll in the leaf. When the leaf has lost its green, place it in water for a minute or so to soften it before placing it in a dish and adding a few drops of iodine solution. The leaf should turn blue/black, indicating the presence of starch.

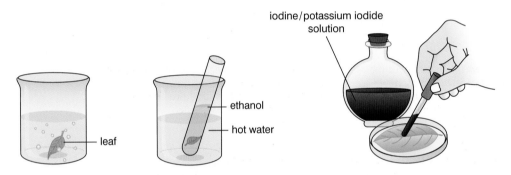

△ Fig. 2.42 Preparing and testing a leaf for starch.

- The investigation above can be adjusted to show the need for light by covering part of the leaf before the destarched plant is brought into the light. Only the part of the leaf that received light should test positive for the presence of starch, showing that photosynthesis is linked to the production of starch.
- This investigation can also be adjusted to show the need for chlorophyll by using variegated leaves. Variegated leaves are partly green (where the cells contain chlorophyll) and partly white (where there is no chlorophyll). A variegated leaf after this investigation will show the presence of starch where there was chlorophyll and not in the parts of the leaf that had no chlorophyll.

- A simple test to show the need for carbon dioxide can be carried out by setting up two bell jars on glass sheets. Sodium hydroxide or potassium hydroxide reacts with any carbon dioxide, removing it from the air. So a dish of one of the hydroxides is placed in one bell jar. Carbon dioxide is added to the other bell jar by burning a candle in it, which also removes some of the oxygen. Similar destarched plants are placed in each bell jar, and the base of the jar sealed to the glass sheet, such as with petroleum jelly. After a few hours in light, a leaf from each plant is tested for starch, which should show that the plant with less carbon dioxide produces less starch.

△ Fig. 2.43 Light was excluded from all of the leaf on the left except an L-shaped window. After exposure to light, only the L-shape tests positive for starch. The whole of the leaf on the right was exposed to light.

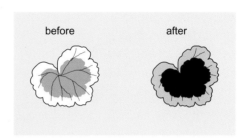

△ Fig. 2.44 Only the green parts of a variegated leaf can photosynthesise, as shown by the diagram on the right which shows the leaf after it has been tested for starch.

Measuring starch production is an indirect measurement of photosynthesis, because starch is made from the glucose produced in photosynthesis. You can investigate photosynthesis more directly by measuring the amount of oxygen produced. The oxygen is usually collected over water, and these investigations are most simply done using aquatic plants (plants that grow in water), such as pondweed, using the apparatus shown in the following box.

- To prove that photosynthesis produces oxygen, simply use the glowing splint test on the gas collected. The splint should re-ignite showing that the gas is oxygen.
- The investigation can be adjusted to test for the effect of light intensity on the rate of photosynthesis as described in the 'Developing investigative skills' box below.
- The investigation can also be adjusted to test for the effect of carbon dioxide concentration by adding different amounts of sodium hydrogen-carbonate (which effectively adds carbon dioxide) to the water and testing the rate at which bubbles of oxygen are produced.

Developing investigative skills

You can investigate the effect of light on photosynthesis by shining a light on a water plant and measuring how quickly bubbles are given off, as shown in Fig. 2.45.

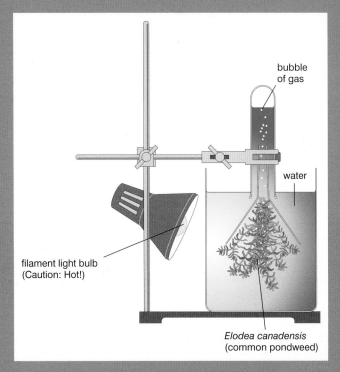

△ Fig. 2.45 The results below were gathered using this apparatus.

	Distance to lamp in cm				
	5	10	15	20	25
Gas bubbles given off in 5 minutes	67	57	40	20	4

Devise and plan investigations

❶ a) Explain why the rate of producing bubbles can be used as a measure of the rate of photosynthesis.

b) Explain how you would identify the gas produced by the plant.

Analyse and interpret data

❷ a) Use the data in the table to draw a suitable graph.

b) Describe and explain the shape of the graph.

Evaluate data and methods

❸ Light is not the only factor that can affect the rate of photosynthesis.

a) Which other factor might have had an effect on these measurements?

b) Suggest how the method could be changed to avoid this problem.

1. List three factors that affect the rate of photosynthesis.

2. Explain how each of these factors affects the rate of photosynthesis.

3. Explain as fully as possible why a variegated leaf tested for starch only causes the iodine/potassium iodide solution to turn from brown to blue-black where the leaf was green.

MINERAL IONS IN PLANT GROWTH

Photosynthesis produces carbohydrates, but plants contain many other types of chemical. Carbohydrates contain just the elements carbon, hydrogen and oxygen, but the amino acids that make up proteins also contain nitrogen. So plants need a source of nitrogen, which they take in in the form of nitrate ions.

Other chemicals in plants contain other elements: for example, chlorophyll molecules contain magnesium and nitrogen. Without a source of magnesium and nitrogen, a plant cannot produce chlorophyll and so cannot photosynthesise.

△ Fig. 2.46 A plant with nitrogen deficiency.

These additional elements are dissolved in water in the soil as **mineral ions**. Plants absorb the mineral ions through their roots, using active transport because the concentration of the ions in the soil is lower than in the plant cells.

Plants that are not absorbing enough mineral ions show symptoms of deficiency. For example, a plant with a nitrogen deficiency has stunted growth, and a plant with magnesium deficiency has leaves that are yellow between the veins, particularly in older leaves as the magnesium ions are transported in the plant to the new leaves.

△ Fig. 2.47 A plant with magnesium deficiency.

QUESTIONS

1. Explain why plants need a supply of mineral ions.

2. Describe the deficiency symptoms in a plant for the following mineral ions

   a) nitrogen

   b) magnesium.

3. Explain why plants show the deficiency symptoms for
   a) nitrogen and **b)** magnesium that you described in Question **2**.

Students set up an investigation into the effect of nitrates on plant growth. They chose two plants that were as similar as possible. Over two months, the plants received the same amount of heat, light, water and carbon dioxide. However, one plant was given a liquid nitrogen feed (liquid containing nitrate ions) in the water and the other only received distilled water. The images show the results at the end of two months.

1. Describe the differences between the two plants as fully as you can in terms of:

   a) the leaves

   b) the overall growth of the plants.

△ Fig. 2.48 Two plants raised with equal amounts of heat, light, water and carbon dioxide. The bigger one also got liquid nitrogen feed.

2. Explain the effect of nitrogen deficiency on the cells in the leaves.

3. Using your knowledge of plant nutrition, explain as fully as you can why these two plants grew differently.

4. Millions of tonnes of nitrogen-containing fertiliser are added to crop fields each year. Explain as fully as you can what would happen if this was not done.

Nutrition in humans

A HEALTHY DIET

To keep us healthy, humans need a diet that includes all the nutrients we need, such as:

- **proteins** – these are broken down to make amino acids, which are themselves used to form enzymes and other proteins needed by cells. Protein sources include eggs, milk and milk products (cheese, yoghurt, etc.), meat, fish, legumes (peas and beans), nuts and seeds.
- **carbohydrates** – these are needed to release energy in our cells, to enable all the life processes to take place. Good sources of carbohydrate include rice, bread, potatoes, pasta and yams.
- **lipids** – as fat deposited just below the skin, these form insulation to maintain body temperature. They are also used as a store of energy for times when the diet does not contain enough energy for daily needs. Fat is present in meat and lipids can also come in oils, milk products (butter, cheese), nuts, avocados and oily fish.
- **vitamins and minerals** – these substances are needed in tiny amounts for the correct functioning of the body. Vitamins and minerals cannot be produced by the body and cooking food destroys some vitamins. For example, vitamin C is best supplied by eating raw fruit and vegetables.

- **dietary fibre** – which is made up of the cell walls of plants. Good sources are leafy vegetables, such as cabbage, and the unrefined grains such as brown rice and wholegrain wheat. It adds bulk to food so that it can be easily moved along the digestive system by a process called peristalsis. This is important in preventing constipation. Fibre is thought to help prevent bowel cancer.
- **water** – the major constituent of the body of living organisms and is necessary for all life processes. Water is continually being lost through excretion and sweating, and must be replaced regularly through food and drink in order to maintain health. Most foods contain some water, but most fruit contain a lot of water.

Vitamins and mineral ions	Job	Good food source	Deficiency disease
Vitamin A	Helps cells to grow and keeps skin healthy, helps eyes to see in poor light	Liver, red and orange vegetables (e.g. carrots), butter and fish oil	Night blindness
Vitamin C	For healthy skin, teeth and gums, and keeps lining of blood vessels healthy	Citrus fruit and green vegetables, potatoes	Scurvy (bleeding gums and wounds do not heal properly)
Vitamin D	For strong bones and teeth	Fish, eggs, liver, cheese and milk	Rickets (softening of the bones)
Calcium	Needed for strong teeth and bones, and involved in the clotting of blood	Milk and eggs	Rickets (softening of the bones)
Iron	Needed to make haemoglobin in red blood cells	Red meats, liver and kidneys, leafy green vegetables such as spinach	Anaemia (reduction in number of red blood cells, person soon becomes tired and short of breath)

△ Table 2.2 Some vitamins and minerals.

◁ Fig. 2.49 A balanced diet contains appropriate amounts of all the essential nutrients. The best way to have a balanced diet is to eat a wide variety of foods including protein, fruit and vegetables and whole grains.

Vitamin D is not only taken in from the diet, it is also made naturally in the skin that is exposed to sunlight. It has been suggested that the paler skin of northern peoples is an adaptation to gathering more light in order to produce more vitamin D.

In the body, vitamin D is used to produce a hormone that controls the uptake of calcium from food. Lack of vitamin D results in softened bones – in children this produces rickets where the long bones of the legs, which take most of the body weight, bend outwards and create a characteristic bowed leg effect. Lack of vitamin D causes softened bones in adults too, and may have other effects although these are uncertain. Too much vitamin D in the diet can cause problems after several months.

△ Fig. 2.50 Rickets – vitamin D deficiency.

1. Explain why rickets is less common in countries nearer the Equator than it used to be in higher latitude countries such as the UK.

2. Rickets used to be common in higher latitude countries among children in poorer families. Suggest a reason for this relationship.

3. Children in higher latitude countries were given regular doses of vitamin D, in the form of fish liver oil, during the winter. Explain why this helped to prevent rickets.

4. Today, more people in higher latitude countries have diets that are less dependent on carbohydrates. Explain how this has reduced the risk of rickets.

5. Doctors in higher latitude countries are now finding vitamin D deficiency among women who wear clothing that covers all of their skin, and in people who eat a vegetarian diet. Describe and explain as fully as you can the reason and the treatment for this.

KWASHIORKOR

Kwashiorkor is a condition found in young children in areas where the diet contains very little protein. It typically occurs in children that were breast-fed, but quickly weaned after the birth of another baby in the family. Breast milk contains proteins, but after weaning, the child may be given a carbohydrate-based diet with few proteins. This diet also often lacks in some vitamins and minerals. Typical symptoms of kwashiorkor include swelling of the feet and abdomen, wasting muscles, thinning hair and loss of teeth. The liver is often damaged, and treatment requires careful adjustment of the diet so as not to damage the liver even more.

Δ Fig. 2.51 A child suffering from kwashiorkor, a protein deficiency.

A **balanced diet** contains all of these nutrients in the right proportions to stay healthy because we need more of some nutrients than of others. Since most foods contain more than one kind of nutrient, trying to work out what a balanced diet looks like can be difficult. Governments use images like this one, of food on a plate, to guide people on what proportions of food to eat.

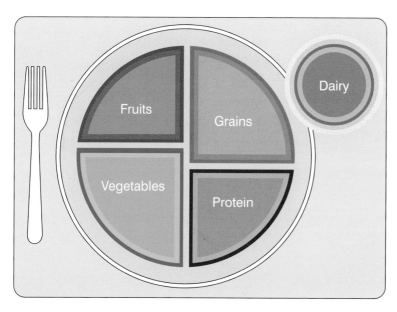

Δ Fig. 2.52 Guidance from the USDA (United States Department of Agriculture) on the proportions of different nutrients in a balanced diet.

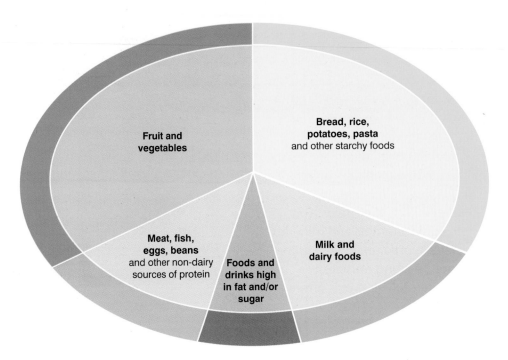

△ Fig. 2.53 Guidance from the UK government on the proportions of different nutrients in a balanced diet.

Different groups of people need different nutrients at different times in their lives, so this balance can change. For example, children need proportionately more protein than adults because they are still growing rapidly. Also, some groups of people have a greater need for a specific nutrient. During pregnancy, for example, women need more iron than usual, to supply what the growing baby needs for making blood cells.

Even with the right proportions of nutrients in our foods, we can still be eating an unhealthy diet. This is because many of our foods, particularly carbohydrates but also fats and proteins, can contribute to the energy our bodies need. If we eat food that supplies more energy than we use, the extra will be deposited as energy stores of fat. This can lead to obesity, which is related to many health problems such as heart disease and diabetes. Controlling the portion size at each meal, keeping between-meal snacks to a minimum, and increasing levels of exercise can help to reduce the risk of becoming overweight.

Energy requirements depend on body size, age, activity levels and pregnancy, as shown in the table.

| | Energy used in a day (kJ) | |
	Male	Female
6-year-old child	7 500	7 500
12–15-year-old teenager	12 500	9 700
Adult manual worker	15 000	12 500
Adult office worker	11 000	9 800
Pregnant woman	–	10 000

△ Table 2.3 Energy requirements for different people.

Developing investigative skills

Combustion (burning) of foods releases heat energy. The word equation for combustion is:

food + oxygen → carbon dioxide + water
(+ heat energy)

This reaction is similar to respiration inside cells, so we can use a combustion experiment to model the energy that is released from foods during respiration.

A food (potato chip/crisp) and leaf of a plant were tested in an investigation to see which released the most energy by combustion. Here are the results.

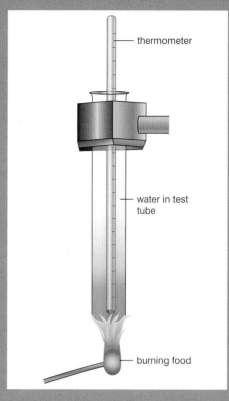

△ Fig. 2.54 Apparatus for measuring heat given off by burning food.

	Plant sample	
	Food	Leaf
Mass of sample in grams	22	12
Temperature of water after burning in °C	27	16
Temperature of water before burning in °C	15	15
Temperature rise in °C	12	1
Energy released by the sample in joules	1260	105

△ Table 2.4 Energy released by combustion.

Demonstrate and describe techniques

❶ a) Look at Fig. 2.54 and describe what happens during the experiment.

b) Identify any areas of safety that should have been considered and suggest how risks could be controlled.

Analyse and interpret data

❷ a) Use the results to calculate the energy released per gram of each sample.

b) Explain why you need to do this.

❸ Which sample released the most energy per gram?

❹ Suggest why some animals that eat the leaves of plants for most of the year change to eating seeds (nuts) when they are available.

Evaluate data and methods

❺ The apparatus shown in Fig. 2.54 does not give accurate results for the amount of energy in the burning material. Explain why, and suggest a method that would increase the accuracy of the results.

QUESTIONS

1. In a balanced diet, apart from water, which three groups of food molecules do we need most of?

2. Look at the advice on a balanced diet in Figs. 2.52 and 2.53.

   a) According to each country's advice, which two types of foods are needed in the greatest amounts in a healthy balanced diet?

   b) Which nutrients do these types of food provide?

   c) Which group in Fig. 2.53 is the smallest?

   d) Suggest why you are recommended to eat less of this group.

3. Explain why different groups of people need different amounts of nutrients. Give examples in your answer.

4. Explain why a healthy diet needs to consider energy as well as nutrients.

THE HUMAN DIGESTIVE SYSTEM

The digestive system

Eating food involves several different processes:

- **ingestion** – taking food into the body (through the mouth in humans)
- **digestion** – breaking down of large food molecules into smaller molecules
- **absorption** of digested food molecules into the blood
- **egestion** – removal of undigested material (**faeces**) from the body.

All these different processes take place in different parts of the **alimentary canal**.

The alimentary canal is a continuous tube through the body, from the mouth, where food is ingested, through the **oesophagus**, stomach, small intestine (**duodenum** and **ileum**), large intestine (**colon** and **rectum**), to the anus, where faeces is egested. You could say that materials in the alimentary canal are not truly in the body. Not until food molecules are absorbed do they cross cell membranes into body tissue. Then they can be used and waste products *excreted* through other organs.

The digestive system includes the alimentary canal and the other organs that contribute to digestion, such as the liver, **pancreas** and gall bladder. The table describes the functions of each of the organs in the digestive system.

Part of digestive system	What happens there
Mouth	Teeth and tongue break down food into smaller pieces. Saliva from salivary glands moistens food so it is easily swallowed and contains the enzyme amylase to begin breakdown of starch.
Oesophagus (or gullet)	Each lump of swallowed food, called a bolus, is moved from the mouth to the stomach by waves of muscle contraction called **peristalsis**.
Stomach	Food enters through a ring of muscle known as a sphincter. Acid and protease enzymes are secreted to start protein digestion. Movements of the muscular wall churn up food into a liquid known as chyme (pronounced 'kime'). The partly digested food passes a little at a time through another sphincter into the small intestine.
Liver	Cells in the liver make bile. Amino acids not used for making proteins are broken down to form urea which passes to the kidneys for excretion. Excess glucose is removed from the blood and stored as glycogen in liver cells.
Gall bladder	Stores bile from the liver. The bile is passed along the bile duct into the small intestine where it neutralises the stomach acid in the chyme and emulsifies lipids.
Pancreas	Secretes amylase, lipase and protease enzymes, as well as sodium hydrogen-carbonate, into the small intestine.
Small intestine (made up of duodenum and ileum)	Secretions from the gall bladder and pancreas as well as further enzymes from the wall of the small intestine complete digestion. Digested food is absorbed into the blood through the villi.
Large intestine (made up of colon and rectum)	In the colon, water and some vitamins are absorbed from the remaining material. In the rectum the remaining material (**faeces**), made up of indigestible food, dead cells from the lining of the alimentary canal and bacteria, is compacted and stored.
Anus	Faeces is egested through a sphincter.

Δ Table 2.5 What happens in the digestive system.

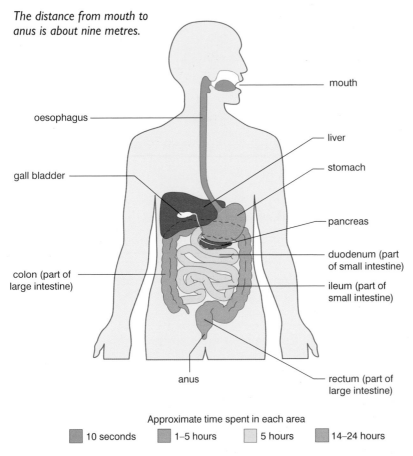

The distance from mouth to anus is about nine metres.

mouth

oesophagus

liver

stomach

gall bladder

pancreas

duodenum (part of small intestine)

colon (part of large intestine)

ileum (part of small intestine)

anus

rectum (part of large intestine)

Approximate time spent in each area

■ 10 seconds   ■ 1–5 hours   ■ 5 hours   ■ 14–24 hours

△ Fig. 2.55 The human digestive system.

Food moves along the alimentary canal, or gut, because of the contractions of the muscles in the walls of the alimentary canal. This is called peristalsis. Fibre in the food keeps the bolus bulky and soft, making peristalsis easier.

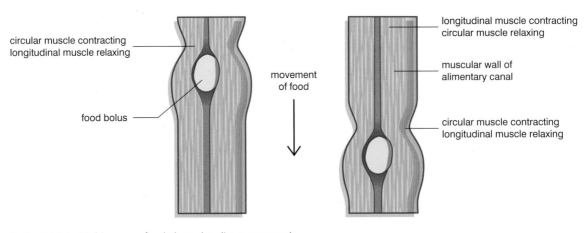

circular muscle contracting longitudinal muscle relaxing

movement of food

longitudinal muscle contracting circular muscle relaxing

muscular wall of alimentary canal

food bolus

circular muscle contracting longitudinal muscle relaxing

△ Fig. 2.56 Peristalsis moves food along the alimentary canal.

1. Sketch the diagram of the digestive system shown in Fig. 2.55. Label the organs and add notes to each organ to explain its function in the system.

2. Explain the difference between egestion and excretion.

3. Explain how the muscles of the alimentary canal wall move food.

DIGESTION

If food is to be of any use to us, it must enter the blood so that it can travel to every part of the body.

Many of the foods we eat are made up of large, insoluble molecules that cannot cross the wall of the alimentary canal and the cell membranes of cells lining the blood vessels. This means they have to be broken down into small, soluble molecules that can easily cross cell membranes and enter the blood. Breaking down the molecules is called digestion.

There are two types of digestion.

1. **Physical (or mechanical) digestion** occurs mainly in the mouth, where food is broken down physically into smaller pieces by the teeth and tongue, and in the stomach, where food is churned by the muscular action of the stomach wall.

2. **Chemical digestion** is the breakdown of large food molecules into smaller ones using digestive enzymes.

Some molecules, such as glucose, vitamins, minerals and water, are already small enough to pass through the gut wall and do not need to be digested.

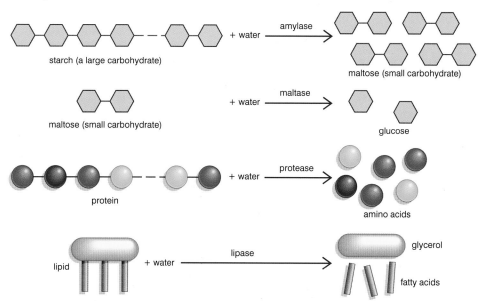

Δ Fig. 2.57 Each food group is digested by specific digestive enzymes.

Chemical digestion happens because of chemicals called enzymes. Enzymes are a type of catalyst found in living things, which speed up the rate of reactions in the body.

Every cell contains many enzymes, which control the many chemical reactions that happen inside it. Digestive enzymes are a group of enzymes that are produced in the cells lining parts of the digestive system and are **secreted** (released) into the alimentary canal to mix with the food.

The digestive enzymes include carbohydrases that break down carbohydrates, proteases that break down proteins and lipases that break down lipids. (Note: the *–ase* at the end of the name means it is an enzyme, and the first part usually names the substrate that the enzyme works on.)

Each of these groups contains enzymes that work on different substrates. For example, the carbohydrases include amylase, which digests starch into the smaller carbohydrate maltose, and maltase, which digests maltose to glucose. Fig. 2.57 shows more details of the digestion of food by digestive enzymes.

SCIENCE IN CONTEXT — LACTOSE

Lactose is the sugar in milk, which is broken down in the alimentary canal by the enzyme lactase to the simple sugars glucose and galactose.

Like all young mammals, human babies produce lactase, which helps them to digest the lactose in breast milk. In most mammals the production of lactase decreases as the young mature, because the adult diet does not include milk. This also happens in adults from many cultures where adults generally do not drink milk, such as in Southeast Asia. However, there are cultures in Europe, India and parts of East Africa, where mammals such as sheep, goats or cattle are kept to supply meat and milk for food. In these human groups the adults continue to produce lactase and are able to digest the lactose in milk. Adults who cannot do this are *lactose intolerant*. In these people, bacteria in the alimentary canal break down the lactose, producing gas which causes great discomfort.

OTHER CHEMICALS IN DIGESTION

As you learned earlier, different enzymes work better in different conditions. Those enzymes that digest food in the stomach work best in acid conditions. Special cells in the lining of the stomach secrete hydrochloric acid into the stomach to create the right conditions for the enzymes. The acid is also helpful in killing any microorganisms taken in with the food.

In the small intestine, the enzymes from the pancreas work best in slightly alkaline conditions. So the acid chyme that enters the intestine from the stomach has to be neutralised. **Bile** is a substance produced by cells in the liver. It is stored in the gall bladder until it is needed and then passes along the bile duct into the small intestine. Bile is highly alkaline, so it neutralises the acid from the stomach and makes the digesting food slightly alkaline.

Bile is also important in the digestion of lipids. Lipids do not mix well with aqueous (water-based) mixtures such as the digesting food, and so remain as large droplets. This produces a small surface area for lipase enzymes to work on, which slows down the rate of digestion. Bile **emulsifies** lipids, breaking them up into much smaller droplets, so that the rate of digestion is much faster.

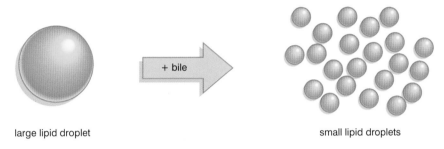

large lipid droplet   small lipid droplets

Δ Fig. 2.58 Bile lowers the surface tension of large lipid droplets so that they break up. This part of the digestive process is called emulsification.

There are many stages in the breakdown of large molecules into their basic units during digestion, each stage requiring a different enzyme; so, although the stomach and the pancreas both produce protease enzymes, these are different proteases. Those in the stomach break down the very large protein molecules into smaller ones that contain only a few amino acids, and those from the pancreas break down the smaller chains into single amino acids.

QUESTIONS

1. Describe the difference between chemical and physical digestion.

2. Explain why enzymes are needed in digestion.

3. a) Which enzyme has starch as its substrate?

   b) Which product is formed by the digestion of starch by this enzyme?

4. Describe the roles of **a)** stomach acid and **b)** bile in digestion.

ABSORPTION OF FOOD

After digestion, small food molecules can diffuse across the intestine wall and be absorbed into the body. The small intestine is over 6 m long in an adult human, but to increase the rate of transport of food molecules across the intestine wall, the surface area of the small intestine wall is increased by millions of finger-like projections called **villi** (singular: villus). The surface area of the cells lining the villi is increased even further by tiny **microvilli**. The combination of all these factors means that the surface area for absorption in the small intestine in an adult human is about 250 m², about the size of a tennis court.

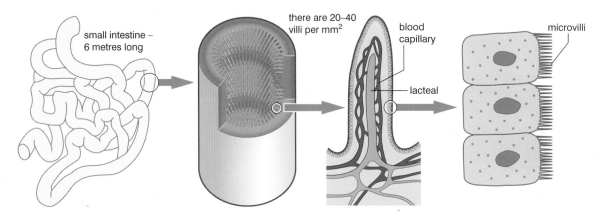

△ Fig. 2.59 The structure of the small intestine.

The villi have other adaptations that help to increase the rate of diffusion.

- They are covered in a thin layer of cells, so that digested food molecules do not have to travel far to be absorbed into the body and into the blood in the capillaries in the villi.
- They are well supplied with blood capillaries, taking absorbed food molecules from the small intestine to the rest of the body and supplying fresh blood – this keeps the concentration gradient between the digested food in the intestine and the cells in the body as high as possible.
- Villi also contain lacteals (part of the lymphatic system) that carry lipid droplets separate from the rest of the food molecules because lipids do not dissolve well in blood.

QUESTIONS

1. Describe all the features of the small intestine that help to maximise the rate of diffusion of digested food molecules into the body.

2. Why is it important to maximise the rate of diffusion of digested food into the body?

End of topic checklist

Bile is a liquid, produced by the liver and stored in the gall bladder, which is highly alkaline and emulsifies lipids.

Chemical digestion is the breakdown of large molecules into smaller ones using enzymes.

Chlorophyll is the green chemical in chloroplasts that captures light energy for photosynthesis.

To **emulsify** is to break up the large droplets of a lipid in an aqueous solution into smaller droplets.

The **epidermis** is the layer of cells on the outer surface of a body or organ, such as a leaf.

Faeces is the undigested material that remains after the digestion and absorption of food in humans.

Fibre is a plant material that is difficult to digest and keeps the food in the alimentary canal soft and bulky, aiding peristalsis.

Ingestion is the taking of food into the alimentary canal.

Microvilli are tiny finger-like extensions of the cell membrane of the surface cells of villi.

Minerals (**mineral ions**) are nutrients that plants and animals need in small amounts, such as nitrates that are needed for making amino acids.

Palisade cell are cells in the upper part of a leaf that contain the most chloroplasts and carry out most of the photosynthesis.

Physical digestion is the breakdown of large food pieces, such as by chewing in the mouth.

Secretion means releasing chemicals that have been made inside a cell into the fluid outside the cell.

The **spongy mesophyll** is the layer of cells in the lower part of the leaf where there are many air spaces, so increasing the internal surface area to volume ratio.

Stomata (singular **stoma**) are tiny holes in the surface of a leaf (mostly the lower epidermis) which allow gases to diffuse in and out of the leaf.

Villus (plural **villi**) is a finger-like projection of the small intestine wall where absorption of digested food molecules occurs.

Vitamins are nutrients needed by the body in tiny amounts to remain healthy, such as vitamins A, C and D.

Waste products are products of a chemical reaction that are not needed, such as oxygen in photosynthesis.

The facts and ideas that you should know and understand by studying this topic:

○ Photosynthesis is the process that takes place in chloroplasts in plant cells, which converts light energy into chemical energy.

○ The equation for photosynthesis is:

$$\text{carbon dioxide} \ + \ \text{water} \ \xrightarrow[\text{light energy}]{\text{chlorophyll}} \ \text{glucose} \ + \ \text{oxygen}$$

It can also be summarised as a balanced symbol equation:

$$6CO_2 \ + \ 6H_2O \ \xrightarrow[\text{light energy}]{\text{chlorophyll}} \ C_6H_{12}O_6 \ + \ 6O_2$$

○ The rate of photosynthesis increases as:

- carbon dioxide concentration increases
- light intensity increases
- temperature increases up to an optimum temperature, after which it decreases as enzymes denature.

○ Leaves are adapted to maximise the rate of photosynthesis in different ways:

- they are broad and thin
- the palisade cells where most photosynthesis takes place are near the upper surface of the leaf
- the epidermis is transparent to let as much light through as possible
- spongy mesophyll maximises the internal surface area for exchange of gases
- stomata allow gases to diffuse into and out of the leaf
- xylem vessels transport water to the leaf
- phloem tissue transports sugars away from the leaf.

○ Plants need mineral ions, such as magnesium and nitrate ions, to convert the sugars from photosynthesis into other essential substances such as chlorophyll, which contains magnesium, and amino acids, which contain nitrogen.

○ The main components of a healthy human diet are: carbohydrates, proteins, lipids, vitamins (such as A, C and D), minerals (such as calcium and iron), water and dietary fibre.

○ A balanced diet includes all the components needed for health in the right proportions.

○ The diet provides energy as well as nutrients, and different groups of people have different energy requirements: for example, very active people need more energy in their food than people who are not active.

○ The human alimentary canal is made up of the mouth, oesophagus, stomach, small intestine (duodenum and ileum) and large intestine (colon and rectum).

○ Ingestion takes place in the mouth.

○ Physical digestion (by chewing) and chemical digestion (by amylase) starts the process of digestion (breakdown) of food in the alimentary canal.

○ Food is moved through the alimentary canal (gut) by peristalsis.

○ Food is further digested in the stomach where acid and protease enzymes are added.

○ Digestion is completed in the small intestine where enzymes from the pancreas and wall of the intestine digest the food into small enough molecules for absorption.

○ Starch is broken down to maltose by amylase, and then maltose is broken down to glucose by maltase; proteins are broken down to amino acids by proteases; lipids are broken down to glycerol and fatty acids by lipase.

○ Bile from the liver, stored in the gall bladder until needed, is added to the small intestine to neutralise the acid from the stomach and emulsify lipids.

○ Digested food molecules are absorbed in the small intestine where the villi and microvilli provide an enormous area for absorption.

○ Undigested food remaining in the alimentary canal passes to the large intestine, where water is reabsorbed into the body and the rest egested through the anus.

End of topic questions

1. **a)** Describe the process of photosynthesis. **(2 marks)**

   b) Explain the role of photosynthesis in the nutrition of plants. **(2 marks)**

   c) Explain the importance of photosynthesis for animal life on Earth. **(2 marks)**

2. Identify the organs of the digestive system involved, and their roles, in each of the following processes:

   a) ingestion **(2 marks)**

   b) digestion **(2 marks)**

   c) absorption **(2 marks)**

   d) egestion. **(2 marks)**

3. Using what have you learned about the effect of concentration gradient and surface area to volume ratio, explain the adaptations of a leaf for photosynthesis. **(4 marks)**

4. State sources in the diet, and describe the importance of the following, in a healthy diet:

   a) vitamins A, C and D **(6 marks)**

   b) the minerals calcium and iron **(4 marks)**

   c) water **(2 marks)**

   d) dietary fibre. **(2 marks)**

5. **a)** Why is the presence of starch in a leaf an indicator that photosynthesis has been taking place? **(1 mark)**

   b) Describe how to destarch a plant before carrying out investigations into photosynthesis in plants, and explain why this is important for the reliability of the investigation. **(3 marks)**

6. There is an old saying that you should chew your food 100 times before swallowing to help look after your stomach. Explain why chewing food well helps digestion. **(1 mark)**

7. Use what you learnt about the effect of different factors on the rate of diffusion to explain the following adaptations in the small intestine:

   a) presence of villi and microvilli **(2 marks)**

   b) single layer of epithelial cells on villi **(2 marks)**

   c) extensive blood supply to villi. **(2 marks)**

8. This is the diet schedule for a male Olympic athlete training for a competition, not including drinks during training.

Breakfast	large bowl of cereal, such as porridge or muesli
	250 ml semi-skimmed milk plus chopped banana
	1–2 thick slices wholegrain bread with olive oil or sunflower spread and honey or jam
	glass of fruit juice plus 1 litre fruit squash
Post-training 2nd breakfast	portion of scrambled eggs
	portion of baked beans
	1–2 rashers grilled lean bacon
	portion of grilled mushrooms or tomatoes
	2 thick slices wholegrain bread with olive oil spread
	1 litre fruit squash
Lunch	pasta with bolognese or chicken and mushroom sauce
	mixed side salad
	fruit
	1 litre fruit squash
Post-training snack	4 slices toast with olive oil or sunflower spread and jam
	large glass of semi-skimmed milk
	fruit
	500 ml water
Dinner	grilled lean meat or fish
	6–7 boiled new potatoes, large sweet potato or boiled rice
	large portion of vegetables, e.g. broccoli, carrots, corn or peas
	1 bagel
	1 low-fat yoghurt and 1 banana or other fruit
	750 ml water and squash
Bedtime snack	low-fat hot chocolate with 1 cereal bar

a) Identify the foods that contribute to each of these food types: **i)** carbohydrates, **ii)** proteins, **iii)** lipids, **iv)** vitamins and minerals, **v)** dietary fibre. **(5 marks)**

b) Which food type is most represented in this diet? **(1 mark)**

c) Explain why this food type is so important in this diet. **(1 mark)**

d) Which food group would you expect to be more represented in an athlete's diet in the early stages of training? Explain your answer. **(2 marks)**

e) Explain why this diet is not suitable for most people. **(2 marks)**

9. Sketch the axes of a graph with time of day along the *x* axis and rate of photosynthesis on the *y* axis. The units on the *x* axis should start at midnight on one day and end at midnight on the following day. Add an arrowhead at the top of the *y* axis to show that the units are arbitrary (have no values) but increase as you go up the axis.

a) Draw a line on your axes to show how the rate of photosynthesis might change during the day for a large tree. **(2 marks)**

b) Annotate your graph to explain which factor (or factors) may be limiting the rate of photosynthesis at different times of the day. **(2 marks)**

10. A farmer is growing tomato plants in a glasshouse. Explain as fully as you can why she might do the following:

a) leave lights on in the glasshouse all night **(1 mark)**

b) close the glasshouse windows at night but open them during the day **(1 mark)**

c) add a liquid feed containing nitrate and magnesium ions to the water for the plants. **(1 mark)**

11. Look at Figs 2.52 and 2.53.

a) Outline the main differences between the diagrams. **(2 marks)**

b) Suggest why the diagrams are different. **(1 mark)**

c) Which diagram do you think is more useful? Explain your answer. **(2 marks)**

Respiration

INTRODUCTION

In general, when we talk about respiration, we often mean breathing (or ventilation). However, in this topic we will look at **cellular respiration**, which is the release of energy from the chemical bonds in food molecules such as glucose. This happens inside all living cells, and most cellular respiration takes place in tiny structures called mitochondria. Cells where a lot of respiration is carried out, such as muscle cells, have many more mitochondria than cells where only a little respiration is carried out, such as in skin cells.

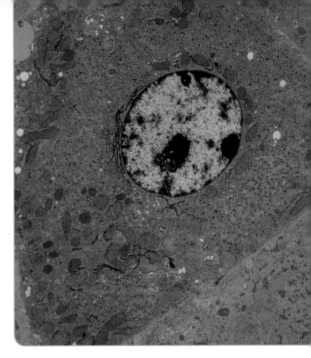

△ Fig. 2.60 This cell contains many mitochondria, which have been coloured red in the image.

KNOWLEDGE CHECK

✓ Organisms need energy for all the life processes that keep them alive.
✓ Plants get this energy from the sugars they make in photosynthesis.
✓ Animals get this energy from their food.

LEARNING OBJECTIVES

✓ Understand how the process of respiration produces ATP in living organisms.
✓ Know that ATP provides energy for cells.
✓ Describe the differences between aerobic and anaerobic respiration.
✓ Know the word equation and the balanced chemical symbol equation for aerobic respiration in living organisms.
✓ Know the word equation for anaerobic respiration in plants and in animals.
✓ Practical: Investigate the evolution of carbon dioxide and heat from respiring seeds or other suitable living organisms.

REMEMBER

Be clear in your answers to distinguish between *cellular respiration* and the word *respiration* when used to mean breathing (ventilation).

AEROBIC RESPIRATION

Cellular respiration releases energy from digested and absorbed food molecules, such as glucose, for all the life processes in the body. For example, energy from respiration is used:

- to keep warm
- to enable muscles to contract
- to build up large molecules from small ones
- in the active transport of substances across cell membranes.

The food is usually glucose (sugar), but other kinds of food molecule can be used if there is not enough glucose available. When food is broken down in respiration a substance called **ATP** (adenosine triphosphate) is made. It is the ATP that directly provides the energy that cells need.

Aerobic respiration is cellular respiration using oxygen. The oxygen needed for aerobic respiration comes from the air (except for a small proportion in photosynthesising plants, which comes from photosynthesis). The carbon dioxide from cellular respiration is released into the air, and the water is either used in the body or excreted through the kidneys.

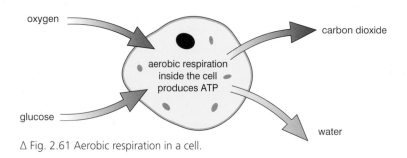

oxygen

carbon dioxide

aerobic respiration inside the cell produces ATP

glucose

water

Δ Fig. 2.61 Aerobic respiration in a cell.

SCIENCE IN CONTEXT

WATER FROM RESPIRATION

A camel can survive for many days without drinking liquid water, which means it survives well in desert conditions. The camel's hump is not a store of water, but a store of fat. Over a long period without food, the fat is broken down by aerobic respiration. Since water is one of the products of aerobic respiration, this also helps the camel to survive longer without drinking water.

Going for days without food and drinking water would kill a human in a few days, because we do not metabolise fat as well as the camel does, or retain water as well. So, before setting out into the desert for a long trip, make sure your camel has a large hump (and you have plenty of food and water).

Δ Fig. 2.62 An Arabian camel (dromedary) can survive for many days in the desert with little food, because fat is stored in its hump.

Most plant and animal cells use aerobic respiration. Water and carbon dioxide are produced as waste products. This is very similar to burning fuel, except that enzymes control the process in our bodies.

Aerobic respiration can be summarised by a word equation:

glucose + oxygen → water + carbon dioxide (+ ATP)

It can also be written as a symbol equation:

$C_6H_{12}O_6 + 6O_2 \rightarrow 6H_2O + 6CO_2$ (+ ATP)

During aerobic respiration, many of the chemical bonds in the glucose molecule are broken down. The ATP produced releases a lot of energy: around 2900 kJ of energy are released for each mole of glucose molecules used in aerobic respiration.

Developing investigative skills

We can use the apparatus in Fig. 2.63 to investigate what happens to seeds as they germinate.

Demonstrate and describe techniques

❶ What changes will this apparatus measure? Explain your answer.

Analyse and interpret data

❷ The results from this investigation are shown in the table.

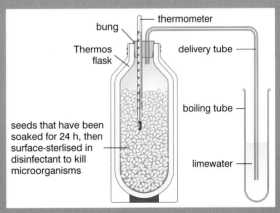
△ Fig. 2.63 Investigating germinating seeds.

Start	Appearance of limewater	Clear
	Temperature in flask in °C	15
After 2 days	Appearance of limewater	Milky
	Temperature in flask in °C	19

a) Describe the changes over 2 days.

b) Explain, as fully as you can, the changes over 2 days.

Evaluate data and methods

❸ What control would you need to carry out to show that it was the seeds that caused these changes? Explain your answer.

QUESTIONS

1. Where does respiration take place in the body?

2. a) Write out the word equation for aerobic respiration.

b) Annotate your equation to show where the reactants come from.

c) Annotate your equation to show what happens to the products of the reaction in a human.

d) Describe how your answer to part **c)** might differ for a camel on a long journey without water, and explain your answer.

3. Explain why it is called *aerobic* respiration.

Anaerobic respiration

Aerobic respiration supplies most of the energy that plant cells and animal cells need most of the time. However, there are times when not enough oxygen is available for aerobic respiration to be carried out fast enough to deliver all the ATP and energy that is needed, for example:

△ Fig. 2.64 Diving animals use anaerobic respiration to help stay under water for a long time.

- in diving animals, such as whales and seals
- inside parts of plants where diffusion of oxygen is too slow for aerobic respiration, such as inside seeds or in root cells in waterlogged ground
- in muscle cells, when vigorous exercise requires more energy than can be provided by an increased supply of oxygen from deeper faster breathing and a faster heart rate.

In these cases, the additional energy needed is supplied by **anaerobic respiration**. This kind of respiration also releases energy from glucose molecules, but without the need for oxygen.

In anaerobic respiration, the glucose molecule is only partly broken down, so far less ATP and therefore far less energy is released from each glucose molecule in anaerobic respiration compared with aerobic respiration. Only about 150 kJ is produced from every mole of glucose molecules respired anaerobically in an animal cell.

Anaerobic respiration in animals

When animal cells respire anaerobically, such as muscle cells during vigorous exercise, the glucose is broken down to lactic acid:

glucose → lactic acid (+ ATP)

Note that, even when a muscle cell is respiring anaerobically, aerobic respiration is also taking place and using all the oxygen that is available. Where aerobic respiration cannot supply all the energy needed, only the additional energy needed comes from anaerobic respiration.

RESPIRATION IN ATHLETICS

If you watch carefully, you may see that a sprint athlete does not breathe during a race. At the start of the race there will be some oxygen in their muscle cells, but this is rapidly used up as they start running. Anaerobic respiration provides virtually all of the energy used in a 100 m sprint by a well-trained athlete.

Sprinting cannot be maintained for long, because the muscle cells also need a rapid supply of glucose for respiration, and the build up of lactic acid causes muscle fatigue and pain. Longer distance races are managed using a combination of aerobic and anaerobic respiration. In marathons, most of the race is run aerobically, with only the last stretch being managed as a sprint using anaerobic respiration.

△ Fig. 2.65 A fit athlete in the middle of a sprint is using almost entirely anaerobic respiration.

Anaerobic respiration in plants

Sometimes plant cells cannot get enough oxygen for aerobic respiration to proceed fast enough, for example, inside a germinating seed. They use a different reaction for anaerobic respiration compared with animals, where the product is ethanol:

glucose → ethanol + carbon dioxide (+ ATP)

Fungi also respire anaerobically like this. Breadmaking makes use of this when we use yeast: the carbon dioxide forms bubbles in the dough, making the bread light and spongy.

▽ Fig. 2.66 Deep inside the germinating seed, energy is needed to start all the reactions that will result in cell division and growth. But there is little oxygen here, so anaerobic respiration is important.

Aerobic respiration	Anaerobic respiration
Oxygen needed	No oxygen needed
Glucose completely broken down	Incomplete breakdown of glucose
End products: carbon dioxide and water	End products: Animal cells: lactic acid Plant cells and yeast: carbon dioxide and ethanol
Lots of ATP/energy released	Relatively small amount of ATP / energy released

△ Table 2.6 Comparing aerobic and anaerobic respiration.

SCIENCE IN CONTEXT

EPOC

During anaerobic respiration, the concentration of lactic acid builds up in the cells until oxygen is available again (which is what happens at the end of exercise). It is then converted back to glucose, for use later, or broken down fully using oxygen to carbon dioxide and water. The additional oxygen needed after exercise used to be called the **oxygen debt**, but it is now understood that additional oxygen is needed after any prolonged exercise, even if completely aerobic, to return many processes in the body back to their resting state. This need, which you can feel when you breathe more deeply after exercise, is now called *EPOC* (excess post-exercise oxygen consumption).

△ Fig. 2.67 Runner recovering from post-exercise oxygen consumption.

QUESTIONS

1. Describe the differences between aerobic respiration and anaerobic respiration.

2. Explain why muscle cells sometimes need to respire anaerobically.

3. Describe the similarities and differences between anaerobic respiration in plant cells and in animal cells.

4. Give one example of where a plant cell respires anaerobically, and explain why it uses this form of respiration.

End of topic checklist

Aerobic respiration is respiration (the breakdown of glucose to release ATP) using oxygen.

Anaerobic respiration is respiration without oxygen. In animal cells it produces lactic acid; in plant cells it produces ethanol and carbon dioxide.

Respiration produces **ATP**, which then provides the energy that cells need.

The facts and ideas that you should know and understand by studying this topic:

○ Respiration is the process in which energy is released from the chemical energy stored in molecules in the cells of living organisms.

○ Respiration uses energy stored in food molecules to build up molecules of ATP, which provide energy to cells.

○ In aerobic respiration, glucose is broken down using oxygen from the air:

glucose + oxygen → carbon dioxide + water (+ ATP)

$C_6H_{12}O_6 + 6O_2 \rightarrow 6CO_2 + 6H_2O$ (+ ATP)

○ Anaerobic respiration is when energy is needed that cannot be supplied by aerobic respiration, usually when not enough oxygen is available.

○ In anaerobic respiration in animal cells, glucose is broken down to lactic acid:

glucose → lactic acid (+ ATP)

○ In anaerobic respiration in plant cells, glucose is broken down to ethanol:

glucose → ethanol + carbon dioxide (+ ATP)

○ Aerobic respiration produces far more molecules of ATP and so releases much more energy from each glucose molecule than anaerobic respiration.

End of topic questions

1. List the body systems in a human that are involved in supplying the reactants, and removing the products, of cellular respiration. **(5 marks)**

2. Draw up a table to summarise the similarities and differences between aerobic and anaerobic respiration. **(8 marks)**

3. Students were studying the results of respiration in some seeds. They set up two identical sets of apparatus, of a boiling tube fitted with a bung and linked to a tube of limewater through a delivery tube. They placed some germinating peas in one boiling tube, and some peas that had been germinated and then boiled in the other tube. The boiling tubes were fitted with their bungs so that no additional air could enter the apparatus and then left overnight.

   a) Suggest what happened to the limewater in the two sets of apparatus. **(2 marks)**

   b) Explain as fully as you can your answer to part **a)**. **(4 marks)**

4. A whale takes a deep breath of air and then dives for half an hour. Suggest how energy is generated in the whale's muscles over the period of the dive. **(3 marks)**

5. Eight chocolate biscuits contain the equivalent of 1 mole of glucose. One of these biscuits, respired aerobically, would provide the energy needed for about one hour by a human body at rest.

   a) How long would eight biscuits provide energy if you were sitting reading and respiring aerobically? **(1 mark)**

   b) How long would eight biscuits provide energy for sitting reading if you were respiring anaerobically. **(2 marks)**

   c) Explain the difference in your answers for parts **a)** and **b)**. **(4 marks)**

Gas exchange

INTRODUCTION

Respiration and photosynthesis need gases from the air (or the water for aquatic organisms), and they produce gases that need to be returned to the environment. These gases must get into and out of the body fast enough to support the rate at which life processes need to work. For single-celled organisms, this is not a problem. They have a large surface area to volume ratio and diffusion across the cell membrane can supply the gases at a fast enough rate. Larger organisms cannot do this. Not only do they have a much smaller surface area to volume

△ Fig. 2.68 The axolotl is a relative of frogs and newts. All its life, it has external gills through which it exchanges gases with the water.

ratio, which slows the rate of diffusion, many of them live on land where the delicate surface for gas exchange would dry out. Different groups of organisms have different solutions to these problems. Plants exchange gases inside the leaf; insects have internal tubes (tracheal system); fish have gills; and many vertebrates have lungs.

KNOWLEDGE CHECK

- ✓ Animals and plants take in oxygen from the air and give out carbon dioxide from respiration.
- ✓ Humans use lungs for breathing.
- ✓ During the day, plants take in carbon dioxide and give out oxygen as a result of photosynthesis.

LEARNING OBJECTIVES

- ✓ Understand the role of diffusion in gas exchange.
- ✓ Understand gas exchange (of carbon dioxide and oxygen) in relation to respiration and photosynthesis.
- ✓ Understand how the structure of the leaf is adapted for gas exchange.
- ✓ Describe the role of stomata in gas exchange.
- ✓ Understand that respiration continues during the day and night, but that the net exchange of carbon dioxide and oxygen depends on the intensity of light.
- ✓ Practical: Investigate the effect of light on net gas exchange from a leaf, using hydrogen-carbonate indicator.
- ✓ Describe the structure of the thorax, including the ribs, intercostal muscles, diaphragm, trachea, bronchi, bronchioles, alveoli and pleural membranes.
- ✓ Understand the role of the intercostal muscles and diaphragm in ventilation.
- ✓ Explain how alveoli are adapted for gas exchange by diffusion between air in the lungs and blood in capillaries.

✓ Understand the biological consequences of smoking in relation to the lungs and the circulatory system, including coronary heart disease.
✓ Practical: Investigate breathing in humans, including the release of carbon dioxide and the effect of exercise.

DIFFUSION

Gas exchange in both plants and humans depends on diffusion. In plants, gases enter and leave the leaf by diffusion through the stomata. In humans, gases pass between the lungs and the blood by diffusion across the thin walls of the alveoli. Remember the factors that affect the rate of diffusion (see pages 63–64) as you learn about gas exchange.

Gas exchange in flowering plants

PHOTOSYNTHESIS AND RESPIRATION

All organisms use cellular respiration to release the chemical energy in food molecules such as glucose. Cellular (usually aerobic) respiration must continue all the time because organisms need energy all the time for other life processes. So animal and plant cells are continually taking in oxygen and releasing carbon dioxide for respiration.

Plants photosynthesise as well as respire. For photosynthesis, plant cells need to take in carbon dioxide and release oxygen. However, photosynthesis can only take place when there is sufficient light.

At night, plants do not photosynthesise, but they do continue respiring. So plants give out carbon dioxide and take in oxygen. At daybreak, as light intensity increases, the rate of photosynthesis increases.

At a particular light intensity, the amount of oxygen produced by photosynthesis will balance the amount used by the plant in respiration, and the *net production of oxygen* will be zero. In the same way, the net production of carbon dioxide is also zero. The point at which photosynthesis and respiration use and produce equal amounts of oxygen and carbon dioxide is called the **compensation point**.

As daylight increases, oxygen production from photosynthesis exceeds its use in respiration, and the opposite for carbon dioxide. This continues until light intensity decreases, when the sun sets and a second compensation point is reached.

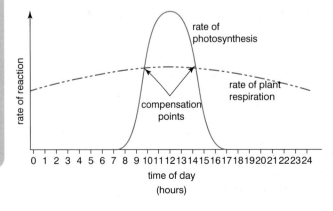

◁ Fig. 2.69 Comparison of the rate of photosynthesis and rate of respiration in a plant over one day.

SCIENCE IN CONTEXT

COMPENSATION POINTS IN A FOREST

The trees in the forest spread their branches wide in the canopy. The leaves on the branches create a lot of shade below them, limiting how much light can reach the forest floor. Plants growing near the forest floor may not only have large leaves, to capture as much as possible of the light that reaches them, they may also have lower compensation points than the leaves on the trees. This means they can start photosynthesising faster than they respire at lower light intensities.

△ Fig. 2.70 Low-growing plants in a forest.

Since plants can only grow when they are photosynthesising faster than they are respiring, this is an important adaptation for living in shaded places. Of course, where there is too little light reaching the forest floor, no plants grow because they would never be able to photosynthesise fast enough to produce new cells and tissues.

QUESTIONS

Use the graph of rate of photosynthesis and respiration (in Fig. 2.69) to help you answer these questions.

1. Within a period of 24 hours, when do plant cells respire? Explain your answer.

2. Within a period of 24 hours, when do plant cells photosynthesise? Explain your answer.

3. What is a compensation point?

4. During which time period is the plant producing more sugars in photosynthesis than it is using in respiration? Explain your answer.

Investigating net gas exchange

You can investigate the effect of light on net gas exchange in a plant using a pH indicator, because carbon dioxide is acidic when dissolved in water. Hydrogen-carbonate solution is often used because it is non-toxic and can be used with living organisms. Care must be taken, however, as the indicator contains two dyes that are harmful and ethanol, which is highly flammable. Before use in an investigation it needs to be *equilibrated*, so that the concentration of carbon dioxide in the solution is the same as the concentration of carbon dioxide in the surrounding air. This is done by drawing air through the solution using a vacuum pump for a few minutes.

Discs can be cut from leaves using a core borer of large diameter. The discs are placed in Petri dishes containing equilibrated hydrogen-carbonate indicator. Placing one dish in bright light and covering the other with dark paper shows a difference in colour of the indicator after 10–15 minutes as a result of the net release or net uptake of carbon dioxide. A similar investigation using an aquatic plant is shown in the 'Developing investigative skills' box below.

Developing investigative skills

Hydrogen-carbonate indicator can be used to indicate the acidity or alkalinity of a solution. With normal atmospheric levels of carbon dioxide it is a red-orange colour. In more acidic solutions it is yellow, and in alkaline solutions it is purple.

Devise and plan investigations

❶ Using the information above, write a plan for testing the effect of light on the net gas exchange from an aquatic plant such as *Elodea* (pondweed).

Analyse and interpret data

The following results were obtained in a similar experiment to the one you have described over 10 minutes after starting the experiment.

Time in minutes	Light dish	Dark dish
0	Red-orange	Red-orange
2	Red	Light orange
4	Reddish-purple	Yellowish-orange
6	Purple	Yellow
8	Purple	Yellow
10	Purple	Yellow

❷ Describe the results shown in the table for
 a) the light dish
 b) the dark dish.

❸ What caused the colour change in the dark dish? What does this suggest is happening in the plant cells?

❹ What caused the colour change in the light dish? What does this suggest is happening in the plant cells?

❺ What other process is happening in the plant cells in the light dish that we cannot show because its effects are masked? Explain your answer as fully as possible.

Evaluate data and methods

❻ What would no change in the colour of the indicator mean?

❼ Explain how you would adapt your method to find the compensation point for this plant.

PLANT STRUCTURE AND GAS EXCHANGE

Many adaptations of the leaf for photosynthesis are also adaptations for the rapid exchange of gases between the photosynthesising cells and the environment. This includes:

- the large surface area for diffusion provided by the internal surfaces of the spongy mesophyll cells
- the short distance that the gases have to diffuse across from the photosynthesising cells to the air spaces in the leaf
- the presence of stomata in the leaf surface (mostly the lower surface).

Carbon dioxide and oxygen enter and leave the plant through the stomata. Normally the stomata open wider when the leaf is exposed to light and tend to close again in the dark. This is because there is a greater demand for gas exchange while the plant is photosynthesising as well as respiring. Closing stomata helps to reduce excessive water loss from the leaf.

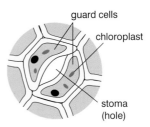

△ Fig. 2.71 Each stoma in the leaf surface is surrounded by two guard cells that control the opening and closing of the hole.

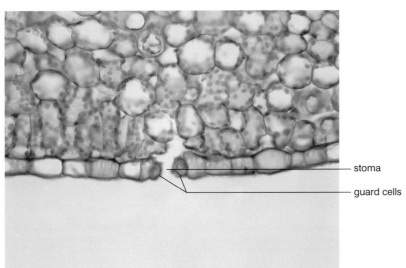

◁ Fig. 2.72 One stoma in the surface of a leaf.

QUESTIONS

1. Explain why a pH indicator can be used to investigate the net exchange of gases in a leaf.

2. Sketch a leaf and label it to show the adaptations for gas exchange.

3. Explain the importance of diffusion in gas exchange in plants.

Gas exchange in humans

Breathing is the way that oxygen is taken into our bodies and carbon dioxide removed. Sometimes it is called **ventilation**.

Do not confuse breathing with respiration. Respiration is a chemical process that happens in every cell in the body. Unfortunately, the confusion is not helped by the parts of the body responsible for breathing being known as the **respiratory system**.

HUMAN RESPIRATORY SYSTEM

When we breathe, air is moved into and out of our lungs. This involves different parts of the respiratory system inside the **thorax** (chest cavity).

When we breathe in, air enters though the nose and mouth. In the nose the air is moistened and warmed.

The air travels down the **trachea** (windpipe) to the **lungs**. Tiny hairs called **cilia** help to remove dirt and microorganisms.

The air enters the lungs through the **bronchi** (singular: bronchus), which branch and divide to form a network of **bronchioles**.

At the end of the bronchioles are air sacs called **alveoli** (singular: alveolus). The alveoli are covered in tiny blood capillaries. This is where oxygen and carbon dioxide are exchanged between the blood and the air in the lungs. This is called **gas exchange**.

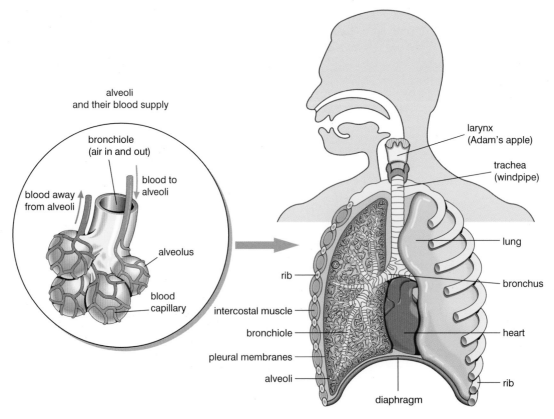

△ Fig. 2.73 The respiratory system.

Surrounding the lungs are the **pleural membranes**, an inner one attached to the lung surface and an outer one attached to the thoracic wall and diaphragm. Between the membranes is a little fluid that helps them slide smoothly over each other. Outside the pleural membranes are ribs, between which are found the **intercostal muscles**, and below the lungs is the **diaphragm**, which is a large sheet of muscle tissue that attaches to the thorax.

The pleural membranes, ribs, intercostal muscles and diaphragm work together to allow breathing.

QUESTIONS

1. List the structures of the human respiratory system and, for each structure, explain its role in breathing.

2. Describe the difference between the terms *gas exchange* and *ventilation*.

INHALATION AND EXHALATION

Breathing in is known as **inhalation** and breathing out as **exhalation** (sometimes they are called inspiration and expiration).

Both happen because of changes in the volume of the thorax. The changes cause pressure changes, which in turn cause air to enter or leave the lungs.

The changes in thorax volume are caused by the diaphragm and the intercostal muscles. In gentle breathing, only the diaphragm may be involved. In deeper breathing, such as during exercise, or when you breathe deliberately, both the ribs and intercostal muscles become involved.

Inhalation

Air is breathed into the lungs as follows.

1. The diaphragm contracts and flattens in shape.

2. The external intercostal muscles contract, making the ribs move upwards and outwards.

3. These changes cause the volume of the thorax to increase.

4. This causes the air pressure in the thorax to decrease.

5. This decrease in pressure causes air to enter the lungs.

Rings of cartilage in the trachea and bronchi keep the air passages open and prevent them from collapsing when the air pressure decreases and bursting when air pressure increases.

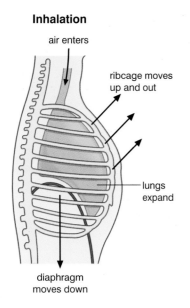
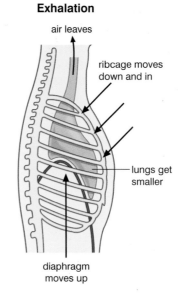

Inhalation

air enters

ribcage moves
up and out

lungs
expand

diaphragm
moves down

Exhalation

air leaves

ribcage moves
down and in

lungs get
smaller

diaphragm
moves up

Δ Fig. 2.74 Left: inhalation. Right: exhalation.

Exhalation

Air is breathed out from the lungs as follows.

1. The diaphragm relaxes and returns to its domed shape, pushed up by the liver and stomach. This means it pushes up on the lungs.

2. The external intercostal muscles relax, allowing the ribs to drop back down. This also presses on the lungs. If you are breathing hard the internal intercostal muscles also contract, helping the ribs to move down.

3. These changes cause the volume of the thorax to decrease.

4. This causes the air pressure in the thorax to increase.

5. This causes air to be forced out of the lungs.

Air contains many gases. Oxygen is taken into the blood from the air we breathe in. Carbon dioxide and water vapour are added to the air we breathe out. The other gases in the air we breathe in are breathed out almost unchanged, except for being warmer.

	In inhaled air	In exhaled air
Oxygen	21%	16%
Carbon dioxide	0.04%	4.5%
Nitrogen and other gases	79%	79%
Water	Variable	High
Temperature	Variable	High

Δ Table 2.7 Contents of inhaled and exhaled air.

Investigating the release of carbon dioxide during breathing

One way to compare the amounts of carbon dioxide in inhaled and exhaled air is to use the apparatus shown in fig 2.75. After cleaning the mouthpiece with antiseptic solution you breathe slowly in and out through it. Exhaled air bubbles through the limewater in the right-hand test tube. Inhaled air bubbles through the limewater in the left-hand test tube. After just a few breaths, the exhaled air will start to turn the limewater in the right-hand test tube cloudy, showing the presence of carbon dioxide. If you continue for several minutes, eventually the inhaled air will start to turn the limewater in the left-hand test tube cloudy too, showing that there is some carbon dioxide present, but far less than in exhaled air.

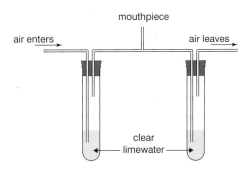

△ Fig. 2.75 Exhaled air leaves through the right-hand test tube. Inhaled air enters through the left-hand test tube.

REMEMBER

Be prepared to answer questions on breathing in terms of comparing the pressure inside the lungs and external air pressure. Air moves from an area of higher pressure to an area of lower pressure.

During inhalation, air enters because the air pressure inside the lungs is lower than the air pressure outside the body.

During exhalation, air leaves the lungs because the air pressure inside is higher than the air pressure outside the body.

SCIENCE IN CONTEXT

ARTIFICIAL VENTILATION

Sometimes an accident or illness can damage a person's ability to breathe. Since exchange of gases is essential for respiration, and so for life, this process must be continued artificially until the patient is able to do it independently again.

In the past, the patient was placed inside a large machine, sometimes called an iron lung, which was sealed from the air. Changes in pressure inside the machine caused changes in thoracic volume, which resulted in air being forced in and out of the patient's lungs. Today, a sealed mask is placed over the patient's mouth and nose, and air is forced into the lungs by increasing the air pressure. The air leaves when the external air pressure is reduced.

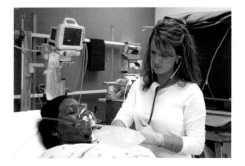

△ Fig. 2.76 A ventilator mask forces air into the patient's lungs by increasing air pressure, and allows air out of the patient's lungs by decreasing the air pressure in the mask.

Alveoli

The alveoli are where oxygen and carbon dioxide diffuse into and out of the blood. For this reason the alveoli are described as the site of gas exchange or the respiratory surface.

The alveoli are adapted for efficient gas exchange by diffusion by having:

- thin permeable walls which keeps the distance over which diffusion of gases takes place, between the air and blood, to a minimum
- a moist lining, in which the gases dissolve before they diffuse across the cell membranes
- a large surface area – there are hundreds of millions of alveoli in a human lung, giving a surface area of around 70 m² for diffusion
- high concentration gradients for the gases, because blood is continually flowing through the capillaries around the alveoli, delivering excess carbon dioxide and taking on additional oxygen, and because of ventilation of the lungs which refreshes the air in the air sacs.

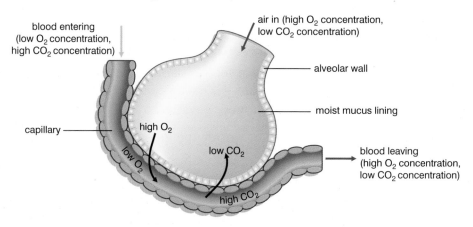

Δ Fig. 2.77 Gas exchange in an air-filled alveolus.

Investigating the effect of exercise on breathing

There are two aspects of breathing that can change during exercise, the rate of breathing and the volume of breath.

Rate of breathing is usually counted as number of breaths per minute.

The volume of a breath can be measured in dm³ using a spirometer. A simple spirometer can be made using a 2-litre plastic bottle that has been marked down the side with volumes of water. (This can be done by adding 500 cm³ of water at a time, and marking the volume on the side of the bottle with a waterproof marker.)

When the bottle is full of water, turn it upside-down into a water trough without allowing any air into the bottle. Insert a flexible plastic tube into the neck of the bottle and secure the bottle and tube in position. Clean the other end of the tubing with antiseptic solution. (Alternatively, add a mouthpiece to the end of the tubing that can easily be removed and sterilised after each test.)

To measure the volume of a breath, ask the person to wear a noseclip and then breathe out a normal breath into the tube. The scale on the bottle can be used to measure the volume of air breathed out.

Developing investigative skills

Devise and plan investigation

❶ Design an investigation into the effect of exercise on breathing. (Hint: think carefully about how many people to test, and how to test them, in order to get reliable results.)

Demonstrate and describe techniques

❷ This investigation could involve vigorous exercise. What risks will you need to prepare for, and how should they be minimised?

Analyse and interpret data

The data in the table are the results from an investigation into the effect of exercise on breathing in 4 people. They were first tested at rest and then after 2 minutes of running on a treadmill set at the same speed.

Person	A		B		C		D	
	Rate (breaths/ minute	Breath volume (dm^3)	Rate (breaths/ minute	Breath volume (dm^3)	Rate (breaths/ minute	Breath volume (dm^3)	Rate (breaths/ minute	Breath volume (dm^3)
At rest	13	0.5	15	0.4	12	1.2	18	0.6
After exercise	19	1.3	23	0.9	18	1.3	26	1.5

❸ Explain how these data should be adjusted before they can give a reliable answer to the question 'How does exercise affect breathing?'

QUESTIONS

1. Describe as fully as you can what happens during
   a) inhalation and
   b) exhalation.

2. a) How many cells does oxygen pass through on its way from the alveoli to the red blood cells?

   b) Why is it important for there to be a large concentration gradient for oxygen between the inside of the alveoli and the blood?

3. Sketch a diagram of an alveolus and annotate it to show how it is adapted for efficient gas exchange. (Hint: remember to refer to diffusion.)

THE EFFECTS OF SMOKING

When a person smokes tobacco, the chemicals in the smoke are taken into the lungs. Those chemicals that have small enough molecules can then diffuse into the blood and be carried around the body. Many of the chemicals in tobacco smoke have damaging effects, not only on the respiratory system, but also on other parts of the body.

Addiction

Nicotine in tobacco smoke is highly addictive, which makes it difficult for smokers to give up smoking. It also alters people's moods because it is a stimulant and a relaxant – smokers often say they feel more relaxed but alert after smoking. These feelings can also become addictive.

Bronchitis

The tar in tobacco smoke is a mixture of chemicals that form a black sticky substance in the lungs. This can coat the tiny hair-like cilia lining the tubes of the lungs, making it more difficult for them to clear out dust and microorganisms. As a result, sticky mucus builds up in the smoker's lungs; this can result in many lung infections and a persistent cough. The irritation and infection can cause a disease called bronchitis.

Emphysema

Over a long time, continued coughing, in order to clear the tar from lungs, damages the alveoli. This breaks down the divisions between them, reducing their surface area. This causes a disease called emphysema where the patient has difficulty getting enough oxygen into their blood. Patients may have to breathe air containing a high concentration of oxygen through a tube or mask.

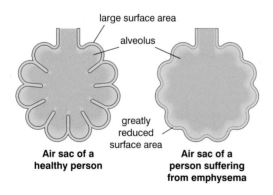

large surface area

alveolus

greatly reduced surface area

Air sac of a healthy person

Air sac of a person suffering from emphysema

◁ Fig. 2.78 Repeated coughing over a long period breaks down the surface of each alveolus, reducing the surface area for exchange of gases. This condition is called emphysema.

◁ Fig. 2.79 People with emphysema may have to breathe air containing a high concentration of oxygen, to make sure their damaged lungs can absorb enough oxygen into their bodies. Breathing masks attached to oxygen tanks such as this one shown can be used.

Cancer

Some chemicals in tar are **carcinogenic**, which means they cause cells to start dividing uncontrollably and so cause cancer. Cancers due to smoking occur most commonly in the mouth, throat and lungs. But many other cancers in the body are also more common in people who smoke than in non-smokers.

Diseases of the circulatory system

Nicotine's stimulant properties can be a problem because stimulants increase blood pressure. Blood pressure is also increased because of the carbon monoxide in smoke. Carbon monoxide replaces oxygen in red blood cells, reducing the amount of oxygen reaching cells. If too little oxygen reaches cells, they can start to die off. This can happen particularly at the ends of fingers and toes. Pregnant women who smoke pass the carbon monoxide to their developing fetus (baby). This can slow down the fetus' rate of growth and result in a baby with a low birth weight, which can have harmful consequences for the baby's health.

CARBON MONOXIDE POISONING

Carbon monoxide is a poisonous (toxic) gas. It is also colourless and has no smell, which makes it a very dangerous gas as it is possible to breathe it without realising it.

During combustion (burning), carbon dioxide is normally produced. In conditions where there is not enough oxygen reaching the flames, carbon monoxide is produced instead. This can happen in faulty gas fires and other appliances, and in the exhaust fumes from vehicles.

In open spaces this is not usually a problem, because there is plenty of oxygen in the air. Also carbon dioxide helps to remove any carbon monoxide that attaches to haemoglobin in red blood cells. However, in enclosed spaces, breathing carbon monoxide-rich air can be fatal because not enough oxygen can get to cells, and organs such as the heart and brain stop working.

△ Fig. 2.80 Exhaust fumes from vehicles contain carbon monoxide.

Some of the chemicals in tobacco smoke cause cholesterol to be released into the blood. Where the linings of blood vessels have been irritated by smoke chemicals, the cholesterol can attach to the lining and cause thickening and clots, which can block the blood flow. This stops oxygen getting to the tissues beyond the blockage. If the clot breaks free and travels to the blood vessels in the brain, it can cause a stroke. In a coronary blood vessel (which supplies the heart muscle with oxygen), a clot can cause **coronary heart disease**, such as a heart attack.

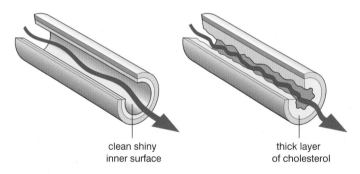

clean shiny
inner surface

thick layer
of cholesterol

△ Fig. 2.81 Smoking can cause thickening of blood vessel linings with cholesterol, which can cause circulatory diseases.

This chart in Fig. 2.82 shows the percentage of low birth weight babies born to Canadian women in 1998 and 1999, in relation to how much the mother smoked during pregnancy.

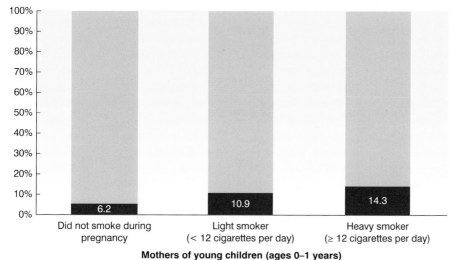

△ Fig. 2.82 Percentages of low birth weight babies born to mothers who smoked and mothers who did not.

1. Before the chart was drawn, the actual number of babies with a low birth weight in each category was calculated as a percentage of the total number of babies born in that category. Explain why this was done.

2. Describe fully the results shown in the chart.

3. A student wrote the following conclusion using the data in this chart:

   'The chart shows that smoking in pregnancy causes low birth weight.'

   Comment on this conclusion and write a better one.

4. Data from many other similar studies in different countries show similar results. Explain why drawing a conclusion from a range of studies gives a more reliable conclusion than just from one study.

5. Name one gas in tobacco smoke that could cause this relationship between maternal smoking and birth weight. Explain as fully as you can how this gas might have this effect.

QUESTIONS

1. List the diseases that smoking can cause, and identify the body systems that they affect.

2. What does the term *carcinogenic* mean?

3. Explain why someone with emphysema may need to breathe additional oxygen.

4. Explain why smoking can cause diseases in other parts of the body, when the tobacco smoke is only breathed into the lungs.

End of topic checklist

Alveoli are tiny air sacs in the lungs where gases diffuse between the air in the lungs and the blood.

Bronchi (single: bronchus) are the two divisions of the trachea as it joins to the lungs.

Bronchioles are the tiny tubes in the lungs that carry air to the alveoli.

Carcinogenic means something that causes cancer, which is the uncontrolled division of cells.

The **compensation point** is the point at which the rate of photosynthesis and respiration in a plant are equal, so that there is no net production or uptake of carbon dioxide or oxygen.

The **diaphragm** is the sheet of muscle at the bottom of the lungs that controls breathing.

Exhalation means breathing out.

Inhalation means breathing in.

Intercostal muscles are the muscles between the ribs that move them during breathing.

Pleural membranes are the two membranes surrounding the lungs.

The **respiratory system** is the body system that includes the lungs, diaphragm and other organs involved in breathing.

The **thorax** is the centre part of the body protected by the ribs, which contains the lungs and heart.

The **trachea** is the tube leading from the mouth to the bronchi, sometimes called the windpipe.

Ventilation is another word for breathing.

The facts and ideas that you should know and understand by studying this topic:

○ Diffusion is important in gas exchange because it controls the rate at which gases enter and leave an organism.

○ As a result of respiration a plant takes in oxygen and releases carbon dioxide.

- As a result of photosynthesis a plant takes in carbon dioxide and releases oxygen.

- Respiration continues while photosynthesis takes place, but when light intensity is high enough, the net exchange of gases in a photosynthesising plant is to take in carbon dioxide and release oxygen.

- The structure of the leaf is adapted for gas exchange by being thin, by having a large surface area of cell surfaces in the mesophyll for gas exchange, and by having stomata.

- The stomata allow gases to diffuse into and out of the leaf, to and from the photosynthesising cells.

- Net gas exchange of a leaf can be shown using hydrogen-carbonate indicator because carbon dioxide is an acidic gas.

- The ribs, intercostal muscles, diaphragm, trachea, bronchi, bronchioles, alveoli and pleural membranes are structures in the thorax concerned with ventilation and gas exchange in humans.

- The intercostal muscles and diaphragm cause ventilation (breathing) by changing the volume of the thorax, which changes the pressure inside the lungs compared with air pressure outside the body.

- Air is drawn into the lungs when the pressure inside the lungs is lower than the outside air pressure, and air is forced out of the lungs when the pressure inside the lungs is greater than the outside air pressure.

- Alveoli have a large surface area, a single layer of cells and an extensive supply of blood capillaries, to maximise the rate of diffusion of gases between the air in the alveoli and the blood in the capillaries.

- Exercise increases the rate and depth of breathing in order to supply more oxygen to muscle cells and remove carbon dioxide from the body more quickly.

- Tobacco smoke contains many chemicals which can harm the respiratory system by damaging tissues, such as cilia in the trachea and the surface of the alveoli.

- Tobacco smoke can also damage other parts of the body, because some of the chemicals enter the blood and are transported in the circulatory system.

- Smoking is associated with the diseases bronchitis, emphysema, lung and other cancers, and low birth weight of babies born to mothers who smoke.

End of topic questions

1. Some students used an oxygen sensor to measure the amount of dissolved oxygen in pond water containing pondweed, when a light was on and when it was switched off. The graph shows their results.

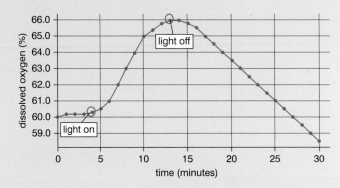

a) The light was switched on after 4 minutes. Describe and explain what happened until the light was switched off again. **(2 marks)**

b) Describe and explain what happened after the light was switched off. **(2 marks)**

2. Look at the table of gases in inhaled and exhaled air on page 112. Describe and explain the difference in proportion of each of these gases in inhaled air and exhaled air:

a) oxygen **(2 marks)**

b) carbon dioxide **(2 marks)**

c) nitrogen. **(2 marks)**

3. The diagram shows a model that can be used to demonstrate the role of the diaphragm in breathing.

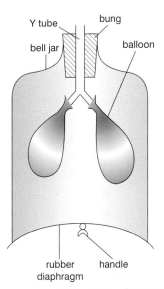

Y tube   bung
bell jar   balloon
rubber diaphragm   handle

   a) Describe and explain what will happen to the balloon 'lungs' when the rubber diaphragm is:

   i) pulled down   **(4 marks)**

   ii) pushed up.   **(4 marks)**

   b) Which parts of the body that can be involved in breathing are not included in this model? Explain their role in breathing.   **(6 marks)**

4. It is commonly stated that 'Plants produce oxygen during the day and carbon dioxide at night'. Explain fully the limits of this statement.   **(5 marks)**

5. a) Describe the terms *diffusion* and *gas exchange*.   **(2 marks)**

   b) Using a suitable example, describe the role of diffusion in gas exchange in organisms, and explain how the tissues and organs of the gas exchange surfaces are adapted to maximise the rate of gas exchange.   **(5 marks)**

6. It took many years to convince people that smoking tobacco could cause circulatory diseases. Explain why this link was difficult to understand, and what we now know about how smoking causes these diseases.   **(4 marks)**

7. Describe the relationship between the compensation points for a plant and the amount that the plant can grow. Explain your answer as fully as possible.   **(4 marks)**

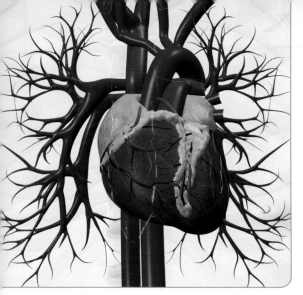

△ Fig. 2.83 Network of arteries and veins in the lungs. The arteries branch from the pulmonary artery, which supplies blood to the lungs. The veins join up to form the pulmonary vein, which returns oxygenated blood to the heart.

Transport

INTRODUCTION

Almost no cell in your body is more than 20 μm (0.02 mm) from a blood vessel. This is necessary because the blood delivers a constant supply of oxygen and glucose for respiration, without which the cells will rapidly die. So it is not surprising that, almost no matter where you cut yourself, you will bleed. Many of the blood vessels that penetrate the tissues are extremely narrow – about 5 to 10 μm wide, which is about the width of one cell. It has been calculated that if you placed the blood vessels of an adult in a line it could wrap four times around the Equator of the Earth.

KNOWLEDGE CHECK

✓ Cells in a plant leaf make glucose by photosynthesis, which is converted to sucrose and transported to other parts of the plant.
✓ The heart and blood vessels form the human circulatory system.
✓ Cells need a continuous supply of oxygen and glucose for respiration, which are supplied by the blood in a human body.

LEARNING OBJECTIVES

✓ Understand why simple, unicellular organisms can rely on diffusion for movement of substances in and out of the cell.
✓ Understand the need for a transport system in multicellular organisms.
✓ Describe the role of phloem in transporting sucrose and amino acids between the leaves and other parts of the plant.
✓ Describe the role of xylem in transporting water and mineral ions from the roots to other parts of the plant.
✓ Understand how water is absorbed by root hair cells.
✓ Understand that transpiration is the evaporation of water from the surface of a plant.
✓ Understand how the rate of transpiration is affected by changes in humidity, wind speed, temperature and light intensity.
✓ Practical: Investigate the role of environmental factors in determining the rate of transpiration from a leafy shoot.
✓ Describe the composition of the blood: red blood cells, white blood cells, platelets and plasma.
✓ Understand the role of plasma in the transport of carbon dioxide, digested food, urea, hormones and heat energy.
✓ Understand how adaptations of red blood cells make them suitable for the transport of oxygen, including shape, the absence of a nucleus and the presence of haemoglobin.

✓ Understand how the immune system responds to disease using white blood cells, illustrated by phagocytes ingesting pathogens and lymphocytes releasing antibodies specific to the pathogen.

✓ Understand how vaccination results in the manufacture of memory cells, which enable future antibody production to the pathogen to occur sooner, faster and in greater quantity.

✓ Understand how platelets are involved in blood clotting, which prevents blood loss and the entry of microorganisms.

✓ Describe the structure of the heart and how it functions.

✓ Explain how the heart rate changes during exercise and under the influence of adrenaline.

✓ Understand how factors may increase the risk of developing coronary heart disease.

✓ Understand how the structure of arteries, veins and capillaries relate to their function.

✓ Understand the general structure of the circulation system, including the blood vessels to and from the heart and lungs, liver and kidneys.

THE NEED FOR TRANSPORT SYSTEMS

Simple unicellular organisms, or those made up of just a few cells, do not need transport systems because they can simply absorb the substances they need, and excrete waste products, across their cell membrane. Their surface area, over which they exchange nutrients and waste with the external environment, is large compared to their total volume, so that diffusion occurs quickly.

In a multicellular organism big enough that many of its cells are more than several cells away from the external world, diffusion becomes too slow to support the cells furthest away from the surface. The larger the organism, and the more active it is, the greater this problem becomes. This is why larger organisms have transport systems, to carry the essential substances around the body to every cell that needs them.

Transport in flowering plants

TRANSPORT TISSUES IN PLANTS

In plants, water and dissolved substances are transported throughout the plant in a series of tubes or vessels. There are two types of transport vessel in plants, called xylem and phloem.

Xylem

Xylem tissue contains hollow xylem cells that form long tubes through the plant. The tubes are the hollow remains of dead cells. The thick strong cell walls help to support the plant.

Xylem tubes are important for carrying water and dissolved mineral ions, which have entered the plant through the roots, to all the parts that need them.

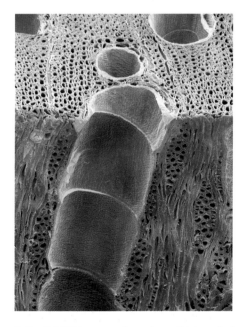

△ Fig. 2.84 Xylem vessels are long, thick-walled tubes that run through the veins of a plant.

They are particularly important for supplying the water that the leaf cells need for photosynthesis.

Phloem

Phloem cells are living cells that are linked together to form continuous phloem tissue. Dissolved food materials, particularly sucrose and amino acids that have been formed in the leaf, are transported all over the plant from the leaves. For example, sucrose is carried to any cell that needs glucose for respiration.

Sucrose is less reactive than glucose and is easier to transport without causing problems for other cells. Sucrose may also be carried to parts of the plant where it is stored, often as another carbohydrate such as starch, which is stored in seeds and root tubers. This transport of sucrose and other materials through the phloem is called **translocation**.

In roots the xylem and phloem vessels are usually in separate groups, but in the stem and leaves they are found together as **vascular bundles** or **veins**.

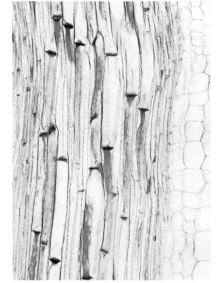

▷ Fig. 2.85 Phloem cells in a longitudinal section of a stem. Some cells still show the green-stained cytoplasm of the phloem cells.

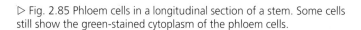

SCIENCE IN CONTEXT

TREE RINGS

The wood of a tree is mostly xylem tissue. Every year, new xylem cells are produced from a ring of cells just inside the bark of the tree. When the tree is growing rapidly, the new xylem cells are large. In temperate regions, such as the UK, the rate of growth, and the size of new cells, decreases as autumn approaches, and stops during winter. The difference in the size of cells produced over one year gives the tree its 'rings' and makes it possible to work out the age of the tree.

▷ Fig. 2.86 Growth rings occur in temperate climates when new xylem cells alternately grow (in spring and summer) and stop growing (in winter).

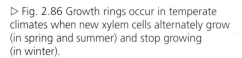

1. Explain why organisms that are more than a few cells thick need transport systems.

2. Describe the function of xylem tissue.

3. Describe the function of phloem tissue.

ABSORBING WATER

Plants absorb water and mineral ions from the soil through **root hair cells**. These are specially adapted for absorption of substances, because they have a fine extension that sticks out into the soil. This greatly increases the surface area for absorption.

△ Fig. 2.87 The root of this germinating seed has many fine root hair cells that greatly increase its surface area.

Root hair cells are found in a short region just behind the growing tip of every root. They are very delicate and easily damaged. As the root grows, the hairs of the cells are lost, and new root hair cells are produced near the tip of the root.

Soil water is a very dilute solution of water and solutes. The concentration of water in the soil is usually greater than the concentration of water in the root hair cell, so water enters the root hair cells by osmosis. This increases the concentration of water (dilutes the cytoplasm) in the root hair cell compared with the cells nearby, so water moves into neighbouring cells by osmosis. This increases the concentration of water in those cells compared with the cells further in, and so on, until the water reaches the vascular bundle in the root. Water molecules enter the xylem tubes by osmosis and are carried to the leaves.

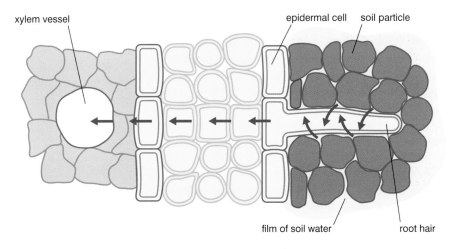

△ Fig. 2.88 The passage of water across a root.

QUESTIONS

1. Which process is used in a root to absorb water from soil water?

2. Copy the diagram of water movement across a root in Fig. 2.88 and annotate it to explain how water enters the root and moves across it to the xylem.

LOSING WATER: TRANSPIRATION

Water is a small molecule that easily crosses cell membranes. Inside the leaf, water molecules pass out of cells into the air spaces. This process is called **evaporation** because the liquid water in the cells becomes water vapour in the air spaces.

Whenever the stomata in a leaf are open, water molecules continually diffuse from the air spaces out into the air (where there is a lower concentration of water molecules). So, in addition to using water in the process of photosynthesis, plants lose water by evaporation from the leaf. This loss of water from the leaves is called **transpiration**.

The loss of water from spongy mesophyll cells to the air spaces decreases the concentration of water molecules in their cytoplasm. So water molecules move from surrounding cells into them by osmosis. That creates a concentration gradient between those cells and the ones further into the leaf, so water molecules move into them by osmosis, and so on, all the way back to the xylem in the vascular bundles.

Water molecules move out of the xylem into surrounding cells by osmosis. This continual movement of water molecules, caused by the transpiration of water from the leaf, is called the transpiration stream.

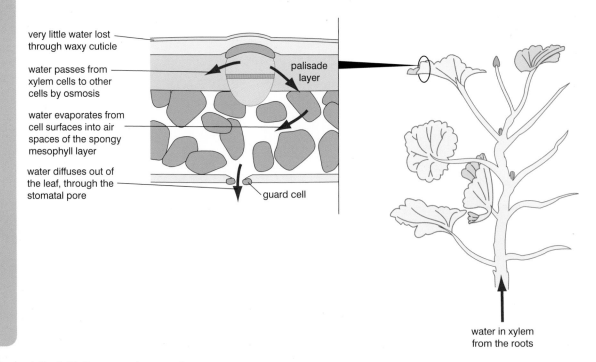

△ Fig. 2.89 How water leaves a plant.

Factors that affect the rate of transpiration

The rate of transpiration from a leaf is affected by anything that changes the concentration gradient of water molecules between the leaf and the air. The steeper the concentration gradient, the faster the rate of transpiration. Several factors can affect the rate of transpiration.

- **Humidity** – This is a measure of the concentration of water vapour in the air. When the air is very humid, it feels damp because there is a high concentration of water vapour in the air. When the air feels dry, the humidity is low. The concentration of water molecules inside the air spaces in the leaf is high. The higher the humidity of the air, the lower the concentration gradient between the air outside and inside the leaf and the slower the rate of transpiration.
- **Wind speed** – The faster the air is moving outside the leaf, the faster any water molecules that diffuse out of the leaf are moved away. This constant removal of water molecules from around the leaf maintains a steeper concentration gradient between the air inside and outside the leaf, so the rate of transpiration is faster.
- **Temperature** – Increased temperature gives particles more heat energy, which results in faster movement of the particles. The faster particles move, the quicker it is for them to evaporate from cell surfaces into the air spaces, diffuse out of the leaf and move away. So increased temperature increases the rate of transpiration.
- **Light intensity** – The higher the light intensity, the more photosynthesis is taking place in the palisade cells. So the stomata are usually opened at their widest in order to exchange carbon dioxide and oxygen as quickly as possible with the cells inside the leaf. Open stomata also make it possible for water molecules to diffuse out of the air spaces into the air more quickly. So a higher light intensity usually increases the rate of transpiration, up to a maximum when the stomata are fully open.

Developing investigative skills

Fig. 2.90 shows an apparatus called a potometer that can be used to investigate the effect of a range of factors on the rate of transpiration. As water evaporates from the leaf surface, the bubble of air in the potometer moves nearer to the leafy twig.

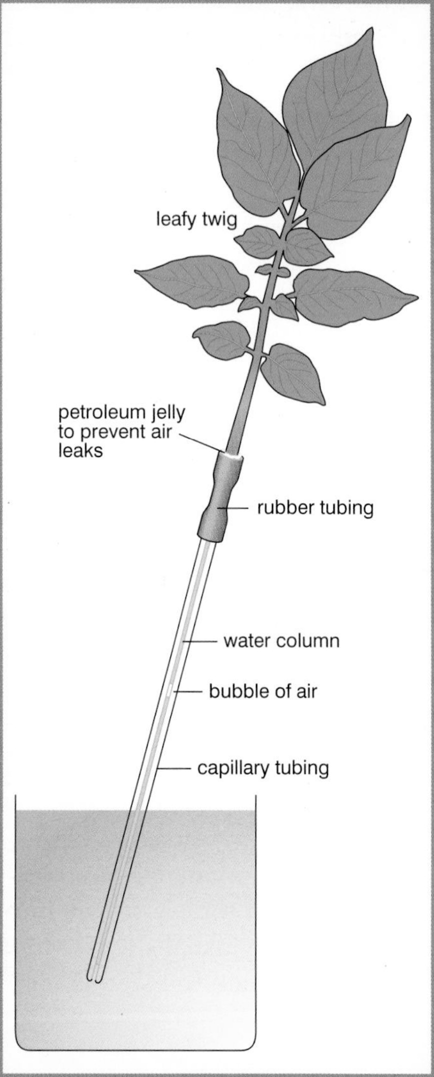

Devise and plan investigations

❶ Suggest how you could use a potometer to measure the effect of the following factors on transpiration: (a) temperature, (b) light intensity, (c) wind speed, (d) humidity.

Analyse and interpret data

The table below shows the results of an investigation, using a potometer in five different sets of conditions.

Conditions	Time for water bubble to move 5 cm (s)
Still air, sunlight	135
Moving air, sunlight	75
Still air, dark cupboard	257
Moving air, dark cupboard	122
Hot, moving air, sunlight	54

❷ For each of the following factors, identify which data in the table should be compared to show the effect of the factor, and explain why those are the right data to compare.

a) temperature

b) light intensity

c) wind speed.

❸ Using the data you have identified, draw a conclusion about the effect of the following on the rate of transpiration.

a) temperature

b) light intensity

c) wind speed.

△ Fig. 2.90 A potometer.

Evaluate data and methods

❹ Explain why the time taken for the bubble to move 5 cm is a measure of transpiration.

❺ What else could make the bubble move?

❻ Explain how could you improve the reliability of the conclusions you drew in Question 3.

Transport in humans

BLOOD

The human circulatory system carries substances around the body. The table shows some of the important substances transported around the human body. These are carried within the blood in different forms.

Substance	Carried from	Carried to
Food (e.g. glucose, amino acids, lipids)	Small intestine	All parts of the body
Water	Intestines	All parts of the body
Oxygen	Lungs	All parts of the body
Carbon dioxide	All parts of the body	Lungs
Urea (waste)	Liver	Kidneys
Hormones	Glands	All around the body to the different target organs affected by the different hormones

△ Table 2.8 What the human circulatory system carries.

Because heat energy transfers from hotter objects to cooler objects, blood also plays an important role in distributing heat energy around the body. It carries heat energy away from cells that are respiring rapidly, such as exercising muscle cells, and prevents them overheating. It also provides heat energy to regions of the body where respiration takes place more slowly. Also, by increasing the diameter of blood vessels in the skin, heat energy can be transferred more rapidly to the environment during exercise to prevent the whole body overheating.

Blood is made from plasma, red blood cells, white blood cells and platelets. Each of these has a particular function.

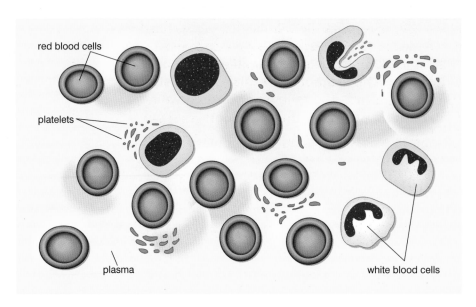

△ Fig. 2.91 Blood is mostly water, containing cells and many dissolved substances.

Plasma

Plasma is the straw-coloured liquid part of blood. It mainly consists of water, which makes it a good solvent. Digested food molecules, such as glucose and amino acids, easily dissolve in plasma. **Urea**, which is formed by the liver from excess amino acids, is also soluble in plasma. Many hormones (see page 169) are also soluble and are carried round the body dissolved in plasma. Carbon dioxide dissolves in water to form carbonic acid (H_2CO_3), and most carbon dioxide is carried in the blood in this form.

Red blood cells

Red blood cells are the most common cell in blood. They have several adaptations that make them efficient at transporting oxygen.

- They have a **biconcave** disc shape (a disc that is thinner in the middle than at the edges) that increases their surface area to volume ratio compared with other cells. This increases the rate at which diffusion into and out of the cell can take place.
- They contain **haemoglobin**, the red chemical that carries oxygen in the cell. In areas of high oxygen concentration (such as in a capillary next to alveoli), haemoglobin binds with oxygen. This reaction is reversible, because in areas of low oxygen concentration (such as in a capillary running through respiring tissue) the oxygen is released from the haemoglobin.

△ Fig. 2.92 Red blood cells (shown in red) and white blood cells (yellow).

- They have no nucleus. This maximises the volume that can be filled with haemoglobin, but means that the cells cannot divide to make new cells as they get older. They are cleaned out of the blood by the liver, and new ones are released into the blood from the bone marrow in long bones such as the femur.
- They are small and flexible, which makes it possible for them to get through the smallest blood vessels (capillaries) that are sometimes no wider than a single red blood cell.

REMEMBER

- In the lungs, haemoglobin combines with oxygen to form oxyhaemoglobin.
- In other organs and tissues, oxyhaemoglobin releases the oxygen to form haemoglobin again.

EXTENSION

One of the factors measured in a blood test is red blood cell count. This is the number of red blood cells in a given volume of blood. In a normal adult male at around sea level it is about 4.7 to 6.1 million cells per mm^3.

If someone who lives at around sea level travels to a place at high altitude, their red blood cell count gradually increases by up to 50% over about two weeks. At high altitude the air is often described as being 'thinner' meaning that the oxygen concentration is lower than nearer sea level.

1. Describe how haemoglobin is adapted for exchanging oxygen with the air and with respiring cells.

2. Suggest how a person's breathing might change if they travelled from low to high altitude quickly. Explain your answer.

3. What advantage is there to the body of producing more red blood cells at high altitude? Explain your answer as fully as you can.

4. Some elite athletes train at high altitude for a few weeks just before a competition even if the race is held at low altitude. Explain as fully as you can what advantage high-altitude training would give over low-altitude training.

White blood cells

There are several different types of white blood cell, but they all play an important role in defending the body against disease. They are part of the **immune system** that responds to infection by trying to kill the pathogen (the disease-causing organism).

- **Phagocytes** – Several types of white blood cell belong to this group, but they all kill pathogens by ingesting them. They engulf pathogens by flowing around them until they are completely enclosed, and then digest them. Different types of phagocytes target different pathogens, such as bacteria, fungi and protoctist parasites.

1 A phagocyte moves towards a bacterium.

2 The phagocyte pushes a sleeve of cytoplasm outwards to surround the bacterium.

3 The bacterium is now enclosed in a vacuole inside the cell. It is then killed and digested by enzymes.

△ Fig. 2.93 Phagocytosis of a bacterium by a phagocyte, a type of white blood cell.

- **Lymphocytes** – This type of white blood cell has a very large nucleus and is responsible for producing chemicals called **antibodies**. When a pathogen infects the body, lymphocytes produce antibodies that specifically match that pathogen. The antibodies attach to the pathogen and either attract phagocytes to engulf the pathogen or cause the pathogen to break open and die.

VACCINATION

Vaccination is a way of preventing disease by making the body respond as if it has already been infected. The vaccine is prepared from small amounts of material from the pathogen, which is either dead or weakened so that it cannot cause infection. The vaccine is put into the body, either through the mouth (for polio vaccination) or by injection. Lymphocytes in the immune system respond by making antibodies to the pathogen, and also **memory cells**. These memory cells remain in the blood after the pathogen has been destroyed, sometimes for the rest of a person's life.

If you are ever infected by the live pathogen, the memory cells recognise it very rapidly and stimulate lymphocytes to produce huge quantities of antibodies very quickly. This response kills off the pathogen rapidly, often before you develop symptoms and realise you have been infected. This rapid response makes you immune to that pathogen.

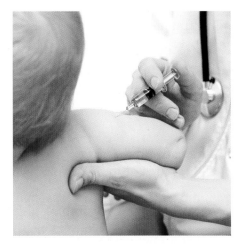

△ Fig. 2.94 An injection of vaccine can be a painful experience, but childhood vaccinations can give life-long protection from dangerous infections.

EXTENSION

Antibodies are another group of chemicals that work by shape, like enzymes. Molecules on the surface of a pathogen have a particular shape, which differs depending on the pathogen. The antibody for a particular pathogen matches this shape, so that it can attach to the surface of the pathogen and either attract phagocytes or cause the pathogen to break up. This is why vaccination for one infection, such as measles, cannot protect you from other infections, such as chicken pox.

This also explains why you can catch a cold or influenza more than once in a lifetime: the pathogens that cause these infections can mutate, changing the shape of their surface molecules frequently, so memory cells do not recognise them when the pathogen infects you the next time.

Platelets

Platelets are small fragments of much larger cells that are also important in protecting from infection. When there is damage to a blood vessel, such as from a cut, the platelets respond by releasing an enzyme that causes the formation of a fibrous protein called fibrin. Fibrin traps blood cells and forms a blood clot. This is essential for staying healthy, as it seals the cut and prevents blood from leaking out and pathogens from getting in.

QUESTIONS

1. Draw up a table to show the components of blood and the roles that they play in the body.

2. Explain how the structure of a red blood cell is adapted to its function.

3. Describe the role of white blood cells in the immune system.

4. Explain how platelets can protect us from infection.

THE CIRCULATION SYSTEM

The human circulatory system consists of the heart and blood vessels. The blood vessels carry blood to the organs, tissues and cells, and the heart is the pump that forces the blood all round the body and back again to the heart. Fig. 2.95 (on the next page) shows a simplified layout of the human circulatory system.

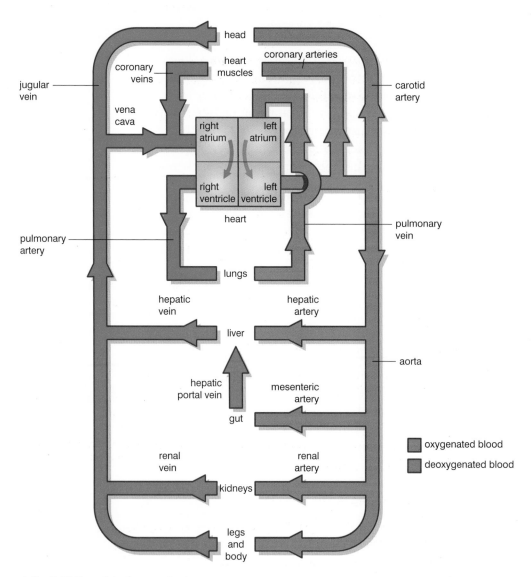

△ Fig. 2.95 Plan of the human circulatory system.

The heart muscles have their own blood supply: the coronary arteries that branch from the aorta and link to the coronary veins that drain into the right atrium.

The name of a major blood vessel is often related to the organ it supplies: *coronary* for heart (from the Latin *corona* for 'crown' because the blood vessels surround the top of the heart like a crown), *hepatic* for liver (from the Greek *hepatos* meaning 'liver'), *renal* for kidneys (from the Latin *renes* meaning 'kidneys'), *pulmonary* for lungs (from the Latin *pulmonis*, 'lungs'). Learn the names of these blood vessels that are associated with the heart, the lungs, liver and kidneys.

The circulatory system in mammals such as humans is a *double circulatory system*. This means that the blood flows twice through the heart for every one time it flows through the body tissues. The advantage of this is that the blood pressure in the circulation through the body can be kept higher than the blood pressure in the circulation through the lungs. A lot of force is needed to pump the blood down to the legs and back, but this force could damage the tiny capillaries in the lungs, which are much closer to the heart.

BLOOD VESSELS

The blood vessels are grouped into three different types: arteries, capillaries and veins.

REMEMBER

Remember: **a** for **a**rteries that carry blood **a**way from the heart. **V**e**in**s carry blood **in**to the heart and contain **v**alves.

Arteries

Arteries are large blood vessels that carry blood flowing away from the heart. Blood in the arteries is at higher pressure than in the other vessels. The highest pressure is in the aorta, the blood vessel that leaves the left ventricle.

Arteries have thick muscular and elastic walls, with a narrow lumen (centre) through which the blood flows. The thick walls protect the arteries from bursting when the pressure increases as the pulse of blood enters them. The recoil of the elastic wall after the pulse of blood has passed through the artery helps to maintain the blood pressure and even out the pulses.

By the time the blood enters the fine capillaries, the change in pressure during and after a pulse has been greatly reduced.

Capillaries

Capillaries are the tiny blood vessels that flow through every tissue and connect arteries to veins. Capillaries have very thin walls, which helps to increase the rate of diffusion of substances. All exchange of substances between the blood and tissues happens in the capillaries.

artery:
thick-walled, carrying blood at high pressure

△ Fig. 2.96 Arteries vary in diameter from about 10 to 25 mm.

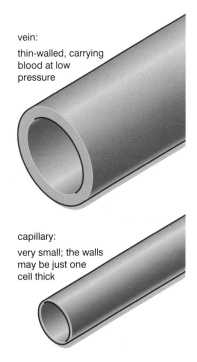

vein:
thin-walled, carrying blood at low pressure

capillary:
very small; the walls may be just one cell thick

△ Fig. 2.97 Veins vary in diameter from about 5 to 15 mm. Capillaries are very small, with a diameter of around 0.01 mm.

Veins

Veins are large blood vessels that carry blood that is flowing back to the heart. By the time blood leaves the capillaries and enters the veins, there is no pulse and the blood pressure is very low.

normal blood flow

veins have valves to stop the blood flowing backwards

 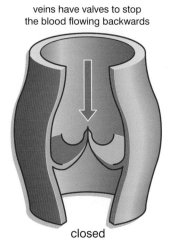

open   closed

△ Fig. 2.98 Valves in the veins make sure that blood can only move in one direction, towards the heart.

Veins have a large lumen to allow blood to flow easily back to the heart. The contraction of body muscles, such as in the legs, helps to push the blood back toward the heart against the force of gravity. **Valves** in the veins prevent backflow and make sure that blood can flow only in the right direction.

QUESTIONS

1. Name the following blood vessels:

   a) the vessels that carry blood to the kidneys

   b) the vessel that carries blood from the heart towards the body

   c) the vessel that carries blood from the liver back towards the heart.

2. Describe the differences in structure of arteries, capillaries and veins.

3. Explain how the structure of arteries helps to even out the blood pulses from the heart.

THE HEART

The heart is a muscular organ that pumps blood by expanding in size as it fills with blood, and then contracting, forcing the blood on its way.

The heart is two pumps in one. The right side pumps blood to the lungs to collect oxygen. The left side then pumps the oxygenated blood around the rest of the body. The deoxygenated (low in oxygen) blood then returns to the right side to be sent to the lungs again.

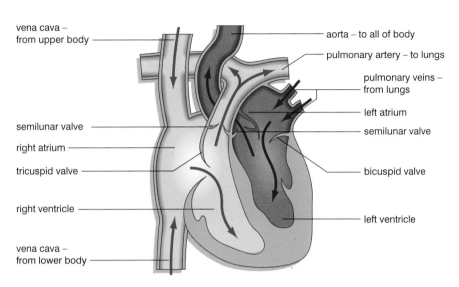

vena cava – from upper body

aorta – to all of body

pulmonary artery – to lungs

pulmonary veins – from lungs

left atrium

semilunar valve

right atrium

tricuspid valve

semilunar valve

bicuspid valve

right ventricle

left ventricle

vena cava – from lower body

△ Fig. 2.99 Section through the heart showing oxygenated blood (red) and deoxygenated blood (blue). The diagram is shown as though you are looking at someone's heart from the front, so the left side of the heart is on the right side of the diagram.

The heart consists of four chambers: two **atria** (single: atrium) and two **ventricles**. The walls of the chambers are formed from thick muscle. Blood passes through the chambers of the heart in a particular sequence as the walls of the chambers contract. First the atria contract at the same time, then the ventricles both contract at the same time, to move the blood through the heart.

- Blood from the body arrives at the heart via the vena cava, and enters the right atrium.
- Contraction of the right atrium passes blood to the right ventricle.
- Contraction of the right ventricle forces blood out through the pulmonary artery to the lungs.
- Blood enters the left atrium from the lungs through the pulmonary vein.
- Contraction of the left atrium passes blood to the left ventricle.
- Contraction of the left ventricle forces blood out through the aorta towards the rest of the body. (Note the muscular wall of the left ventricle is thicker than that of the right ventricle as it has to produce a greater force.)

To make sure that blood only flows in one direction through the heart, there are valves at the points where blood vessels leave the heart, and between the atria and ventricles.

Changing heart rate

Heart rate is the measure of how frequently the heart beats, usually given as beats per minute. We usually take measurements of heart rate by feeling for a pulse point, where the blood flows through an artery near to the skin, such as in the wrist or at the temple.

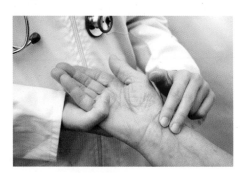

▷ Fig. 2.100 Taking a pulse is actually measuring the expansion and contraction of the artery wall as the blood passes through it. However, as each pulse of blood is created by one contraction of the ventricles, we say that we are measuring heart beats.

Resting heart rate is the rate at which the heart beats when the person is at rest. On average it is between 60 and 80 beats per minute for an adult human, but this range is very variable. Resting heart rate may vary as a result of:

- age – children usually have a faster average than adults
- fitness – a trained athlete may have a resting rate as low as 40 beats per minute because their heart is stronger and can pump out more blood on each contraction
- illness – infection can raise resting heart rate, but some diseases of the circulatory system can slow resting heart rate.

Heart rate increases during activity in order to pump blood more rapidly around the body. This supplies oxygen and glucose more rapidly to respiring cells particularly in the muscles, and removes waste products more rapidly.

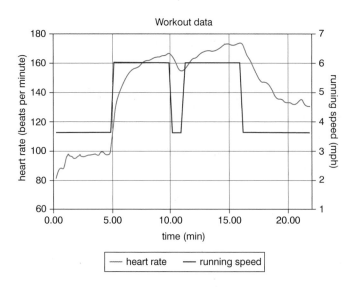

△ Fig. 2.101 How heart rate changes with exercise.

Heart rate is controlled mainly by nerves, but the hormone **adrenaline** also increases it. Adrenaline is produced in the adrenal glands just above the kidneys. It is released into the blood when the brain detects a situation of threat or excitement. Adrenaline has many effects on the body, only one of which is increased heart rate. The full range of effects is often called the 'fight or flight' response because it prepares the body for action – either to fight the threat or to run away from it.

◁ Fig. 2.102 The people on this rollercoaster will have a large amount of adrenaline in their blood at this moment.

Coronary heart disease

Factors such as smoking, a lack of exercise, a diet high in fatty foods and being overweight, can increase the risk of developing **coronary heart disease** (CHD). In CHD the coronary arteries that supply blood to the heart muscle itself become narrowed by a build-up of cholesterol. If this happens, the supply of oxygen to the heart muscle is reduced, which can cause pain called angina. The build-up of cholesterol can also lead to a clot forming. If a clot blocks a coronary artery, stopping the supply of blood and oxygen to part of the heart muscle, then that part of the heart may stop working, causing a heart attack.

QUESTIONS

1. Starting in the vena cava, list the chambers and blood vessels in the order that blood passes through them until it reaches the aorta.

2. Explain why blood can only flow in one direction through the heart.

3. Describe and explain the effect of exercise on heart rate.

Developing investigative skills

The effect of exercise on heart rate can be measured by taking pulse measurements after different levels of exercise.

Devise and plan investigations

In an investigation of the effect of exercise on the heart rate, a student was asked to exercise at different levels for 2 minutes, at which point the pulse rate was measured. The student was then allowed to rest for 5 minutes and then continue to exercise at the next level of activity.

❶ a) Explain why the pulse rate was taken after 2 minutes of exercise and not sooner.

b) Explain why they rested for 5 minutes before starting the next level of exercise.

Analyse and interpret data

The table below shows the results of an investigation into the effect of exercise on heart rate of one student.

	Resting	Walking	Jogging	Running
Heart rate (beats per minute)	72	81	96	122

❷ Describe the pattern shown in the data.

❸ Use the data to draw a conclusion for the investigation.

❹ Explain why heart rate responds like this to different levels of exercise.

Evaluate data and methods

❺ How reliable is this conclusion, and what could have been done during the investigation to improve the reliability?

End of topic checklist

Adrenaline is a hormone that stimulates heart rate as preparation for 'fight or flight'.

An **atrium** (plural atria) is one of two chambers of the heart that receive blood from veins and pumps it into the ventricles.

Biconcave describes the shape of a red blood cell in which the cell is thinner in the middle than at the edges.

Coronary heart disease occurs when the coronary arteries to the heart have become narrowed by a build-up of cholesterol, caused for example by smoking.

Evaporation is when particles in a liquid (such as water) gain enough energy to move fast enough and become a gas (as in water vapour).

Haemoglobin is the red chemical in red blood cells that combines reversibly with oxygen.

Heart rate is the number of heart beats in a given time, for example, beats per minute.

Humidity is a measure of the concentration of water molecules in the air.

The **immune system** protects the body against infection and includes white blood cells.

Lymphocytes are a type of white blood cell that makes antibodies to attack a pathogen.

Memory cells are white blood cells that respond when a pathogen re-infects the body, quickly producing large quantities of antibodies to destroy it.

Phagocytes are a type of white blood cell that ingests and destroys pathogens.

Phloem tissue is formed from many living cells that carry dissolved substances, such as sucrose and amino acids, from the leaves to other parts of a plant.

Plasma is the liquid, watery part of blood which carries dissolved food molecules, urea, hormones, carbon dioxide and other substances around the body and also helps to distribute heat energy.

Platelets are fragments of much larger cells that cause blood clots to form at sites of damage in blood vessels.

Root hair cells are cells in the epidermis of roots that have a long extension of cytoplasm, where uptake of substances from soil water occurs.

Transpiration is the evaporation of water from the leaf surface.

Vaccination is a technique for giving a person a vaccine to stimulate the immune system and protect them from infection.

Valves are flaps in the heart, and in veins, that prevent the flow of blood in the wrong direction.

Ventricles are the two chambers of the heart that receive blood from the atria and pump it out through arteries.

Xylem tubes are formed from dead cells in the vascular bundles of a plant which carry water and dissolved substances from the roots to the leaves and other parts of the plant.

The facts and ideas that you should know and understand by studying this topic:

○ Unicellular organisms can rely on diffusion for the movement of substances between the cell and the environment, but multicellular organisms need transport systems to supply materials to cells.

○ Xylem tissue transports water and mineral ions from plant roots to other parts of the plant.

○ Phloem tissue carries sucrose and amino acids from the leaves to other parts of the plant.

○ Water is absorbed from the soil by osmosis through root hair cells and crosses the root cells by osmosis to the xylem.

○ Transpiration is the evaporation of water from the leaf surface, mostly through the stomata.

○ The rate of transpiration increases with increased temperature, increased wind speed, increased light intensity and decreased humidity.

○ Human blood is formed from liquid plasma that carries red blood cells, white blood cells and platelets around the body.

○ Plasma is mostly water in which many substances dissolve, such as carbon dioxide, food, urea and hormones.

○ Blood flow around the body carries heat energy from warmer tissues to cooler tissues, and to the skin surface where the energy is transferred to the environment.

○ Red blood cells have a biconcave disc shape, contain large amounts of haemoglobin, have no nucleus, and are small and flexible so that they can carry oxygen efficiently to all the cells in the body.

○ White blood cells include phagocytes that ingest pathogens, and lymphocytes which produce antibodies specific to a pathogen, to help protect the body from infection.

○ Vaccination encourages lymphocytes to produce memory cells that respond more rapidly to another infection by the same pathogen, so making the body immune.

○ Platelets cause blood to clot where there is damage to a blood vessel, preventing further blood loss and entry by pathogens.

○ The human circulation system consists of the heart, which pumps blood through arteries, then through capillaries, and finally through the veins back to the heart.

○ Arteries have thick, elastic, muscular walls to resist the pressure of blood as it enters and even out the change in pressure as blood flows through.

○ Capillaries have very thin walls, through which substances such as carbon dioxide, oxygen and glucose are exchanged with cells.

○ Veins have a large lumen so that blood can flow easily through them, and valves to prevent the backflow of blood so it only flows in one direction.

○ The heart is formed from four chambers, two atria and two ventricles, which have muscular walls to push blood through the heart. Valves prevent backflow of blood so that it only flows in one direction.

○ Heart rate increases during exercise, and under the effect of the hormone adrenaline.

○ Various factors, such as smoking, or being overweight, may increase the risk of developing coronary heart disease.

End of topic questions

1. The flatworm *Planaria* is a multicellular organism that is about 10 mm long, a few millimetres wide and less than 1 mm thick. It is one of the largest organisms that does not have a transport system. Explain how it manages without one. **(2 marks)**

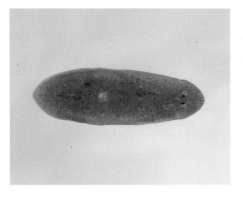

△ A *Planaria* flatworm seen from underneath.

2. **a)** What is the role of arteries in the circulatory system? **(1 mark)**

   b) What is the role of the veins in the circulatory system? **(1 mark)**

   c) Explain as fully as you can why the blood pressure is higher in arteries than in veins. **(2 marks)**

   d) How is the structure of the arteries adapted to cope with the higher pressure? **(2 marks)**

   e) How is the structure of veins adapted to deal with the lower pressure? **(2 marks)**

3. If you place a stalk of celery into water that contains colouring, after a few days the leaves of the celery will start to turn the colour of the water. This is evidence for transpiration.

△ A celery stem that has been sitting for a day in water containing red dye.

   a) What is transpiration? **(1 mark)**

   b) Which tissue in the celery makes this change in the leaves possible? **(1 mark)**

   c) Describe how this tissue is adapted for its function. **(2 marks)**

   d) Explain how this experiment demonstrates transpiration. **(4 marks)**

4. People who have suffered a thrombosis, where the blood unexpectedly clots and blocks a blood vessel, may be given aspirin every day to help protect them from it happening again. Aspirin interferes with the way platelets function.

   a) Describe the role of platelets in the blood. **(2 marks)**

   b) Explain why this normal function helps us to remain healthy. **(2 marks)**

   c) What damage might be caused if a blood vessel that supplies heart muscle is blocked by a blood clot? **(3 marks)**

   d) Explain why aspirin is effective in reducing the risk of another thrombosis. **(1 mark)**

5. People who suffer from anaemia often have a low red blood cell count (fewer blood cells per mm^3 blood) than usual. One of the symptoms of anaemia is becoming tired more easily than usual. Explain why these symptoms occur. **(4 marks)**

6. Describe the factors that increase the risk of coronary heart disease. **(4 marks)**

7. The blood pressure of blood leaving the right ventricle of the human heart is 3 kPa, and from the left ventricle is around 13 kPa. Explain how the heart can produce these different pressures, and why this difference is important for the body. **(4 marks)**

Excretion

△ Fig. 2.103 Shedding leaves is a way of getting rid of waste.

INTRODUCTION

The shedding of leaves by trees, either all together in the autumn by deciduous trees or a few at a time by evergreens, is a form of excretion. Trees store metabolic waste substances in cells in the leaves, out of the way so that they do not interfere with other life processes. When the leaves are shed, this waste is shed also – we say it has been excreted because it has been removed from the body of the tree.

KNOWLEDGE CHECK

✓ Plants produce oxygen from photosynthesis, and plants and animals release carbon dioxide from respiration – these are waste substances if they are not used in other processes.
✓ Excess amino acids from digestion are broken down to form urea in the liver.

LEARNING OBJECTIVES

✓ In plants, understand the origin of carbon dioxide and oxygen as waste products of metabolism and their loss from the stomata of a leaf.
✓ In humans, know the excretory products of the lungs, kidneys and skin (organs of excretion).
✓ Understand how the kidney carries out its roles of excretion and osmoregulation.
✓ Describe the structure of the urinary system, including the kidneys, ureters, bladder and urethra.
✓ Describe the structure of a nephron, to include Bowman's capsule and glomerulus, convoluted tubules, loop of Henle and collecting duct.
✓ Describe ultrafiltration in the Bowman's capsule and the composition of the glomerular filtrate.
✓ Understand how water is reabsorbed into the blood from the collecting duct.
✓ Understand why selective reabsorption of glucose occurs at the proximal convoluted tubule.
✓ Describe the role of ADH in regulating the water content of the blood.
✓ Understand that urine contains water, urea and ions.

EXCRETION IN FLOWERING PLANTS

Excretion is defined as the process or processes by which an organism eliminates the waste products of its metabolic activities. (Remember that excretion is different from egestion.) In flowering plants two waste products that need to be excreted are carbon dioxide and oxygen. Carbon dioxide is produced in respiration while oxygen is a product of photosynthesis. Excess amounts of these gases (not needed for other processes) are excreted through the stomata of the leaves.

EXCRETION IN HUMANS

The metabolic activities in human cells produce many waste products that need to be excreted.

Carbon dioxide is the waste product from respiration. If it remained in cells, it would change their pH and affect the activity of enzymes. It diffuses from respiring cells into the plasma of the blood and is carried around the body until it reaches the lungs. There it diffuses through the capillary and alveoli walls and is breathed out.

△ Fig. 2.104 Sweat is water and ions that have been secreted from the body via the skin.

The skin plays a minor part in excretion. Sweat, which is secreted on to the skin surface from special cells in the skin, contains water and some minerals such as sodium and chloride ions (salt).

Waste products of many cell processes dissolve in the blood and are carried to the **kidneys**, where they are excreted. These products include **urea**, produced from the breakdown of excess amino acids by the liver.

THE URINARY SYSTEM

Humans have two kidneys situated just under the rib cage at the back of the body, about halfway down the spine. The kidneys are well supplied with blood, which enters through the renal arteries and leaves through the renal veins. Inside the kidneys, the blood is filtered to remove waste substances no longer needed by the body. These include excess water, urea and mineral ions, which together form **urine**. Urine flows out of the kidneys down the **ureters** and into the **bladder**. The urine is stored in the bladder until a ring of muscle at the base is relaxed (usually when you go to the toilet). The urine then flows out of the bladder, through the **urethra** to the environment.

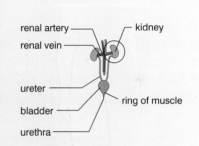

△ Fig. 2.105 The human urinary system.

QUESTIONS

1. Which is the main organ of excretion in plants? Explain your choice.

2. Which are the main organs of excretion in humans? Explain your choices.

3. Draw up a table to list the main structures of the urinary system and their functions.

STRUCTURE OF THE KIDNEY

Each kidney consists of around one million tiny tubules called **nephrons**. Each nephron is associated with a blood capillary. The exchange of substances between the nephron and capillary forms urine, which passes into the ureter.

Each nephron has a particular structure.

- It begins with a small cup-shaped structure called a **Bowman's capsule**. This surrounds a knot of tiny capillaries called a **glomerulus**.
- The Bowman's capsule leads into the **proximal convoluted tubule**. This part of the nephron is very closely associated with the capillary.
- The proximal convoluted tubule leads into a long loop called the **loop of Henle**.
- This loop leads into the **distal convoluted tubule**.
- The distal convoluted tubules from several nephrons open into a **collecting duct**.
- All the collecting ducts empty into the ureter.

Different parts of the nephron are involved in urine formation and excretion, and osmoregulation (water control).

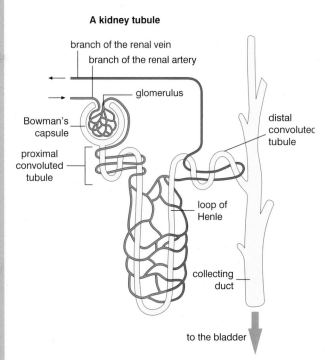

◁ Fig. 2.106 The structure of a nephron.

EXCRETION BY THE KIDNEYS

The formation and excretion of urine by the kidneys involves two separate processes: ultrafiltration and reabsorption.

Ultrafiltration

Ultrafiltration is filtration using pressure. The walls of the Bowman's capsule and glomerulus are only one cell thick, and there are tiny gaps between the cells. Blood enters the glomerulus from the renal artery. The blood vessel (afferent arteriole) narrows, increasing the blood pressure. The pressure of the blood in the glomerulus squeezes many small molecules out from the blood and into the capsule. These molecules include water, glucose, mineral ions (such as sodium and chloride), hormones, vitamins and urea, which together form the **glomerular filtrate**. Large molecules such as blood proteins and blood cells are too large to be filtered through into the capsule, so they remain in the blood.

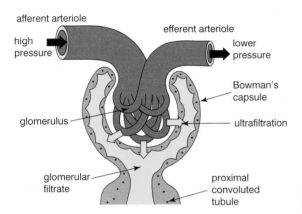

◁ Fig. 2.107 The glomerulus and Bowman's capsule.

Selective reabsorption

As the filtrate passes through the proximal convoluted tubule, some substances are reabsorbed from the filtrate back into the blood in the capillary that runs close by. Some substances are reabsorbed more than others, so this is called **selective reabsorption**. For example, a lot of the sodium ions, and all of the glucose, that were filtered into the tubule are reabsorbed by active transport here. These substances are very important in the body, so it is essential they are not lost in urine.

A lot of water is also reabsorbed from the convoluted tubule by osmosis, because the concentration of water molecules is higher in the filtrate than in the blood. Many other substances that the body needs, such as vitamins, other mineral ions and amino acids, are also reabsorbed here. Other substances that the body needs to remove, such as much of the urea, are not reabsorbed.

Additional water is reabsorbed into the blood from the collecting duct, but the amount that is reabsorbed depends on the state of hydration of the body as you will see below.

QUESTIONS

1. List the structures in the order that a urea molecule would pass through from the blood to the urethra.

2. In which parts of the nephron do the following take place?

a) ultrafiltration

b) reabsorption

3. a) Name one substance that is selectively reabsorbed by the nephron.

b) Which process is used to reabsorb this substance?

OSMOREGULATION AND THE KIDNEYS

Osmoregulation is the control of the concentration of water in the blood. Water is essential for many cell reactions and as a solvent for many substances both within cells and in blood.

- If the concentration of water in the blood falls too low, the body is in a state of dehydration. The blood will draw water out of cells by osmosis, which may damage the processes going on in the cells.
- Too much water in the blood is called overhydration. This can be caused by drinking a lot of water very quickly. It is also dangerous, for example because too much water in animal cells can cause them to burst.

The body controls the amount of water in the blood by adjusting the amount that is reabsorbed from the collecting ducts in the nephrons. Sensors in the **hypothalamus** in the brain detect the concentration of water in the blood.

- If the concentration of water in the blood is too low (not enough water), a hormone called **ADH** (antidiuretic hormone) is released into the blood. When it reaches the kidneys, the ADH makes the walls of the collecting ducts more permeable to water molecules, so more water molecules pass out of the collecting ducts and into the blood capillaries. This also results in a smaller volume of darker urine, more concentrated in solute molecules.
- If the concentration of water in the blood is too high (too much water), less ADH is produced. So the walls of the collecting ducts become less permeable to water, and fewer water molecules are reabsorbed into the blood. This also results in a larger volume of paler urine, less concentrated in solute molecules.

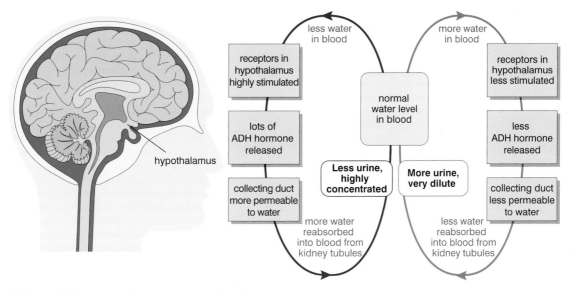

Δ Fig. 2.108 The control of water content of the blood.

QUESTIONS

1. Use your own words to define the term *osmoregulation*.

2. a) Which hormone controls the concentration of water in the blood?

   b) Where is this hormone produced?

   c) Where are the target cells for this hormone?

3. Describe the effect of the hormone you identified in Question 2.

SCIENCE IN CONTEXT

SURVIVING WITHOUT WATER

Australian hopping mice live in the deserts of Australia. These mice can survive without ever drinking liquid water. This is partly because they lose very little water in their breath, but also because most of the water that is released from food molecules during respiration is reabsorbed in their kidneys and kept within the body. The mice produce very small amounts of thick, syrupy urine.

Δ Fig. 2.109 An Australian hopping mouse.

End of topic checklist

ADH (antidiuretic hormone) is the hormone involved in regulating water content in the blood by changing the permeability of the collecting duct of nephrons.

The **bladder** is where urine is held in the body before it is released into the environment.

The **Bowman's capsule** is the cup-shaped structure at the start of a nephron where ultrafiltration occurs.

A **collecting duct** is the tube at the end of a nephron, where additional water may be reabsorbed from the urine.

Convoluted tubules are two sections of the nephron closely associated with a capillary; selective reabsorption of glucose takes place in the proximal convoluted tubule.

Glomerular filtrate is the liquid in the Bowman's capsule produced by ultrafiltration; it contains water, glucose, ions, urea, hormones and vitamins.

The **glomerulus** is a small knot of capillaries associated with a Bowman's capsule.

The **hypothalamus** is the part of the brain that monitors water content of the blood, and is where ADH is produced.

The **loop of Henle** is a large loop of the nephron.

A **nephron** is a tiny kidney tubule where ultrafiltration and reabsorption take place to produce urine.

Osmoregulation is the regulation of the concentration of water in the blood.

Selective reabsorption is the reabsorption of more of some substances (such as glucose from filtrate) than others.

Ultrafiltration is filtration using pressure, as happens between the glomerulus and Bowman's capsule.

Urea is a substance produced in the liver from the breakdown of amino acids not needed in the body.

Ureters are tubes that connect the kidneys to the bladder.

The **urethra** is a tube that connects the bladder to the outside of the body.

The **urinary system** is the body system that includes the kidneys, ureters, bladder and urethra.

Urine is a liquid waste produced by kidneys, containing water, urea and ions.

The facts and ideas that you should know and understand by studying this topic:

- ◯ Plants excrete the waste products of respiration (carbon dioxide) and photosynthesis (oxygen) into the environment through the stomata in their leaves.

- ◯ In humans, the lungs excrete carbon dioxide, the skin excretes water and ions, and the kidneys excrete urea and other substances that the body does not need.

- ◯ The urinary system includes the kidneys, ureters, bladder and urethra.

- ◯ Each kidney is made of millions of nephrons (tubules) that carry out the processes of ultrafiltration and selective reabsorption.

- ◯ Each nephron consists of a Bowman's capsule surrounding a glomerulus of capillaries, the proximal and distal convoluted tubules, the loop of Henle, and the collecting duct into which urine drains.

- ◯ Ultrafiltration occurs between the glomerulus and Bowman's capsule, where small molecules in the blood are forced through the walls from the blood into the nephron to form the glomerular filtrate.

- ◯ Selective reabsorption of glucose occurs in the proximal convoluted tubule, where all the glucose in the filtrate is reabsorbed back into the blood in the capillary.

- ◯ The liquid that flows out of the kidneys is urine, which contains water, urea and excess ions that the body does not need.

- ◯ ADH (antidiuretic hormone) increases the amount of water reabsorbed from the filtrate into the blood by making the walls of the collecting ducts more permeable.

End of topic questions

1. **a)** Describe the term *excretion* in your own words. (1 mark)

   b) Give examples of excretion from **i)** flowering plants, **ii)** humans. (4 marks)

2. Explain how the structure of the glomerulus and Bowman's capsule are adapted to their function in ultrafiltration. (3 marks)

3. **a)** Explain what is meant by *selective reabsorption*. (1 mark)

   b) By which process is **i)** glucose and **ii)** water reabsorbed in the nephron? (2 marks)

   c) If some nephrons were treated with a poison that kills cells, would either of these processes still be possible? Explain your answer. (3 marks)

   d) Explain as fully as you can why glucose needs to be selectively reabsorbed in the kidneys. (3 marks)

4. In which set of conditions would there be the most ADH in a person's blood:

   a) after several hours without drink on a hot sunny day

   b) after drinking a large glass of lemonade?

   Explain your answer. (3 marks)

5. Explain as fully as you can why osmoregulation is essential for health. (4 marks)

Coordination and response

INTRODUCTION

US tennis star Andy Roddick can hit a tennis ball so hard that it travels at around 250 km per hour. In order to return the ball successfully, his opponents have only a fraction of a second to work out where to stand and how best to return the ball. Their response is built on years of training, so that they can respond without consciously thinking.

△ Fig. 2.110 Professional tennis players serve so quickly that a radar gun is used to measure the speed of the ball.

KNOWLEDGE CHECK

✓ Plants and animals detect the environment with specialised sense organs.
✓ Animals respond to changes in the environment using nervous and hormonal systems.
✓ Plants respond to changes in the environment using their hormonal system.
✓ Nerve cells are specialised cells adapted to the function of carrying electrical impulses.

LEARNING OBJECTIVES

✓ Understand how organisms are able to respond to changes in their environment.
✓ Understand that homeostasis is the maintenance of a constant internal environment, and that body water content and body temperature are both examples of homeostasis.
✓ Understand that a coordinated response requires a stimulus, a receptor and an effector.
✓ Understand that plants respond to stimuli.
✓ Describe the geotropic and phototropic responses of roots and stems.
✓ Understand the role of auxin in the phototropic response of stems.
✓ Describe how nervous and hormonal communication control responses and understand the differences between the two systems.
✓ Understand that the central nervous system consists of the brain and spinal cord and is linked to sense organs by nerves.
✓ Understand that stimulation of receptors in the sense organs sends electrical impulses along nerves into and out of the central nervous system, resulting in rapid responses.
✓ Understand the role of neurotransmitters at synapses.
✓ Describe the structure and functioning of a simple reflex arc illustrated by the withdrawal of a finger from a hot object.
✓ Describe the structure and function of the eye as a receptor.
✓ Understand the function of the eye in focusing on near and distant objects, and in responding to changes in light intensity.
✓ Describe the role of the skin in temperature regulation, with reference to sweating, vasoconstriction and vasodilation.
✓ Understand the sources, roles and effects of the following hormones: adrenaline, insulin, testosterone, progesterone and oestrogen.
✓ Understand the sources, roles and effects of the following hormones: ADH, FSH and LH.

SENSITIVITY

Sensitivity, the ability to recognise and respond to changes in external and internal conditions, is recognised as one of the characteristics of living organisms.

A change in conditions is called a **stimulus**. To produce a coordinated response to that stimulus, there must be a **receptor** that can recognise the stimulus and an **effector**, a mechanism to carry out the response. **Coordination** means detecting and responding appropriately to a particular stimulus.

Coordination and response in flowering plants

TROPISMS

Plants generally respond to changes in the environment by a change in the way they grow. For example, a shoot grows towards light, and in the opposite direction to the force of gravity, while a root grows away from light, but towards moisture and in the direction of the force of gravity. These growth responses to a stimulus in plants are called **tropisms**. These responses help the plant produce leaves where there is the most light, and roots that can supply the water that the plant needs.

◁ Fig. 2.111 The response of growing towards light helps a plant get more light for photosynthesis.

Growth in response to the direction of light is called **phototropism**. If the growth is *towards* light, it is called *positive* phototropism, as in shoots. Roots grow *away* from light so they are *negatively* phototropic. Growth in response to gravity is called **geotropism**. Roots show *positive* geotropism, but shoots are *negatively* geotropic.

Tropisms are controlled by plant hormones called **auxins**. These hormones are made in the tips of shoots (and roots) and diffuse away from the tips. Further back along a shoot, the hormones *stimulate* cells to elongate (grow longer) so that the shoot grows longer.

SCIENCE IN CONTEXT   **GARDENER'S TIP**

One effect of the hormone auxin is to inhibit the growth of side shoots. This is why a gardener who wants a plant to stop growing taller and encourage it to become more bushy will take off the shoot tip, so removing a source of auxin.

The growth of shoots towards light can be explained by the response of auxin to light.

- When all sides of a shoot receive the same amount of light, equal amounts of auxin diffuse down all sides of the shoot. So cells all around the shoot are stimulated equally to grow longer. This means the shoot will grow straight up.
- When the light on the shoot comes mainly from the side, auxin on that side of the shoot moves across the shoot to the shaded side. The cells on the shaded side of the shoot will receive more auxin, and so grow longer, than those on the bright side. This causes the shoot to curve as it grows, so that it grows towards the light.

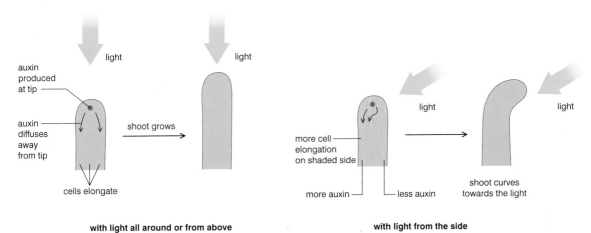

Δ Fig. 2.112 The effect of light and auxin on the growth of shoots.

The hormone is also made in root tips. However, it has the opposite effect on root cells compared with shoot cells, because it *reduces* how much the cells elongate.

- When roots are pointing straight down, all sides of the root receive the same amount of hormone, so all cells elongate by the same amount.
- When the root is growing at an angle to the force of gravity, gravity causes the hormone to collect on the lower side. This reduces the amount of elongation of cells on the lower side of the root, so that the root starts to curve as it grows until it is in line with the force of gravity.

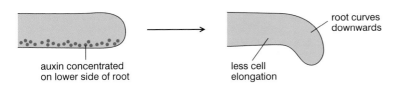

Δ Fig. 2.113 The effect of gravity on the growth of roots.

REMEMBER

A full understanding of the phototropic responses in stems is needed to gain higher marks. Remember that auxin causes shoots to curve by the elongation of existing cells, not by the production of more cells.

Developing investigative skills

Fig. 2.114 shows apparatus that can be used to investigate the effect of light on the growth of seedlings.

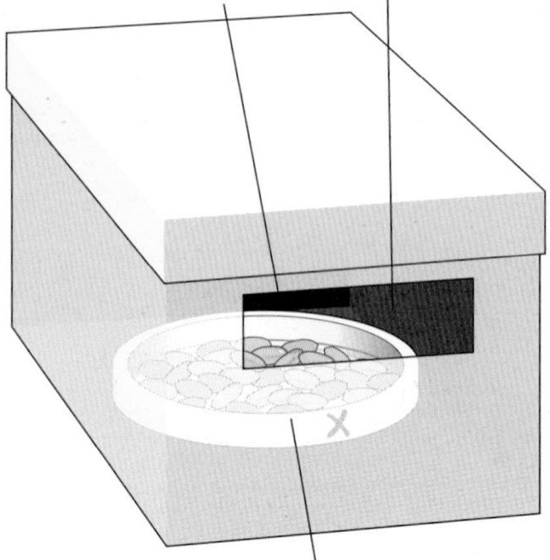

Devise and plan investigations

❶ Describe how the apparatus could be used for this investigation.

❷ Describe how you would set up a control for this investigation.

Make observations and measurements

❸ a) If this investigation were set up correctly, what result would you expect to see in the seedlings from the windowed box, compared with your control?

b) Explain your answer.

Evaluate data and methods

❹ Suggest how this investigation could be extended to investigate whether roots also show a phototropic response.

inner walls painted black

window to allow light in

Petri dish with damp paper towel and seeds

△ Fig. 2.114 Apparatus for investigating the effect of light on growing seeds.

QUESTIONS

1. Describe the term *tropism* in your own words.

2. Give one example of:

a) positive phototropism

b) positive geotropism.

3. Describe the action of auxin in a shoot growing in one-sided light.

Coordination and response in humans

There are two systems involved in coordination and response in humans.

• One is the **nervous system**, which includes the brain, the spinal cord, the peripheral nerves and specialist sense organs such as the eye and the ear. Communication in the nervous system is in the form of **electrical impulses** and responses may be very rapid.

- The other is the **hormonal** (or endocrine) **system**, which uses chemical communication by means of hormones. Hormones are secreted by **endocrine glands** and act upon target cells in other tissues and organs. The hormonal system helps to maintain basic body functions including metabolism and growth. It often has more long-term effects than the nervous system and usually its response is less rapid.

THE NERVOUS SYSTEM

In the human nervous system:

- Specialised **sense organs** that contain receptor cells sense stimuli (changes in conditions).
- Information about these stimuli is sent as electrical impulses from the receptor cells through nerves to the central nervous system.
- The **central nervous system** (**brain** and **spinal cord**) processes the electrical impulses and coordinates the response.
- The response passes as electrical impulses along nerves to effectors, which are often muscles but may be endocrine glands.
- The effectors produce the response to the stimulus.

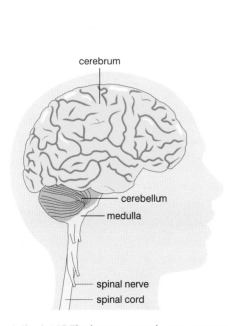

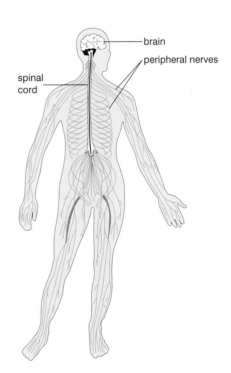

△ Fig. 2.115 The human central nervous system.　　　△ Fig. 2.116 The nervous system.

Nerve cells

Nerves connect the sense organs to the central nervous system, and the central nervous system to effectors. Nerves, the brain and the spinal cord are all made of specialised cells called **neurones**.

Neurones are specially adapted for their function because they have many endings that connect with other neurones for passing electrical impulses, and long cell extensions, called axons and dendrons, that carry the electrical impulses.

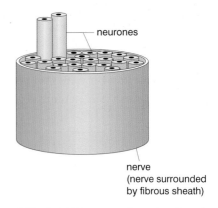

△ Fig. 2.117 Nerves are large bundles of many neurones.

- Neurones that link sense organs to the central nervous system are called sensory neurones.
- Neurones within the central nervous system may be very short, and are called relay (sometimes intermediate or connecting) neurones.
- Neurones that connect the central nervous system to an effector, such as a muscle, are called motor neurones.

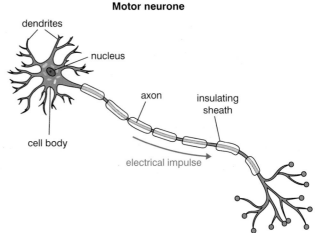

Synapses

Sometimes impulses may have to pass along several neurones. The place where an impulse passes from one neurone to another is called a **synapse**. This is a very small gap between the ends of two neurones. The electrical impulse itself does **not** travel across the gap, but instead causes chemicals called **neurotransmitters** to be released from the end of one neurone which diffuse across the synapse. When the neurotransmitters reach the next neurone they cause a new electrical impulse to be sent. Although the neurotransmitters travel by diffusion, because the synapse is so narrow (about 1/50,000th of a mm), the whole process happens very quickly.

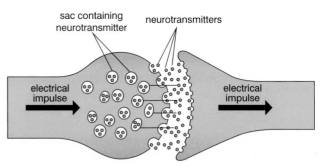

△ Fig. 2.118 When an electrical impulse arrives at a synapse, neurotransmitters are released which start a new impulse in the next neurone.

Synapses mean that impulses can only travel in one direction from one neurone to another. They also allow one neurone to pass impulses to several others at the same time, or to receive impulses from several others.

SENSE ORGANS

Different sense organs contain different specialised receptor cells that respond to different stimuli. The table shows the different sense organs in humans.

Sense organ	Sense	Stimulus
Skin	Touch	Pressure, pain, hot/cold temperatures
Tongue	Taste	Chemicals in food and drink
Nose	Smell	Chemicals in the air
Eyes	Sight	Light
Ears	Hearing	Sound (vibrations in air)
	Balance	Movement/position of head

△ Table 2.9 Sense organs in humans.

The eye

In humans, the eye is the sense organ that responds to changes in light. The specialised light-sensitive receptor cells are found in the **retina** at the back of the eye. Light passing through the eye and reaching the retina causes changes in these cells, sending electrical impulses to the brain along the optic nerve. Other structures of the eye support this process.

- The **cornea**, lens, aqueous and vitreous humour are all transparent, to let light pass through. The cornea and lens refract (bend) light to form an image on the retina. The jelly-like aqueous and vitreous humour maintain the shape of the eyeball.
- The pupil is a hole in the iris that controls how much light enters the eye.
- The inside of the eye is very dark, to stop light reflecting and so cause multiple images.

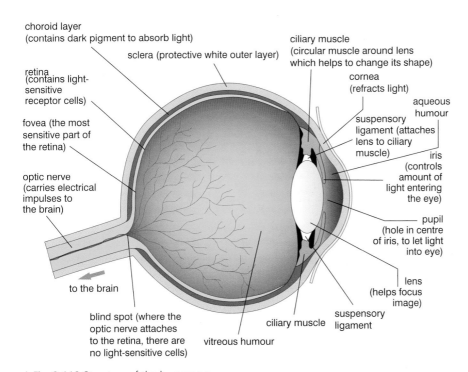

△ Fig. 2.119 Structure of the human eye.

There are two different kinds of light-sensitive cells in the retina.

- **Cone cells** respond to light of different *wavelengths*, and therefore respond to different colours. They only work well in bright light, so we only see colour images when the light is bright enough. The cone cells are mostly clustered at the fovea on the retina, where most light falls.
- **Rod cells** respond to differences in light *intensity*, not wavelength. They are more sensitive at low light intensities than cone cells, so we use these mostly in low light conditions. They can not distinguish different colours. Rod cells are found all over the retina.

DARK ADAPTATION

In bright light the rod cells lose their ability to respond. This is because the coloured pigment in the cells that responds to light is changed into another form. As the light becomes dimmer, the pigment slowly changes back into the form that detects light. However, it can take up to half an hour for the rods to fully recover. This is known as dark adaptation and explains why, if you move from a bright place to a darker place, it takes a while for your eyes to adjust and see clearly again.

Changing light conditions

The light-sensitive cells in the retina only respond to the stimulus of light above a certain light intensity. When it is so dark that the cells are not stimulated, we cannot see. Since vision is an important sense, at low light intensities our eyes need to gather as much light as possible. However, rod and cone cells are easily damaged by high light intensity – which is why you should NEVER look directly at a bright light source such as the Sun.

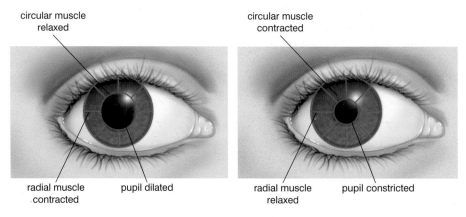

Δ Fig. 2.120 The pupil response to light. Left: in dim light. Right: in bright light.

The iris (ring-shaped, coloured part of the eye) controls the amount of light entering the eye by controlling the size of the hole in the centre, the pupil. The iris contains circular and radial muscles. In bright light the circular muscles contract and the radial muscles relax, making the pupil smaller. This prevents damage to the cells in the retina by reducing the amount of light entering the eye. The reverse happens in dim light, when the eye has to collect as much light as possible to see clearly.

Focusing light

In order to see clear images of our surroundings, the light that enters our eyes needs to be focused properly on the retina.

The thick clear cornea bends light rays as they enter the eye in order to focus them on the retina. The lens provides fine focus to sharpen the image.

Rays of light from distant objects are almost parallel when they enter the eye. They require less refraction (bending) to come to a focus on the retina – the cornea can manage most of this without help from the lens. The ciliary muscles, which are circular, relax and the lens is pulled into a thinner shape by the suspensory ligaments. This provides the correct focusing power.

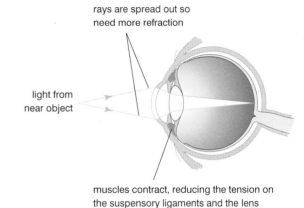

△ Fig. 2.121 How the eye focuses light from distant and near objects.

Rays of light from near objects are diverging when they enter the eye. They need much more powerful refraction to bend them to a focus on the retina. The ciliary muscles contract, which means they pull less on the suspensionary ligaments. This allows the elastic lens to return to a more rounded shape. This refracts light more to achieve a focused image on the retina.

QUESTIONS

1. Explain how the following structures are adapted to support the role of the eye in sensing light:

   a) cornea

   b) pupil

   c) retina.

2. Explain how the eye responds to changing light intensity.

3. Explain how the eye produces a focused image of an object that is near to the person.

REFLEX RESPONSES

The simplest type of response to a stimulus is a **reflex**. Reflexes are rapid, automatic responses to a specific stimulus that usually protect you in some way, for example blinking if something gets in your eye or sneezing if you breathe in dust.

The pathway that nerve impulses travel along during a reflex is called a **reflex arc**:

stimulus → receptor → sensory neurone → relay neurone in CNS →
motor neurone → effector → response

Simple reflexes are usually **spinal reflexes**, which means that the impulses are processed by the spinal cord, not the brain. The spinal cord sends an impulse back to the effector. Effectors are the parts of the body that respond, either muscles or glands. Examples of spinal reflexes include responses to standing on a pin or touching a hot object.

stand on pin → pain receptors → sensory neurone → spinal cord → motor neurone →
leg muscles → leg moves

When the spinal cord sends an impulse to an effector, other impulses are sent on to the brain so that it is aware of what is happening. It also allows the brain to over-ride the reflex response. For example, if you were holding a large bowl of hot food that you were looking forward to eating, you might look around quickly for somewhere to put it down rather than drop it immediately and risk breaking the bowl.

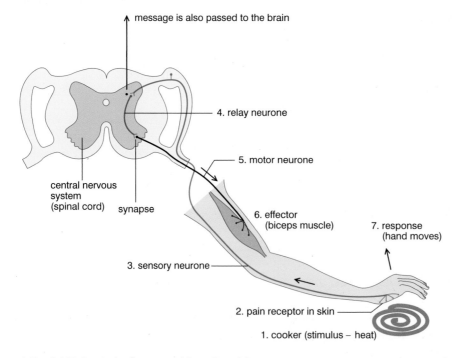

△ Fig. 2.122 A spinal reflex to touching a hot object.

QUESTIONS

1. What is meant by the term *reflex response* and why are these responses important for survival?

2. Describe the reflex response that occurs when you put your finger on something hot.

HOMEOSTASIS

For our cells to carry out all the life processes properly, they need the conditions in and around them, such as the temperature and amount of water and other substances, to stay within acceptable limits. Keeping conditions within these limits – that is, keeping the internal environment constant – is called **homeostasis**.

Temperature control

The temperature in the core of your body is about 37 °C, regardless of how hot or cold you may feel on the outside. This **core temperature** may naturally vary a little, but it never varies a lot unless you are ill.

Heat energy is constantly released by cells as a result of respiration and other chemical reactions, and is transferred to the surroundings outside the body. To maintain a constant body temperature these two processes must balance. The temperature of the blood from the core of the body is monitored by the hypothalamus in the brain. If the temperature varies too much from 37 °C, the hypothalamus causes changes to happen that bring the temperature back to about 37 °C. The hypothalamus also receives electrical impulses from heat sensor cells in the skin surface.

If core temperature rises too far:

- Sweat is released on to the surface of the skin from glands. Sweat is mostly water, and this water evaporates. Evaporation needs heat energy, so heat energy is removed from the skin surface as the sweat evaporates, cooling the skin.
- Blood vessels carrying blood near the surface of the skin dilate (get wider) so more blood flows through them. This is known as **vasodilation**, and it is what makes light-skinned people look pink when they are hot. Vasodilation makes it easier for heat energy to be transferred to the skin surface and from there to the environment by radiation and conduction.

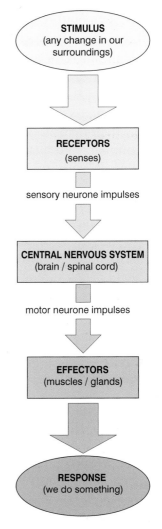

△ Fig. 2.123 A diagrammatic reflex arc.

167

If core temperature falls too far:

- Blood vessels carrying blood near the surface of the skin constrict (get narrower), which reduces the amount of blood flowing through them. This is known as **vasoconstriction**. As the warm blood is kept deeper in the skin, this reduces the rate of heat transfer by conduction to the skin surface and from there to the environment.
- Body hair may be raised by muscles in the skin. This has little effect in humans (often called goose bumps) but is more effective in mammals with fur and in birds, because the fur or feathers trap air next to the skin. Air is not a good conductor of heat energy, so this still layer of air acts as insulation.
- Muscles may start to 'shiver'. This means they produce rapid, small contractions. Cellular respiration is used to produce these contractions, releasing heat energy at the same time which heats the blood flowing through the muscles.

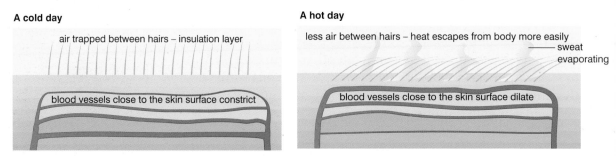

Δ Fig. 2.124 The skin responds to maintain body core temperature.

Control of body water content

Our bodies gain water through:

- food – most foods contain water
- drink
- respiration and other chemical reactions – water is one of the waste products of respiration.

Our bodies lose water through:

- breathing
- sweating
- egestion – faeces contains water
- excretion of urine.

Although we are constantly gaining and losing water, it is important that our overall water content does not vary too much. If we contain too little water, and are in danger of dehydrating, we can drink more, and we produce less, more concentrated urine. (We still have to produce some urine to excrete urea.) If we contain too much water, and are in danger of overhydrating, we produce a larger volume of more dilute urine. The amount and concentration of urine produced is controlled by the hypothalamus in the brain, which monitors the body's water content by monitoring the concentration of the blood.

QUESTIONS

1. Describe the term *homeostasis* in your own words.

2. Give two examples of homeostasis in the human body.

3. Explain the role of skin blood vessels in maintaining core body temperature.

Developing investigative skills

You can use a test tube of warm water wrapped in wet paper towel as a model to investigate whether sweating really does cool the body, measuring how the temperature of the water changes over time.

Devise and plan investigation

❶ Explain how the tube models sweating in a human.

❷ How would you set up the control for this investigation? Explain your answer.

Make observations and measurements

The table shows the results of an investigation like the one described above.

	Time (min)	0	2	4	6	8	10	12	14	16
Temperature of water in tube (°C)	Wet towel	56	50	46	42	39	36	34	32	31
	Dry towel	56	52	49	46	44	44	41	40	39

❸ Use the results to draw a suitable graph.

Analyse and interpret data

❹ Describe any pattern shown in your graph.

❺ Draw a conclusion from the graph.

❻ Explain your conclusion using your scientific knowledge.

HORMONES

Hormones are chemical messengers. They are made in the **endocrine glands**.

Endocrine glands do not have ducts (tubes) to carry away the hormones they make: the hormones are secreted directly into the blood to be carried around the body in the plasma. (There are other types of glands, called exocrine glands, such as salivary and sweat glands, that do have ducts.)

Most hormones affect several parts of the body; others only affect one part of the body, called the **target organ**.

The changes caused by hormones are usually slower and longer-lasting than the changes brought about by the nervous system.

Adrenaline

The **adrenal glands** produce **adrenaline**. This hormone is released in times of excitement, anger, fright or stress, and prepares the body for 'flight or fight': the crucial moments when an animal must instantly decide whether to attack or run for its life.

The effects of adrenaline are:

- increased heart rate
- increased depth of breathing and breathing rate
- increased sweating
- hair standing on end (this makes a furry animal look larger but only gives humans goose bumps)
- glucose released from liver and muscles
- dilated pupils
- paling of the skin as blood is redirected to muscles.

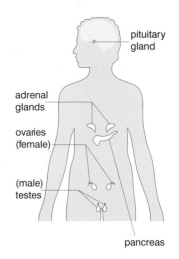

△ Fig. 2.125 The position of some endocrine glands in the human body.

ADH (Antidiuretic hormone)

ADH (antidiuretic hormone) is made by the hypothalamus but then stored and released by the **pituitary gland** in the brain. It helps regulate the body's water content. It causes the kidney tubules to reabsorb more water into the blood by increasing the permeability of the collecting ducts (see page 152).

The pituitary gland also produces many other hormones which control other processes in the body, such as growth.

Insulin

Insulin is secreted by the pancreas, which is also part of the digestive system. Insulin controls the concentration of glucose in the blood, which must remain within a small range. If it rises or falls too much, you can become very ill.

After a meal containing carbohydrates, the blood glucose concentration rises rapidly as glucose is absorbed from digested food in the small intestine. This rise is detected by the pancreas, and causes the pancreas to release insulin. The insulin travels in the blood to the liver. Here it causes any excess glucose to be converted to glycogen, another carbohydrate, which is insoluble and is stored in the liver.

Between meals, glucose in the blood is constantly diffusing into cells for use in cellular respiration. So the blood glucose concentration falls. When a low level of glucose is detected by the pancreas, the pancreas stops secreting insulin and secretes the hormone glucagon instead. Glucagon converts some of the stored glycogen back into glucose, which is released into the blood to raise the blood glucose concentration again.

People with one kind of diabetes (called Type I) are unable to produce insulin. This means that their blood glucose concentration may become so high or so low that it damages cells, which in extreme cases may lead to unconsciousness and death. These people may need to inject insulin to help prevent this damage.

1. Control of blood glucose concentration is an example of homeostasis. Explain what this means.

2. Describe the role of insulin in controlling blood glucose concentration.

3. Patients need to check their blood glucose concentration with a simple blood test before deciding how much insulin to inject. Explain why this is important.

4. The amount of exercise that a person with diabetes does will affect the amount of insulin they inject. Explain why.

△ Fig. 2.126 A portable kit for instant testing of blood glucose. People with Type I diabetes use this to check their blood glucose before and after meals, and throughout the day.

Testosterone

Testosterone is the male sex hormone and is secreted from the **testes**. Testosterone causes secondary sexual characteristics in boys (see page 203) and is needed for the production of sperm.

Progesterone and oestrogen

The **ovaries** produce the female sex hormones **progesterone** and **oestrogen**. Oestrogen is responsible for the development of secondary sexual characteristics in girls (see page 203), and together with progesterone it helps control the menstrual cycle (see page 204).

FSH and LH

FSH (follicle-stimulating hormone) and **LH** (luteinising hormone) are both produced by the pituitary gland in the brain. They work together with progesterone and oestrogen to control the menstrual cycle (see page 204).

QUESTIONS

1. Explain the meaning of the following terms:

    a) hormone

    b) endocrine gland

    c) target organ.

2. Draw up a table to show the following hormones, where they are produced in the body, and what effects they have: adrenaline, insulin, testosterone, progesterone and oestrogen.

End of topic checklist

ADH (antidiuretic hormone) is a hormone produced in the hypothalamus but stored and released by the pituitary gland. It is involved in regulating water content in the blood by changing the permeability of the collecting duct of kidney nephrons.

Adrenaline is a hormone produced by the adrenal glands that is responsible for the 'fight or flight' response.

Auxins are plant hormones that control phototropism and geotropism.

The **central nervous system** is the part of the nervous system that coordinates and controls responses, consisting of brain and spinal cord.

An **effector** carries out the response to a stimulus; in animals these are muscles or glands.

Electrical impulses are the form in which information is sent along nerves.

The **endocrine glands** are a collection of cells that secrete hormones into the blood.

FSH (follicle-stimulating hormone) is a hormone produced by the pituitary gland that helps control the menstrual cycle.

Geotropism is a growth response in plants affected by the direction of the force of gravity.

Homeostasis is the maintenance of a constant internal environment.

A **hormonal system** is a chemical response system in humans where hormones produced by the endocrine glands are carried in the blood to target organs where they affect the cells.

Insulin is a hormone produced by the pancreas that causes muscle and liver cells to take glucose from the blood.

LH (luteinising hormone) is a hormone produced by the pituitary gland that helps control the menstrual cycle.

Nerves are bundles of neurones that connect receptors to the central nervous system, and the central nervous system to effectors.

A **neurone** is a nerve cell, which is specially adapted for carrying electrical nerve impulses.

Neurotransmitters are chemicals that pass from one neurone to another across a synapse.

The **nervous system** is a response system in humans that uses electrical impulses between receptor cells, neurones and effector cells to produce a response to a stimulus.

Oestrogen is a hormone produced by the ovaries that helps to control the menstrual cycle and produces secondary sexual characteristics in girls.

Phototropism is a growth response in plants affected by light.

Progesterone is a hormone produced in the ovaries that helps to control the menstrual cycle.

Receptor cells detect a stimulus, for example cells in the retina detect light.

A **reflex arc** is the pathway that nerve impulses travel along during a reflex.

Sense organs are organs containing receptor cells adapted for the receiving of a particular type of stimulus.

A **stimulus** is a change in the environment that triggers a response in an organism.

A **synapse** is a small gap between two neurones across which neurotransmitters travel.

A **target organ** is an organ of the body containing cells that respond to a particular hormone.

Testosterone is a hormone produced by the testes that produces secondary sexual characteristics in boys.

Vasoconstriction means narrowing of blood vessels.

Vasodilation means widening of blood vessels.

The facts and ideas that you should know and understand by studying this topic:

- ◯ Organisms respond to changes in their environment.
- ◯ A coordinated response requires a stimulus that is sensed by receptor cells, which results in a change in the organism brought about by a receptor (usually muscles or glands in an animal).
- ◯ Plants respond to stimuli by growth responses called tropisms, which are controlled by plant hormones called auxins.
- ◯ Plant shoots show positive phototropism when they grow towards light.
- ◯ Plant roots show positive geotropism when they grow towards the force of gravity. Plant shoots show negative geotropism when they grow in the opposite direction to the force of gravity.
- ◯ Rapid responses in humans are coordinated through nervous impulses in the nervous system, and longer-term responses are coordinated through chemicals (hormones) in the hormonal system.

End of topic checklist continued

○ The central nervous system consists of the brain and spinal cord, which coordinate and control responses.

○ Stimulation of receptor cells in sense organs causes electrical impulses to travel along nerves to the central nervous system.

○ Neurotransmitters travel across synapses allowing an electrical impulse in one neurone to start another impulse along the next neurone.

○ The reflex arc is a connection of sense organ, nerves and spinal cord, and effector, which allows rapid responses without thinking, such as when you rapidly pull your finger away after touching a hot object.

○ The eye has many adaptations for its function in sensing light, including the light-sensitive cells in the retina, and the cornea and the lens for focusing light.

○ Contraction or relaxation of the muscles in the iris change the size of the pupil in response to changes in light intensity, to control the amount of light entering the eye.

○ Contraction or relaxation of the ciliary muscles supporting the suspensory ligaments attached to the lens change the shape of the lens to aid fine focusing of images on the retina.

○ Homeostasis is the control of the internal environment of the body within narrow limits so that cells can function well.

○ Examples of homeostasis in humans includes the regulation of blood water concentration and temperature regulation.

○ Core body temperature is regulated by vasodilation near the skin surface and sweating when too hot, and vasoconstriction near the skin surface and muscle shivering when too cold.

○ ADH is the hormone made in the hypothalamus and released by the pituitary gland, which controls the amount of water reabsorbed from urine passing through the collecting ducts in the kidney.

○ Adrenaline is the hormone that is made in the adrenal glands and prepares the body in many ways for action, including increasing heart rate.

○ Insulin is the hormone made in the pancreas that causes blood glucose concentration to be reduced when it is too high.

○ Testosterone is the male sex hormone made in the testes that controls the development of secondary sexual characteristics in boys and is needed for sperm production.

○ Progesterone and oestrogen are the female sex hormones made in the ovaries that help control the menstrual cycle; oestrogen also controls the production of secondary sexual characteristics in girls.

○ FSH and LH are hormones made in the pituitary gland that help control the menstrual cycle.

End of topic questions

1. Some sprouting tomato seedlings are placed in a dimly lit room near to a brightly lit window.

△ Sprouting tomato seedlings.

  a) Which of these statements best describes the seedlings over the next few weeks? Explain your choice. **(2 marks)**

    i) The seedlings wilt and bend towards the light.

    ii) The seedlings grow towards the brightest light.

    iii) The seedlings grow straight up.

    iv) The seedlings bend towards the light.

  b) What is meant by the term *positive phototropism*? **(1 mark)**

  c) Explain the survival advantage to plants of having shoots that are positively phototropic. **(2 marks)**

  d) What is meant by *positive geotropism*? **(1 mark)**

  e) Explain the survival advantage to plants of having roots that are positively geotropic. **(2 marks)**

2. a) What is the purpose of the nervous and hormonal systems in humans? **(1 mark)**

  b) Use the following table headings to compare the nervous and hormonal systems in humans. **(6 marks)**

System	Cells of system	Method of transmission	Speed of response

  c) Explain as fully as you can why it is advantageous to have both systems. **(5 marks)**

3. a) Describe the sequence of sensing and response in the nervous system of one of Andy Roddick's opponents who returns a serve successfully. **(4 marks)**

  b) Is this a reflex action? Explain your answer. **(2 marks)**

4. Explain the following.

  a) A student visiting a coal mine could not see anything in the mine when the lights were turned off. **(2 marks)**

  b) A cataract is a clouded lens in the eye, caused by many conditions. A patient with cataracts cannot see clear images. **(2 marks)**

  c) A person who is long-sighted needs to wear spectacles with converging lenses in order to read something near to them. **(2 marks)**

5. One example of homeostasis in humans is the control of core body temperature.

  a) Identify the receptors, monitoring area, and effectors in the response to a change in external temperature. **(6 marks)**

  b) Explain why changes in skin blood flow affect the rate of heat loss from the body. **(2 marks)**

  c) Explain why homeostasis of core body temperature is important for survival. **(4 marks)**

6. For each of the following hormones, name the gland that produces it, one target organ, and describe its effect on that target organ:

  a) insulin **(3 marks)**

  b) adrenaline. **(3 marks)**

Exam-style questions
Sample student answer

Question 1

A Biology student sets up an investigation in which three cubes of different sizes are cut from an agar jelly block. The agar jelly contains a pink indicator that turns colourless in the presence of acid.

The cubes are placed in an acid. The student measures the time taken for the cubes to turn completely colourless.

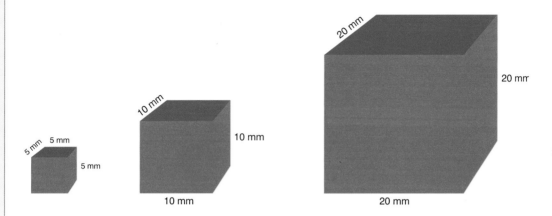

a) What is the name of the process that causes the acid to penetrate the agar jelly? **(1)**

diffusion ✓ ①

b) The student uses three cubes of different dimensions:

Dimensions of cube (mm)

$5 \times 5 \times 5$

$10 \times 10 \times 10$

$20 \times 20 \times 20$

For each cube, calculate its:

- surface area

- volume

- surface area: volume ratio.

Give the surface area: volume ratio in the form 'n : 1'. **(9)**

Exam-style questions continued

Dimensions of cube (mm)	Surface area of cube (mm²)	Volume of cube (mm³)	Surface area: volume ratio
$5 \times 5 \times 5$	150 ✓ ①	125 ✓ ①	150 : 125 ✗
$10 \times 10 \times 10$	600 ✓ ①	1000 ✓ ①	600 : 1000 ✗
$20 \times 20 \times 20$	2400 ✓ ①	8000 ✓ ①	2400 : 8000 ✗

c) Describe the relationship between surface area and volume as the cube increases in size. (1)

As the volume of the cube increases, so does the surface area, but not to the same extent. ✓ ①

d) During the Biology lesson, the acid completely penetrates the two smaller cubes, but by the end of the lesson, the 20 mm × 20 mm × 20 mm cube has still not turned completely colourless.

  i) Explain these results. (3)

The acid diffuses into the cubes over their surface. ✓ ①

In the small and medium cubes, there is sufficient surface area for the acid to penetrate quickly. ✓ ①

For the 20 mm x 20 mm x 20 mm cube, there is insufficient surface area, in relation to the cube's volume, for the acid to completely penetrate quickly. ✓ ①

  ii) Explain what implications this has for organisms of increasing size. (3)

As organisms increase in size, and therefore volume, diffusion of useful substances into their bodies and waste substances out ✓ ① *will not occur quickly enough to meet their needs* ✓ ①

(Total 17 marks)

⑬/⑰

EXAMINER'S COMMENTS

1. a) Correct.

b) The student has:

produced a table, which is a good way of organising the answer, and the correct units for surface area and volume have been used

appreciated that a cube has six faces, so the total area is six times the surface area of one face.

The student has not, however, given the ratio in the form n : 1, but simply used the figures from the table.
The correct version of the table is:

Dimensions of cube (mm)	Surface area of cube (mm²)	Volume of cube (mm³)	Surface area: volume ratio
5 × 5 × 5	150	125	1.2:1
10 × 10 × 10	600	1000	0.6:1
20 × 20 × 20	2400	8000	0.3:1

c) Correct. A better way of expressing this, however, would be to say that as the volume of the cubes increases, so does the surface area, but not at the same rate relative to the volume.

d) i) This is a good answer.

The student has given a full explanation for International GCSE level. They have begun the answer well by explaining that the acid diffuses into the cubes over their surface. The student has appreciated that in the small and medium cubes, there is sufficient surface area, in relation to the cube's volume, for the acid to penetrate quickly and completely.

They have also appreciated that for the 20 mm × 20 mm × 20 mm cube, there is insufficient surface area, in relation to the cube's volume, for the acid to penetrate quickly.

ii) The answer is correct, but the student has missed one of the marking points.

Organisms solve the problem of increasing size by providing additional surface area, i.e. additional absorbing surfaces.

You can see this in plants, for instance in the internal structure of the leaf. Many animals increase their absorbing surfaces with long digestive systems with folded walls with tiny projections, and with the huge surface area of their lungs. You can use examples such as these to illustrate your answer.

Exam-style questions continued

Question 2

Most living organisms are made up of cells.

a) The passage below describes the structure of animal and plant cells.

Use suitable words to complete the sentences in the passage.

The .... holds the cell together and controls substances entering and leaving the cell. The .... is the jelly-like substance contained within the cell. It is where many different chemical processes occur.

The .... is the control centre of the cell and contains genetic material as .... These control how a cell grows and works.

Plant cells also have features that are not found in animal cells. These include the ...., made of ...., which gives the cell extra support and defines its shape.

Many plant cells have a large central, permanent .... that contains cell sap. It is used for storage of some chemicals, and to support the shape of the cell.

Plant cells exposed to the light contain .... These contain the green pigment ...., which absorbs the light energy that plants use for the process of ....

(10)

b) The levels of organisation in multicellular organisms include cells, tissues, organs and systems.

Put ticks in the table to show whether each of the following structures is a cell, tissue or organ.

Structure	Cell	Tissue	Organ
Blood			
Brain			
Liver			
Muscle			
Neurone			
Ovum			
Skin			
Sperm			

(8)

(Total 18 marks)

Exam-style questions continued

Question 3

The table below lists some molecules that are important biologically.

a) Give the units that each one of the following biological molecules is made up of.

Biological molecule	Units that make up the molecule	
Glycogen		(1)
Lipids		(2)
Proteins		(1)
Starch		(1)

b) Describe a test that can be carried out in the laboratory for each of the following carbohydrates.

   i) glucose   (3)

   ii) starch.   (2)

c) The graphs show the effect of pH on the activity of two enzymes (proteases) that break down proteins.

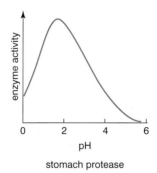

stomach protease

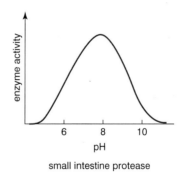

small intestine protease

   i) What are the optimum pHs of stomach and small intestine proteases?   (2)

   ii) Explain why the functioning of enzymes is sensitive to pH.   (4)

d) In humans, proteases are produced by the stomach, pancreas and small intestine. Identify the location of these organs by adding labels to the diagram below.   (3)

(Total 19 marks)

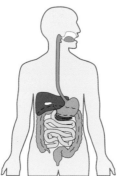

Question 4

A student cut a number of cylinders from a potato and weighed them. These were placed in sucrose solutions of different concentrations.

After one hour, the cylinders were removed, blotted dry and reweighed. The student calculated the percentage change in mass for each cylinder. The results are shown below.

Concentration of sucrose in mol per dm³	Percentage change in mass of potato cylinders				Average percentage change in mass
	Experiment 1	Experiment 2	Experiment 3	Experiment 4	
0.0	+31.4	+33.7	+31.2	+32.5	
0.2	+20.9	+22.2	+22.8	+21.3	
0.4	−2.7	−1.8	−1.9	−2.4	
0.6	−13.9	−12.8	−13.7	−13.6	
0.8	−20.2	−19.7	−19.3	−20.4	
1.0	−19.9	−20.3	−21.1	−20.3	

a) Calculate the average percentage changes in mass for each of the sucrose concentrations. (6)

b) **i)** Draw a graph of these results. Join the points with a line of best fit. (4)

ii) Use the graph to work out at what concentration of sucrose there would be no net movement of water. (2)

iii) Explain the student's results fully. (7)

(Total 19 marks)

Question 5

The diagram shows part of a kidney.

a) Name parts A–H shown in the diagram. (8)

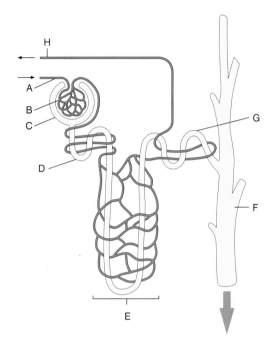

b) The table shows the concentration of certain substances in the plasma, glomerular filtrate and urine of a healthy person.

Substance	Concentration in plasma (g per 100 cm³)	Concentration in glomerular filtrate (g per 100 cm³)	Concentration in urine (g per 100 cm³)
Amino acids	0.05	0.05	None
Glucose	0.10	0.10	None
Ions	0.90	0.90	0.90 – 3.60
Protein	8.00	None	None

Referring to the structures in part **a)**, explain the changes in concentrations of each of the four types of substance. (4)

c) Describe the role of ADH in regulating the water content of the blood. (7)

(Total 19 marks)

Question 6

Plants respond to the light available.

a) The graph below shows the effect of light intensity on photosynthesis in a plant.

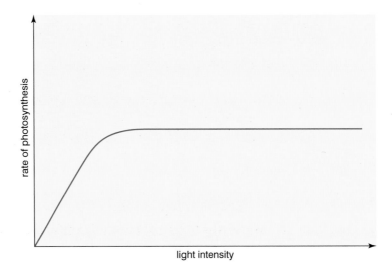

i) Describe and explain the effect of light intensity on the plant. (4)

ii) The investigation was also carried out in a high concentration of carbon dioxide.

Sketch a graph of what you would expect so as to compare it with the graph above. Explain the shape of the graph you have drawn. (3)

b) A plant's response of growing towards light helps it to receive more light for photosynthesis.

i) The passage below describes plants' responses towards stimuli.

Use suitable words to complete the sentences. (8)

Growth in response to the direction of light is called ....
If the growth is towards light, it is called .... ,
as shown by plant ....

Growth in response to gravity is called .... Plant roots are
.... This response helps the plant
.... to grow .... , so the plant can obtain the
.... it needs.

Exam-style questions continued

ii) A scientist investigating the response of plants to light placed:

- one group in the light, given even illumination
- one group of plants in the dark
- one group exposed to light from one side.

The plants were in an atmosphere of radioactive carbon dioxide, and after five hours, the amount of radioactive auxin in the area below the shoot tip was measured. The scientist's results are shown below.

	Plants in the light	Plants in the dark	Plants exposed to light from one side	
			Dark side	Lighted side
Total radioactive auxin in counts per minute	2985	3004	2173	878

Explain fully what these results tell the scientist about the effect of light on auxin in the plants.

(4)

(Total 19 marks)

Question 7

The diagram below shows a section through a leaf.

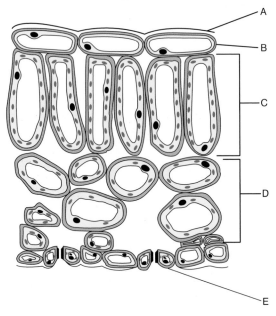

a) **i)** Name parts A–E shown in the diagram. (5)

   ii) Explain how the leaf is adapted to exchanging gases required for photosynthesis. (5)

b) Chemical substances in a plant are transported in the xylem and phloem.

   Copy and complete the table below.

	Phloem	Xylem
Substances transported	(2)	(2)
Substances are transported from	(1)	(1)
Substances are transported to	(1)	(1)

(8)

(Total 18 marks)

Question 8

The nervous system is involved in the body's response to stimuli.

a) **i)** When a person puts her hand on a hot object, she removes it quickly using a reflex action.

   Draw a flow chart to show the reflex arc involved. (6)

   ii) As soon as she removes her hand, she realises that she has touched a hot object. Explain how this occurs. (3)

b) What is the name of the other system involved in the body's coordination? How does the response of this system differ from that of the nervous system? (4)

(Total 13 marks)

Exam-style questions continued

Question 9

The circulatory system has several functions, including the transport of substances, temperature regulation and defence.

a) The diagram below shows the structure of the heart.

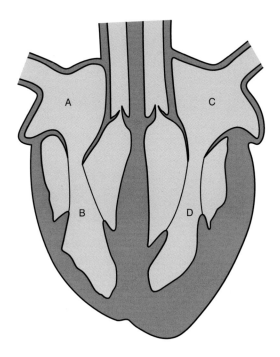

   i)   Name the chambers of the heart, A, B, C and D.   **(4)**

   ii) Explain why the wall of chamber D is much thicker than the wall of chamber B.   **(2)**

b) Describe the functions of each of the different components of the blood:

   i)   red blood cells   **(2)**

   ii) white blood cells   **(2)**

   iii) plasma   **(2)**

   iv) platelets.   **(2)**

c) The blood system is involved in the body's immunity.

   Explain how a vaccination can make a person immune to a particular disease.   **(4)**

(Total 18 marks)

There has been life on Earth for over 3500 million years. The oldest living things we know of were simple bacteria. These bacteria reproduced. Reproduction led to different combinations of characteristics, and mutation produced new characteristics. The environment determined which of these combinations of characteristics were the most successful, and so which individuals survived and which went extinct. Those that survived passed their genes on to their offspring through reproduction. This led to the evolution of new species and eventually to the evolution of animals and plants, and to the millions of species that are alive on Earth today.

STARTING POINTS

1. How can you produce more plants without using flowers?

2. Why do flowers have to be pollinated before they make seed?

3. Why are some flowers large and brightly coloured and others small and inconspicuous?

4. What do seeds need to help them start growing?

5. How do the structures of the male and female human reproductive system support their function?

6. What controls the menstrual cycle every month?

7. What is DNA and what does it do?

8. How can we predict the inheritance of a characteristic controlled by a gene?

9. Why are there two types of cell division?

10. What is natural selection and how can it bring about evolution?

SECTION CONTENTS

a) Reproduction

b) Inheritance

c) Exam-style questions

3
Reproduction and inheritance

△ A simplified model of some DNA strands.

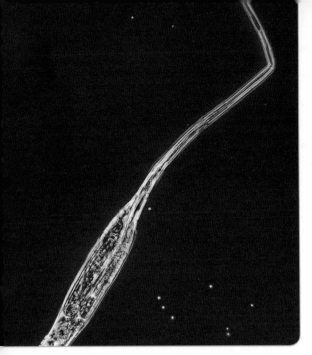

△ Fig. 3.1 Rotifers are not believed to have reproduced sexually for at least 40 million years.

Reproduction

INTRODUCTION

Most multicellular organisms are able to reproduce sexually. This requires the joining of gametes from the male and the female in fertilisation. Many plants and animals can also reproduce *a*sexually. This is where gametes and fertilisation are not needed to produce new individuals. Until recently, scientists believed that asexual reproduction in animals only happened in addition to sexual reproduction. However, we now know that some species of rotifers (microscopic aquatic animals) have not reproduced sexually for at least 40 million years. Males of these species simply do not exist.

KNOWLEDGE CHECK

✓ The flower is the reproductive structure in flowering plants.
✓ The human reproductive system consists of organs, tissues and cells that are specially adapted for their role in reproduction.
✓ Sexual reproduction is the production of new individuals as a result of fertilisation; asexual reproduction is the production of new individuals without fertilisation.

LEARNING OBJECTIVES

✓ Understand the differences between sexual and asexual reproduction.
✓ Understand that fertilisation involves the fusion of a male and female gamete to produce a zygote that undergoes cell division and develops into an embryo.
✓ Describe the structures of an insect-pollinated and a wind-pollinated flower and explain how each is adapted for pollination.
✓ Understand that the growth of the pollen tube followed by fertilisation leads to seed and fruit formation.
✓ Practical: Investigate the conditions needed for seed germination.
✓ Understand how germinating seeds use food reserves until the seedling can carry out photosynthesis.
✓ Understand that plants can reproduce asexually by natural methods (illustrated by runners) and by artificial methods (illustrated by cuttings).
✓ Understand how the structure of the male and female reproductive systems are adapted for their functions.
✓ Understand the roles of oestrogen and progesterone in the menstrual cycle.
✓ Understand the roles of FSH and LH in the menstrual cycle.
✓ Describe the role of the placenta in the nutrition of the developing embryo.
✓ Understand how the developing embryo is protected by amniotic fluid.
✓ Understand the roles of oestrogen and testosterone in the development of secondary sexual characteristics.

SEXUAL REPRODUCTION

Sexual reproduction is the most common method of reproduction for the majority of larger organisms, including almost all animals and plants. To produce a new organism, two **gametes** (sex cells) fuse. This is known as **fertilisation** and the resulting cell is called a **zygote**. The zygote then undergoes cell division to develop into an **embryo**.

Usually sexual reproduction involves two parent organisms of the same species. The zygote that is formed contains some genes from each of the parents, so it is not genetically identical to either parent. Some plants produce male and female gametes in the same flowers, so they are able to reproduce by self-fertilisation. But this is still sexual reproduction because it involves the fusion of a male and female gamete.

There are advantages and disadvantages to sexual reproduction.

Advantages

- Fusion of gametes combines genetic information from two parents, which results in variety in the offspring. This produces individuals that may be better adapted to different conditions than the parents and other offspring.
- The species has a greater chance of survival in changing conditions.

Disadvantages

- Sexual reproduction usually requires a second parent for fertilisation and finding a mate can require the individual to use up energy.
- Sexual reproduction usually takes longer to produce offspring than asexual reproduction.

ASEXUAL REPRODUCTION

Some organisms increase in number by **asexual reproduction**. For this it is not necessary to have two parents. During asexual reproduction, cells from the body of the parent divide to produce the offspring. This means that the offspring are genetically identical to their parent and to each other.

Asexual reproduction also has advantages and disadvantages.

Advantages

- Only one parent is required. There is no need for a parent animal to find a mate or for **pollination** in plants.
- Large numbers of organisms can often be produced in a short time.
- All the offspring produced are genetically identical, so they should survive well in the conditions in which the parent grows well.

Disadvantages

- The lack of variation in the offspring means that any change in the environment that reduces their chance of survival can affect all individuals equally.
- Because the offspring do not vary, they are not suited to moving away and living in environments with different conditions.

▷ Fig. 3.2 Aphids are insects that reproduce asexually during the summer.

QUESTIONS

1. Draw up a table to summarise the differences between sexual and asexual reproduction.

2. Give one example when offspring from sexual reproduction would have a survival advantage over those produced by asexual reproduction.

3. Aphids are common pests that damage crop plants. Explain why it is an advantage for female aphids to reproduce asexually over the few warm summer months in countries where the winter is cold.

Reproduction in flowering plants

Flowering plants are the most successful group of plants. These are the only plants that have true flowers and produce **seeds** with a tough protective coat. During sexual reproduction in flowering plants:

- They produce male and female gametes (sex cells) – some species may produce male and female gametes in the same flowers; other species may have male-only flowers and female-only flowers on the same plant; and in other species male flowers and female flowers are produced on different plants.

- Male pollen is transferred to the female part of the flower so that pollination can take place.
- The male gamete and female gamete fuse during fertilisation to form a zygote.
- The zygote develops to form an embryo within a seed, which protects the embryo and provides food during germination of the seed.
- Seeds are dispersed so that they germinate away from the parent.

STRUCTURE OF FLOWERS

All flowers have a similar basic arrangement. Their structures are stacked on top of each other along a short stem, arranged either in a spiral or in separate rings.

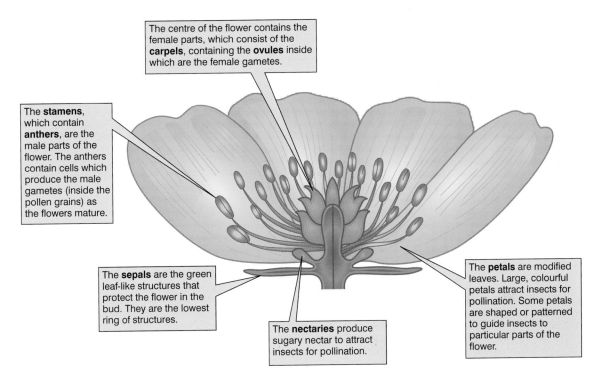

The centre of the flower contains the female parts, which consist of the **carpels**, containing the **ovules** inside which are the female gametes.

The **stamens**, which contain **anthers**, are the male parts of the flower. The anthers contain cells which produce the male gametes (inside the pollen grains) as the flowers mature.

The **sepals** are the green leaf-like structures that protect the flower in the bud. They are the lowest ring of structures.

The **nectaries** produce sugary nectar to attract insects for pollination.

The **petals** are modified leaves. Large, colourful petals attract insects for pollination. Some petals are shaped or patterned to guide insects to particular parts of the flower.

Δ Fig. 3.3 Structure of a flower.

The male parts of the flower

The male part of a flower is the ring of **stamens**. There may be a few stamens, or as many as 100. Each stamen consists of two parts: the **anther** at the top and a stalk called the filament.

The **pollen grains** contain the male gametes in flowering plants. Pollen develops in the pollen sacs of the anthers. Cells lining the inside of the pollen sacs divide by meiosis (see page 231) to give four cells. Each of these cells develops into a pollen grain. As a grain matures, it develops a thick outer wall to protect the delicate male gamete inside. When all the pollen grains in the anther are mature, the anther splits open to release them.

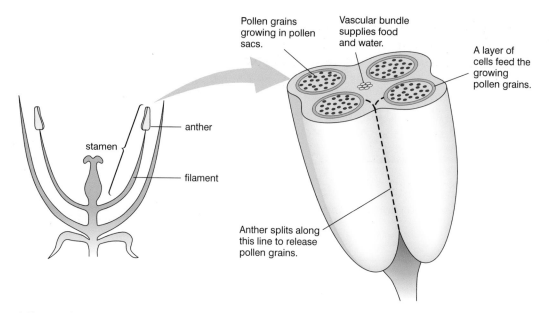

Pollen grains growing in pollen sacs.

Vascular bundle supplies food and water.

A layer of cells feed the growing pollen grains.

anther

stamen

filament

Anther splits along this line to release pollen grains.

△ Fig. 3.4 The structure of a stamen.

△ Fig. 3.5 Alder catkins shedding pollen.

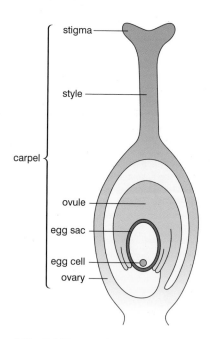

stigma

style

carpel

ovule

egg sac

egg cell

ovary

△ Fig. 3.6 The carpel.

The female parts of the flower

The female part of the flower is the **carpel**. A flower may have more than one carpel, each with its own **style** and **stigma**. The stigma is the part of the carpel where the pollen lands during pollination.

The **ovary** at the base of the carpel protects the female gamete from the dry air outside. The ovary contains one or more **ovules**, and each ovule contains an egg sac that surrounds the **egg cell** (female gamete).

QUESTIONS

1. What is the function of the stamen in a flower?
2. Where are the male gametes of a plant found?
3. Name the main parts of a carpel and explain their role in a flower.

△ Fig. 3.7 Even complex flowers like daisies, which contain many male and female parts, have carpels surrounded by stamens.

POLLINATION

Before fertilisation can take place, the male gametes have to reach the female gametes. This involves transferring the pollen to the stigma, a process known as pollination. In many plants this means transferring the pollen from one flower to another. Some plants use the wind to transfer their pollen between flowers. Others use animals, especially insects, to carry the pollen. Flowers have different features depending on whether they are pollinated by wind or insects.

Wind-pollinated plants	Insect-pollinated plants
Small petals, which do not get in the way when wind blows the pollen	Large petals for insects to land on
Green or tiny petals	Brightly coloured petals to attract insects
No scent	Often scented to attract insects
No nectaries	Nectaries present at the base of the flower produce a sugary liquid to attract insects, such as bees and butterflies
Many anthers which are often large and hang outside the flower so that pollen is easily dispersed	A few small anthers, usually held inside the flower
Pollen grains have smooth outer walls	Pollen grains have sticky or spiky outer walls
Stigmas are large and feathery, often hanging outside the flower to trap pollen	Stigmas are small and held inside the flower
Produce large amounts of pollen	Produce smaller amounts of pollen
Pollen is lightweight	Pollen is heavier

△ Table 3.1 Flowers pollinated by wind and insects have different characteristics.

▷ Fig. 3.8 In insect-pollinated plants, nectaries secrete a sugary liquid to attract insects.

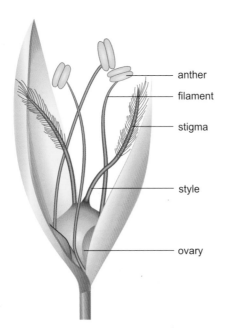

anther

filament

stigma

style

ovary

△ Fig. 3.9 These grass plants have anthers that hang outside the flowers and release large amounts of pollen to the wind. The stigmas also hang outside the flower to collect pollen from other grass plants.

SCIENCE IN CONTEXT

POLLINATOR RELATIONSHIPS

Different pollinators are attracted to different features of animal-pollinated flowers. Tube-shaped flowers attract insects with long tubular mouthparts, such as butterflies, or birds with long beaks, such as hummingbirds. Blue and violet-coloured flowers are more attractive to bees. Butterflies often prefer red. Plants pollinated by moths or bats tend to open at night and may not be brightly coloured but instead produce a strong sweet scent. Plants that rely on flies to pollinate them, like the titan arum, often smell like rotting meat.

Perhaps the most bizarre partnerships between flowers and insects occur where some species of orchid produce flowers that mimic the female of a particular species of wasp. Male wasps are attracted to the flowers to mate with what they think are female wasps. During the 'mating' the flowers deposit pollen on the insect, which is then carried to the next flower that the insect is attracted to.

▷ Fig. 3.10 A male wasp receiving pollen while 'mating' with an orchid flower.

QUESTIONS

1. Describe the main features of a wind-pollinated flower.

2. Explain how wind-pollinated flowers are adapted to distribute their pollen.

3. Describe the main features of an insect-pollinated flower.

4. Explain how insect-pollinated flowers are adapted to distribute their pollen.

FERTILISATION AND SEED FORMATION

Fertilisation occurs when the male nucleus from the pollen grain fuses with the female egg cell in the egg sac. To get from the tip of the stigma to the egg sac, the pollen grain produces a thin tube called the **pollen tube**. Fig. 3.11 (next page) shows how the pollen tube grows down through the style into the ovule and delivers the male gamete to the egg. Once inside the egg, the male gamete fuses with the female egg cell to produce a zygote.

The zygote develops quickly to form an embryo plant, which needs to be protected from drying out. This embryo plant contains tissues that will become roots, stem and leaves when it starts to grow or germinate.

Surrounding the zygote, other tissue grows to form food stores of carbohydrate, lipids and proteins. These food stores are called **cotyledons** – some species of plants have two cotyledons, for example beans. Other plants have only one cotyledon, such as maize and grasses.

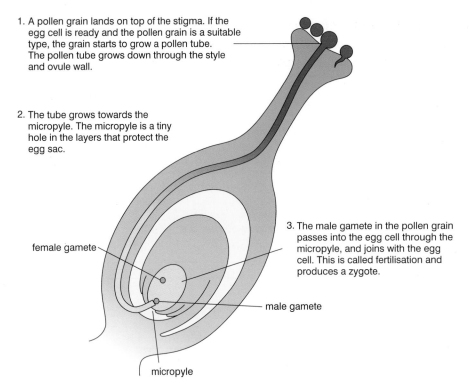

1. A pollen grain lands on top of the stigma. If the egg cell is ready and the pollen grain is a suitable type, the grain starts to grow a pollen tube. The pollen tube grows down through the style and ovule wall.

2. The tube grows towards the micropyle. The micropyle is a tiny hole in the layers that protect the egg sac.

3. The male gamete in the pollen grain passes into the egg cell through the micropyle, and joins with the egg cell. This is called fertilisation and produces a zygote.

female gamete

male gamete

micropyle

△ Fig. 3.11 Pollination leads to fertilisation.

The ovule wall hardens as the zygote grows, to form a tough or hard protective casing around the seed called the seed coat. The ovary tissue surrounding the ovule often develops into a **fruit**. If the ovary contained several ovules, the fruit will contain several seeds, such as in apples and grapes.

Fruits often have soft and sweet flesh to attract animals to eat them.

plumule
(embryo shoot)

seed coat

embryo

cotyledon

micropyle

radicle
(embryo root)

△ Fig. 3.12 The structure inside a bean seed.

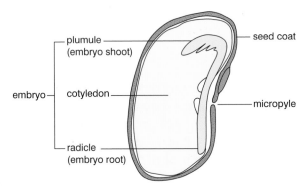

ovary turns
into the fruit

ovule turns
into the seed

fleshy fruit

hard seed

▷ Fig. 3.13 In a plum the ovule turns into a hard seed.

The hard coat of the seed protects the embryo as the seed passes through the animal's gut and egested. The seed is deposited usually at a distance from the parent plant with a useful supply of nutrients in the animal's faeces.

QUESTIONS

1. Distinguish between *pollination* and *fertilisation* in a plant.

2. Describe the process of fertilisation in a flowering plant, from the point of pollination of the flower.

3. Identify what the following structures in the carpel will develop into when fertilisation has taken place: female gamete (egg cell), ovule wall, ovary.

GERMINATION

Germination begins when the seed absorbs large amounts of water. The water acts as a solvent for many chemicals, and provides the right conditions for enzymes to start working. Warmth helps to increase the rate of enzyme action. Oxygen allows cell respiration to increase to release the energy needed for growth.

The presence of light is not usually needed for germination. This is because seeds may be below ground, where they cannot get their food from photosynthesis. The energy they need comes from breaking down food reserves in the cotyledons. Usually the embryo root (the radicle) starts to grow first, so that the developing seedling can get more water. Then the shoot starts to grow. In some plants the cotyledons are lifted above ground as the shoot grows, so they become seedling leaves. These usually turn green and start to photosynthesise. Other plants leave the cotyledons below ground, and photosynthesis does not begin until the first real leaves develop.

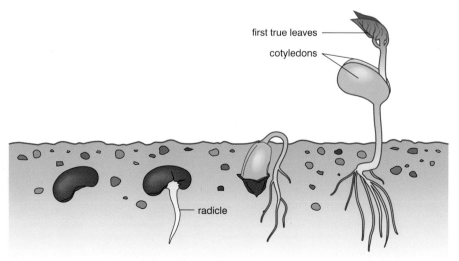

Δ Fig. 3.14 Germination of a bean seed.

CONDITIONS FOR GERMINATION

Some seeds need particular conditions before they will germinate. For example, certain seeds require exposure to very high temperatures before they will germinate. This is because they are adapted to living in dry places where fire is a common occurrence. Germinating after a fire means there is likely to be less competition with other species that usually cover the ground. Also, the ash left from the burning acts as a natural fertiliser for the new plants.

Developing investigative skills

Seeds need particular conditions for germination.

Devise and plan investigations

❶ Using the apparatus in Fig. 3.15, write a plan to investigate the effect of **(a)** light, **(b)** water and **(c)** temperature on the germination of seeds. Think carefully about what controls to use in each case.

Make observations and measurements

An investigation was carried out using two Petri dishes containing 20 seeds of the same species. Both dishes received the same amount of light and moisture, but they were kept at different temperatures. The table shows the number of seeds that germinated over a period of 8 days.

△ Fig. 3.15 A simple set-up for germinating seeds.

Day	Total number germinated seedlings	
	Cool (10 °C)	Warm (20 °C)
1	0	0
2	0	0
3	0	5
4	1	11
5	6	15
6	16	17
7	18	17
8	18	17

❷ Display the results of this investigation in a suitable way.

Analyse and interpret data

❸ Describe the patterns shown by these results.

❹ Draw a conclusion from these results.

❺ Explain the results using your scientific knowledge.

ASEXUAL REPRODUCTION IN PLANTS

Many plants also reproduce asexually to produce new individuals. Some plants produce **runners**, which are stems that grow along the ground. At the point where leaves grow from the stem, new roots can grow as well. Until the new roots are well developed in the ground, the leaves of the new plant get the water they need for growth from the roots of the parent plant, through the xylem tissue in the stem. Once the roots are growing well enough to support the new plant, the stem between the two plants can die off. Since the new plant has grown from the cells of the old plant, the two plants are genetically identical.

New plants can also be grown from bits of old plants, such as bits of leaf, stem or root. Some plants do this naturally, but farmers and plant growers make use of this when they grow new plants artificially from **cuttings**. Pieces of leaf, stem or root are cut from a healthy plant, treated with plant hormones to encourage root formation, and then planted in compost. The cuttings are kept moist, and after a few weeks, if conditions are right, the cuttings will have developed new roots. They will then start growing new leaves and shoots. Again, the new plants will be genetically identical to the parent plant from which the cuttings were taken.

△ Fig. 3.16 New plants can be produced asexually from runners.

△ Fig. 3.17 Some plant cuttings will even grow roots when placed in water. When the roots are large enough, the new plant can be planted in compost.

Be prepared to discuss the advantages and disadvantages of asexual reproduction in terms of growing new plants by taking cuttings. It is an excellent way of growing many new plants with characteristics that a grower wants (such as a particular flower colour), but all the plants will also show the same weaknesses (such as susceptibility to infection by a particular pathogen).

QUESTIONS

1. Explain the role of food reserves in a seed during germination.

2. What effect do the following conditions have on the germination of seeds?

   a) moisture

   b) warmth.

3. Explain why most seeds do not need light for germination.

4. Draw a flow chart to describe the stages in producing new plants from old ones by taking cuttings.

Sexual reproduction in humans

MALE REPRODUCTIVE SYSTEM

In humans, the male has two testes (singular: testis). These are the organs in which **sperm** are produced. It is important that they are kept at a lower temperature than the core of the body, because at this temperature the production of sperm is inhibited. The testes are supported outside the body in the scrotum.

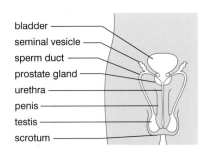

bladder
seminal vesicle
sperm duct
prostate gland
urethra
penis
testis
scrotum

△ Fig. 3.18 The male reproductive system. (Note that the bladder is not part of the reproductive system.)

Sperm ducts carry the sperm from the testes to the penis, passing through the prostate gland and seminal vesicles. The prostate gland and seminal vesicles together produce the liquid in which the sperm are able to swim. Semen is the mixture of sperm cells and fluids.

Semen passes along the sperm duct to the urethra to outside the body. The urethra also carries urine from the bladder to outside of the body. When the man is sexually excited, large blood spaces in the penis fill with blood. This causes the penis to become larger and stiffer in what is called an erection. At the same time a muscle ring (sphincter) at the top of the urethra contracts, preventing urine entering the urethra from the bladder.

QUESTIONS

1. Sketch a diagram of the human male reproductive system.

2. Add labels to your sketch to name the main parts of the system.

3. Describe the role of each of the main parts of the system in human reproduction.

FEMALE REPRODUCTIVE SYSTEM

The two ovaries are the organs that produce eggs in women. They are positioned within the abdominal cavity, either side of the uterus and joined to it by the oviducts.

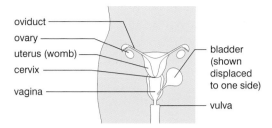

Every month from puberty until menopause, when a woman is around 50, one ovary releases one egg which travels down the oviduct to the **uterus** (womb). If it is not fertilised, the egg will be flushed from the uterus during the monthly period (bleed).

△ Fig. 3.19 The female reproductive system. (Note that the bladder is not part of the reproductive system.)

The lower end of the uterus, at the cervix, leads into the vagina, an elastic, muscular tube. The vagina opens to the outside of the body at the vulva, which is formed by the meeting of folds of skin called the labia.

QUESTIONS

1. Sketch a diagram of the human female reproductive system.

2. Add labels to your sketch to name the main parts of the system.

3. Describe the role of each of the main parts of the system in human reproduction.

SECONDARY SEXUAL CHARACTERISTICS

Secondary sexual characteristics are characteristics that develop at puberty. Some prepare the body for reproduction, others make it clear that the body is now mature enough for reproduction. The development of these characteristics is controlled by the production of sex hormones.

The male sex hormone is testosterone, which is secreted from the testes. At puberty, the increased secretion of testosterone causes the development of the following secondary sexual characteristics in boys:

• an increase in rate of growth until adult size
• hair growth on face and body including pubic hair
• penis, testes and scrotum growth and development
• deepening of voice
• increased muscle development
• sperm production.

The female sex hormones are progesterone and oestrogen, which are both produced in the ovaries. At puberty, increased secretion of oestrogen causes the development of the following secondary sexual characteristics in girls:

• an increase in rate of growth until adult size
• breast development
• vagina, oviducts and uterus development
• start of **menstrual cycle**
• hips widening
• pubic hair and under-arm hair growth.

THE MENSTRUAL CYCLE

The **menstrual cycle** is a sequence of changes that occur in a woman's body every month. The average cycle is 28 days long, but it is normal for it to vary in different women. These changes are controlled by the female sex hormones oestrogen and progesterone.

The first day of the cycle is when the lining of the uterus prepared for the previous egg starts to break down, and another egg starts to develop in one of the ovaries.

- Cells surrounding the developing egg secrete oestrogen.
- The oestrogen stimulates the lining of the uterus to repair and thicken.
- About two weeks through the cycle (around day 14) an egg is released from the ovary into the oviduct. This is called **ovulation**. When this happens, cells in the ovary start to secrete progesterone.
- Progesterone maintains the thickened uterus lining, in preparation to receive a fertilised egg.
- If the egg is not fertilised, the concentrations of progesterone starts to fall.
- The fall in hormone concentration causes the uterus lining to break down – it is lost from the body during menstruation (a period).
- When the hormone concentrations are low enough, another egg starts to develop in one of the ovaries and the cycle begins again.

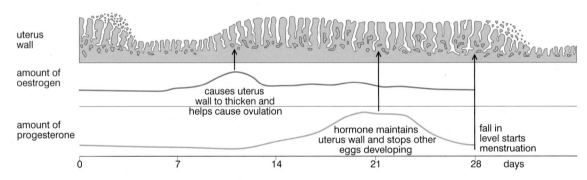

△ Fig. 3.20 Oestrogen and progesterone levels through the menstrual cycle.

If the egg is fertilised, then progesterone continues to be released from the ovary. This maintains the uterus lining during pregnancy and prevents further ovulation.

QUESTIONS

1. Describe the term secondary sexual characteristics.

2. In the menstrual cycle, what changes in the body mark the start of the cycle?

3. Draw the menstrual cycle as a circle of 28 days. On your diagram label:

a) ovulation, b) menstruation, c) increases and decreases in oestrogen secretion, d) increases and decreases in progesterone secretion.

FSH AND LH

The control of the menstrual cycle also involves two other hormones that are produced by the pituitary gland in the brain. FSH (follicle-stimulating hormone) is released from the pituitary at the start of the cycle and stimulates the development of an egg in the ovary. The developing egg produces increasing amounts of oestrogen as it develops. Oestrogen stimulates the release

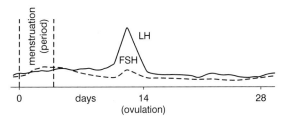

△ Fig. 3.21 FSH and LH levels through the menstrual cycle.

of LH (luteinising hormone) from the pituitary which causes ovulation. Progesterone inhibits the production of FSH and LH. If the egg is not fertilised, concentrations of oestrogen and progesterone fall, which allows the pituitary to start secreting FSH again.

QUESTIONS

1. a) Name the four hormones involved in controlling the menstrual cycle.

   b) For each hormone, identify the site of secretion and the site of any target cells.

2. Describe how the hormones are transported through the body.

3. a) Sketch a diagram to show how the concentrations of the hormones vary through the menstrual cycle.

   b) Annotate your sketch to show the cause of a change in a hormone.

4. Explain why the menstrual cycle is not coordinated by the nervous system.

5. The control of the menstrual cycle could be considered as an example of homeostasis in the body. Explain why.

FERTILISATION AND DEVELOPMENT OF THE FETUS

Stiffening of the penis makes it possible for a man to insert his penis into the vagina of a woman to deliver sperm during sexual intercourse. Rapid contractions of muscles in the penis during ejaculation send the sperm shooting out into the vagina, on their way to the egg cell.

Sperm deposited near the cervix swim up into the uterus, and then along the oviduct to the egg. Many sperm fail to make the journey but some will reach the oviducts at the top end of the uterus. The egg will have been travelling along the oviduct while the sperm have been swimming up from the uterus. Fertilisation takes place in the oviduct. The nucleus of one sperm cell fuses with the nucleus of the egg cell, forming a fertilised egg or zygote.

After fertilisation in the oviduct, the fertilised egg (zygote) travels on towards the uterus. The journey takes about three days, during which time the zygote will divide several times to form a ball of cells, which is now called an embryo.

In the uterus, the embryo embeds in the thickened lining (implantation) and cell division and growth continue. For the first three months, the embryo gets nutrients from the mother by diffusion through the uterus lining.

By the end of three months, the placenta has developed, and the embryo has developed into a **fetus** in which all the main organs of the body can be identified.

Over the next 28 weeks the fetus will increase its mass very rapidly. At no other point in an individual's lifetime will it grow at such a high rate. The period of development in the uterus is known as **gestation**, and lasts about 40 weeks in humans, measured from the time of the woman's last period.

The rapid growth during gestation depends on a good supply of food and oxygen, provided by the mother.

△ Fig. 3.22 Just before fertilisation when a sperm fuses with an egg.

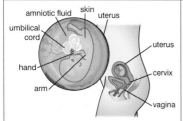

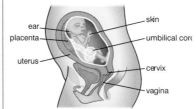

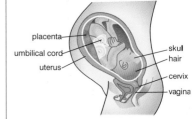

11 weeks:
- fetus about 4 cm long
- most of the main body structures have formed

23 weeks:
- fetus about 29 cm long and weighs about 500 g
- fetus hears sound from outside, and moves

40 weeks:
- fetus is about 51 cm long and weighs about 3.4 kg
- all organs fully developed ready for birth

△ Fig. 3.23 The fetus in the uterus at three times during gestation.

The fetus develops inside a bag of fluid called **amniotic fluid**. This fluid is produced from the membrane (amnion) that forms the inner layer of the bag (amniotic sac). The fluid protects the fetus from damage. It also reduces the effect of large temperature variations which would affect the rate of development of the fetus. One of the signs that birth will be soon is when this bag bursts during labour.

SCIENCE IN CONTEXT — ULTRASOUND SCANS

Ultrasound is very high frequency sound, far above the frequency that can be heard. It is used in medical imaging for showing soft tissues inside the body. It is particularly useful for looking at the developing fetus in the uterus, because it does not harm either the fetus or the mother.

An ultrasound scan is commonly done about halfway through gestation, to make sure that the fetus is developing normally. At about this stage, if the fetus is lying at the right angle, it may even be possible to tell if it is a male fetus because the testes can be distinguished at this age.

Ultrasound scans may be done at other times during gestation if there is any concern about the development of the fetus.

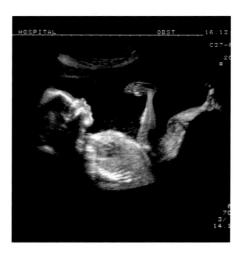

△ Fig. 3.24 This ultrasound scan was taken in the 20th week of gestation and shows that the fetus is developing normally.

THE PLACENTA

The **placenta** is an organ that allows a constant exchange of materials between the mother and fetus. It develops from fetal tissues. The placenta and the uterus wall have a large number of blood vessels that run very close to each other but do not touch. Maternal and fetal blood do not mix. If they did, the higher blood pressure in the mother could damage the fetus. It also helps to prevent pathogens and some chemicals getting into the blood of the fetus.

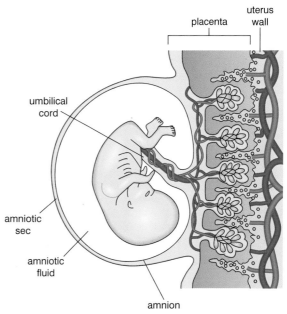

▷ Fig. 3.25 Substances are exchanged between the mother's blood and the fetus's blood across the placenta.

The fetus is joined to the placenta by the umbilical cord, which carries the blood vessels of the fetus. Dissolved food molecules, oxygen and other nutrients that the fetus needs for growth diffuse from the mother's blood into the blood of the fetus. Waste products from metabolism in the fetus diffuse across into the mother's blood.

Once the placenta has formed, until birth, this is the only way that the developing fetus exchanges materials with the outside world. Birth usually occurs when all the organs of the fetus are fully developed and ready to carry out the life processes on their own.

QUESTIONS

1. Explain the following terms in your own words:

   a) zygote

   b) embryo

   c) fetus.

2. Describe the role of the following during the development of a fetus:

   a) amniotic fluid

   b) placenta.

3. Explain as fully as possible why the placenta contains a large number of blood vessels from the mother and from the fetus that run very close together.

End of topic checklist

Amniotic fluid is fluid surrounding the developing fetus in the uterus, which protects the fetus from mechanical damage.

The **anther** is the male part of a flower that produces pollen.

Asexual reproduction is the production of new individuals without fertilisation, from the division of body cells in the parent.

The **carpel** is the female structure in flowers which contains one or more ovaries and their stigmas and styles.

A **cutting** is a part taken from a plant (such as piece of stem, root or leaf) and treated so that it grows into a new plant, a form of artificial asexual reproduction of plants.

An **egg** cell is the female gamete.

An **embryo** is a developing young organism, where cell division and differentiation are taking place rapidly. In plants it develops in the seed. In humans, the embryo is the stage between zygote and fetus.

Fertilisation is when male and female gametes join together.

Fetus is the name given to the developing baby in the uterus.

Fruit is the usually soft, fleshy structure surrounding a seed, formed from the ovary after fertilisation.

FSH (follicle-stimulating hormone) is a hormone produced by the pituitary gland that helps control the menstrual cycle.

A **gamete** is a sex cell.

Germination is when a plant starts to grow from a seed.

LH (luteinising hormone) is a hormone produced by the pituitary gland that helps control the menstrual cycle.

The **menstrual cycle** is the sequence of events that take place each month in a woman's reproductive organs.

Oestrogen is a hormone produced by the ovaries that helps to control the menstrual cycle and produces secondary sexual characteristics in girls.

The **ovary** (in plants and humans) is a structure that contains immature egg cells.

An **ovule** is the female structure in a flower that contains one egg cell.

End of topic checklist continued

The **placenta** is the structure formed by a fetus that attaches to the uterus wall and exchanges essential substances between the mother's blood and the blood of the fetus.

A **pollen grain** is a male structure in plants that contains the male gamete.

A **pollen tube** is a tube that develops from a pollen grain down through the style, carrying the male gamete to the female gamete.

Pollination is the process in which pollen from one flower is transferred to another flower, before fertilisation can take place.

Progesterone is a hormone produced in the ovaries that helps to control the menstrual cycle

A **runner** is a stem produced by some plants that grows along the ground from which new plants develop.

Secondary sexual characteristics are the physical characteristics that develop at puberty.

A **seed** is a hard-shelled structure formed from an ovule that contains the plant embryo and food stores.

Sexual reproduction is the production of new individuals by the fusion of a male and a female gamete.

A **sperm** is the male gamete in humans.

The **stamen** is the male structure in flowers that contains the anther.

The **stigma** is the female structure in flowers where pollen grains attach in pollination.

A **style** is a structure that supports the stigma in a flower.

The **testes** are the site of sperm production in men.

Testosterone is a hormone produced by the testes that produces secondary sexual characteristics in boys.

A **zygote** is a fertilised egg, formed from the fusion of a male gamete and female gamete.

The facts and ideas that you should know and understand by studying this topic:

◯ Asexual reproduction is the production of new individuals without fertilisation. It is the division of the body cells of one parent. It produces offspring that are genetically identical to the parent and each other.

- Sexual reproduction is the production of new individuals from the fusion of a male gamete and a female gamete during fertilisation. It requires two parents, and produces offspring that are genetically different to their parents and each other.
- Fertilisation of a male gamete and female gamete produces a zygote that develops into an embryo by cell division.
- An insect-pollinated flower has features such as coloured petals, scent and nectaries. These attract insects to feed at the flower. The insects pick up pollen which they transfer to other flowers when they feed there.
- A wind-pollinated flower is usually small, without coloured petals, scent or nectaries. It produces a large amount of lightweight pollen that is scattered over a large distance in the wind.
- After pollination, a pollen tube grows down inside the style to deliver the male gamete to the female gamete in the ovule so they can join in fertilisation. After fertilisation, the ovule develops into a seed, and the ovary becomes the fruit.
- Seeds need moisture and warmth and oxygen for successful germination, but not light.
- Germinating seeds use the chemical energy of their food reserves to support the life processes for growth until they produce leaves that can photosynthesise.
- Some plants can reproduce asexually by natural methods (e.g. runners) or by artificial methods (e.g. cuttings).
- The human male reproductive system includes: the testes where sperm are made; the sperm ducts that carry the sperm to the urethra; the prostate gland and seminal vesicles that produce liquid in which the sperm swim; the penis, which when erect delivers sperm into the vagina of the woman; and the urethra, which carries the sperm from the sperm ducts to the outside of the body.
- The human female reproductive system includes: the ovaries, where the egg cells are made; the oviducts, which carry the eggs to the uterus and where fertilisation takes place with sperm cells; the uterus, where the embryo embeds and develops into a fetus; the cervix, where sperm are deposited at the base of the uterus; and the vagina, where the penis is inserted during sexual intercourse.
- The stages of the menstrual cycle are controlled by the hormones oestrogen, progesterone, FSH and LH.
- The hormones oestrogen and testosterone bring about the development of the secondary sexual characteristics.
- Once the embryo has embedded in the uterus wall, it develops the placenta. This is the site of exchange of nutrients and waste materials between the blood of the mother and of the fetus.
- During development, the embryo (and later the fetus) is protected from mechanical damage and temperature fluctuations by amniotic fluid that surrounds it in the amniotic sac.

End of topic questions

1. The photograph shows a catkin on a goat willow tree. A catkin is formed from a group of flowers.

   a) What is the purpose of the flowers on a goat willow tree? **(1 mark)**

   b) Name the yellow parts of the flowers shown in this photograph. **(1 mark)**

   c) Describe their purpose in a flower. **(1 mark)**

   d) Are goat willow flowers pollinated by the wind or by insects? Explain your answer using clues from the photograph. **(2 marks)**

2. A gardener has some packets of seeds for planting. The packets explain how to plant the seeds to get the best germination.

   a) What is meant by *germination*? **(1 mark)**

   b) All the packets say that the seeds need to be planted in moist compost and kept warm. Explain why the seeds need these conditions. **(2 marks)**

   c) Larger seeds can be planted deeper in the compost, but the tiniest seeds need to be scattered on the surface of the compost. Explain why different seeds need to be planted at different depths. (Hint: think about food reserves.) **(3 marks)**

   d) Some seeds that come from plants in high-latitude regions (such as Canada or Russia) need to be placed in the freezer for a few weeks before they will germinate. This makes them respond as if they had been through a cold winter. Explain the survival advantage of this adaptation. **(2 marks)**

3. a) Explain the advantage to a flower of having adaptations for attracting insects rather than relying on wind for pollination. **(1 mark)**

   b) Explain one disadvantage for an insect-pollinated plant that relies on one or just a small number of insect species for pollination. **(1 mark)**

4. a) Where are sperm cells made in the human body? (1 mark)

b) Where are egg cells made in the human body? (1 mark)

c) Where is an egg cell fertilised by a sperm cell? (1 mark)

d) Starting from the point of their formation, explain how a sperm cell reaches the egg cell at fertilisation. (4 marks)

5. a) Describe fully how the hormones oestrogen and progesterone control the menstrual cycle. (4 marks)

b) Describe the roles of FSH and LH in the menstrual cycle. (2 marks)

6. a) What is the placenta? (1 mark)

b) What role does the placenta play in supporting the fetus? (1 mark)

c) How are substances exchanged across the placenta? (2 marks)

d) Explain one advantage of keeping the mother's blood separated from the blood of the fetus. (1 mark)

e) Tobacco smoke contains carbon monoxide. Explain as fully as you can why, if a woman smokes during pregnancy, her baby may weigh less than the average at birth. (4 marks)

Inheritance

INTRODUCTION

Unless a zygote (fertilised egg) divides completely into two, and develops as two identical twins, the baby that develops from that zygote is genetically unique. Each cell in a zygote contains genetic information, half of which came from the father and half from the mother. And during gamete formation, some changes may have occurred in some of that genetic information. So the baby may have some variations in its genes that neither of its parents has. Interestingly, although the baby is genetically unique, virtually all of the cells in its body are genetically identical.

△ Fig. 3.26 This baby will have inherited some characteristics from her mother and some from her father.

KNOWLEDGE CHECK

✓ Organisms show variation in their features.
✓ Variation can be inherited or caused by the environment.
✓ Genes are small parts of the genetic information (DNA) found in the nucleus of a cell.

LEARNING OBJECTIVES

✓ Understand that the genome is the entire DNA of an organism and that a gene is a section of a molecule of DNA that codes for a specific protein.
✓ Understand that the nucleus of a cell contains chromosomes on which genes are located.
✓ Describe a DNA molecule as two strands coiled to form a double helix, the strands being linked by a series of paired bases: adenine (A) with thymine (T), and cytosine (C) with guanine (G).
✓ Understand that an RNA molecule is single stranded and contains uracil (U) instead of thymine (T).
✓ Describe the stages of protein synthesis including transcription and translation, including the role of mRNA, ribosomes, tRNA, codons and anticodons.
✓ Understand how genes exist in alternative forms called alleles which give rise to differences in inherited characteristics.
✓ Understand the meaning of the terms: dominant, recessive, homozygous, heterozygous, phenotype and genotype.
✓ Understand the meaning of the term codominance.
✓ Understand that most phenotypic features are the result of polygenic inheritance rather than single genes.
✓ Describe patterns of monohybrid inheritance using a genetic diagram.
✓ Understand how to interpret family pedigrees.
✓ Predict probabilities of outcomes from monohybrid crosses.

- ✓ Understand how the sex of a person is controlled by one pair of chromosomes, XX in a female and XY in a male.
- ✓ Describe the determination of sex of the offspring at fertilisation, using a genetic diagram.
- ✓ Understand how division of a diploid cell by mitosis produces two cells that contain identical sets of chromosomes.
- ✓ Understand that mitosis occurs during growth, repair, cloning and asexual reproduction.
- ✓ Understand how division of a cell by meiosis produces four cells, each with half the number of chromosomes, and that this results in the formation of genetically different haploid gametes.
- ✓ Understand how random fertilisation produces genetic variation of offspring.
- ✓ Know that in human cells the diploid number of chromosomes is 46 and the haploid number is 23.
- ✓ Understand that variation within a species can be genetic, environmental, or a combination of both.
- ✓ Understand that mutation is a rare, random change in genetic material that can be inherited.
- ✓ Understand how a change in DNA can affect the phenotype by altering the sequence of amino acids in a protein.
- ✓ Understand how most genetic mutations have no effect on the phenotype, some have a small effect and rarely do they have a significant effect.
- ✓ Understand that the rate of mutations can be increased by exposure to ionising radiation and by some chemical mutagens.
- ✓ Explain Darwin's theory of evolution by natural selection.
- ✓ Understand how resistance to antibiotics can increase in bacterial populations, and appreciate how such an increase can lead to infections being difficult to control.

CHROMOSOMES, DNA AND GENES

Inside virtually every cell in the body is a nucleus, which contains long threads called **chromosomes**. These threads are usually stretched out and fill the nucleus but, when the chromosomes condense (gather into bundles) just before cell division, they can be seen through a microscope. The chromosomes are made of a chemical called **DNA** (deoxyribonucleic acid).

Different sections of a DNA molecule carry codes that tell cells how to make different proteins. This is significant because proteins are the basis for the characteristics of an individual. For example, a protein called melanin is responsible for dark hair and skin. The section of DNA that produces a particular characteristic or protein is called a **gene**. Each chromosome consists of many different genes. All the DNA or genes of an organism are called its **genome**.

DNA is a very long molecule that is formed from two parallel strands ('backbones') joined together at regular spaces by pairs of **bases**, like the rungs of a ladder. The whole structure is twisted, forming a **double helix**. There are four different bases in DNA: **adenine (A)**, **thymine (T)**, **cytosine (C)** and **guanine (G)**.

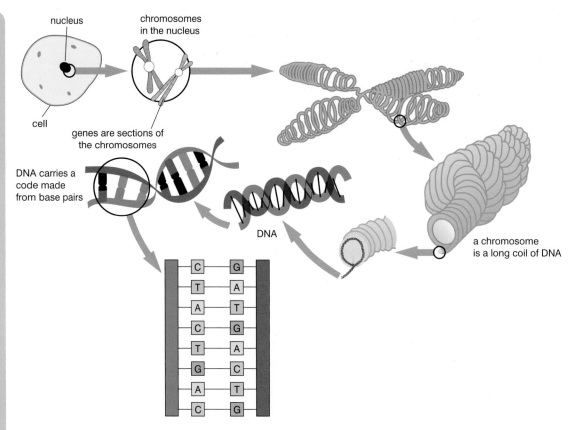

△ Fig. 3.27 The relationship between cell, nucleus, chromosome, DNA, gene and bases.

The bases always form the same pairs: A pairs with T and C pairs with G (see Fig. 3.27). So you can use the order of bases on one strand of DNA to identify the bases that will be on the opposite strand. For example, CTACTGAC on one strand is matched with GATGACTG on the opposite strand.

REMEMBER

To help you remember the base pairing:
- the straight letter A pairs with the straight letter T
- the round letter G pairs with the round letter C.

PROTEIN SYNTHESIS

Each gene is a code that tells the ribosomes in a cell how to make a specific protein. Making proteins is called **protein synthesis**. This involves another type of molecule called **RNA** (ribonucleic acid). RNA is similar to DNA but it is single stranded, and contains the base **uracil** (**U**) instead of thymine (T).

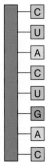

△ Fig. 3.28 RNA molecules are single stranded and contain U instead of T.

Proteins are made of amino acids and each type of protein has a different amino acid sequence. There are 21 different naturally occurring amino acids but a protein molecule may contain hundreds or even thousands of amino acids. (It's a bit like English, where there are 26 letters, but you can make up many different words, of varying lengths, by using different sequences of letters. Letters are like amino acids, and words are like proteins - except in English, words are not hundreds of letters long!)

The amino acid sequence is coded for by the sequence of bases on DNA and RNA, and every three bases codes for one particular amino acid. This is called the **genetic code**. Each group of three bases is called a **codon**.

The stages in protein synthesis are as follows.

1. A section of DNA that makes up a gene 'unzips' forming two single strands.

2. One of the strands acts as a 'template' for the formation of a corresponding strand of RNA. This process is called **transcription**. The bases on the RNA pair with the bases on the DNA, so C and G pair with each other, RNA A bases pair with DNA T bases, and RNA U bases pair with DNA A bases. The new length of RNA formed is called **mRNA** (messenger RNA).

3. The mRNA molecules leave the nucleus and travel to the **ribosomes** (which are also made of RNA).

4. At the ribosomes, amino acids are joined together to make up proteins. This process is called **translation**. The sequence of bases on the RNA determines the sequence of amino acids that make up the protein. Translation also involves another type of RNA called **tRNA** (transfer RNA). There are different types of tRNA molecule. Each one has a particular amino acid attached at one end, and a corresponding sequence of three bases at the other end called an **anticodon**. A ribosome moves along an mRNA molecule 'reading' the bases. Every three bases on the mRNA is called a codon. The ribosome attaches a tRNA molecule, with the corresponding anticodon, to the codon on the mRNA. The ribosome then 'reads' the next mRNA codon and attaches another corresponding tRNA molecule. The amino acids on the two tRNA molecules become attached and so start to build up a protein. The ribosome continues to move along the mRNA until it reaches the end. By this stage a long sequence of amino acids has been put together which makes up a new protein molecule. Once they have 'delivered' their amino acids, the tRNA molecules can collect new amino acids.

These steps are shown in Fig. 3.29 (next page).

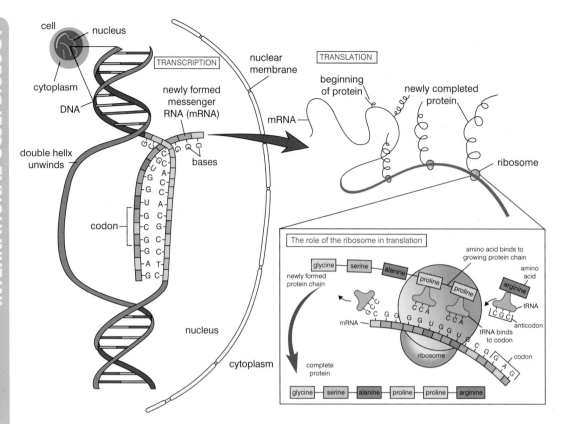

△ Fig. 3.29 Protein synthesis showing the stages of transcription and translation.

Enzymes are proteins, so many of the effects of genes are the result of the way the enzymes that are produced control reactions inside the cell.

ALLELES

Genes code for particular characteristics, such as the colour of an animal's fur. However, variations in a characteristic can be caused by different forms of a gene, called **alleles**. For example, leopards have two alleles controlling whether they have spots or are completely black (see page 219).

If you take all the chromosomes from a body cell, you can arrange them into pairs. This is because you inherit one of each chromosome pair from your father and one from your mother. Both chromosomes in the pair carry the same genes (except for the sex chromosomes – see page 230), but the genes on each chromosome may be in the form of the same allele or different alleles. In humans there are 46 chromosomes in a body cell, which make up 23 pairs.

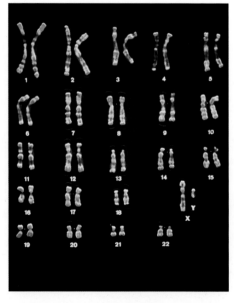

△ Fig. 3.30 The chromosomes of a human male arranged in chromosome pairs.

QUESTIONS

1. Put the following in order of size, starting with the smallest: cell, chromosome, gene, nucleus.

2. In relation to DNA, explain what we mean by:

   a) a *double helix*, b) *paired bases.*

3. What are *alleles*?

4. How is RNA different from DNA?

5. In your own words describe what happens in:

   a) transcription, b) translation.

INHERITING CHARACTERISTICS

Some characteristics, such as the colour of your eyes, are passed down (**inherited**) from your parents, but other characteristics may not be passed down. Sometimes characteristics appear to miss a generation: for instance, you and your grandmother might both have dry earwax (see page 228), but both of your parents may have wet earwax.

Spotted leopard parents occasionally have a cub that has completely black fur instead of the usual spotted pattern. It is known as a black panther but is still the same species as the ordinary leopard.

△ Fig. 3.31 Two spotted leopard parents may produce offspring with spotted coats or black coats.

Just as in humans, leopard chromosomes occur in pairs. One pair carries a gene for fur colour. There are two copies of the gene in a normal body cell (one on each chromosome). The version of the gene (allele) may be identical but sometimes they are different, one being for a spotted coat and the other for a black coat.

Leopard cubs receive half their genes from each parent. Eggs and sperm cells contain only half the number of chromosomes of normal body cells. This means that egg and sperm cells contain only one of each pair of alleles. When an egg and sperm join together at fertilisation, forming a zygote which will develop into the new individual, it now has two alleles of each gene.

Different combinations of alleles will produce different fur colour:

spotted coat allele	+	spotted coat allele	=	spotted coat
spotted coat allele	+	black coat allele	=	spotted coat
black coat allele	+	black coat allele	=	black coat

△ Table 3.2 Combinations of fur colour alleles in leopard cubs.

The black coat only appears when *both* of the alleles for the black coat are present. As long as there is at least one allele for a spotted coat, the coat will be spotted, because the allele for a spotted coat overrides the allele for a black coat. It is the **dominant allele**. Alleles like the one for the black coat are described as **recessive**.

- An individual with two identical alleles for a gene is said to be **homozygous** ('homo' means 'the same') for that gene.
- An individual with two different alleles for a gene is said to be **heterozygous** for that gene ('hetero' means 'different').

A leopard with a spotted coat may be homozygous for the spotted coat allele, or heterozygous. A leopard with a black coat can only be homozygous for the black coat allele.

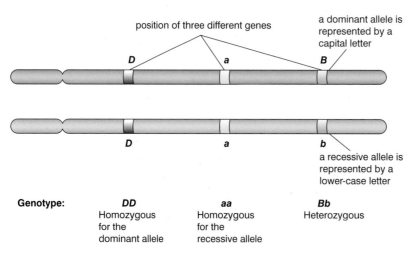

△ Fig. 3.32 A pair of chromosomes showing some definitions in genetics.

If you are female, you have two copies of every gene on all chromosomes. In males, a few genes are present on the X sex chromosome but not on the Y chromosome. These genes can produce sex-linked characteristics, where men are more likely to be affected by a recessive allele than women.

QUESTIONS

1. Explain the following terms in your own words:

   a) *dominant*

   b) *recessive*

   c) *homozygous*

   d) *heterozygous.*

2. How many alleles for a particular gene would be found in:

   a) a body cell

   b) a gamete

   c) a zygote?

MONOHYBRID CROSSES

An individual's combination of genes is his or her **genotype**. An individual's combination of physical features is his or her **phenotype**. Your genotype influences your phenotype.

We can show the influence of the genotype in a **genetic diagram**. This uses a capital letter for the dominant allele and a lower case letter for the recessive allele.

REMEMBER

To avoid the risk of confusion when drawing genetic diagrams, we choose a letter that is easily distinguished in capital and lower-case, especially when written by hand. For example, S or C for the coat gene might seem reasonable to use for the leopard example, but this gives alleles S/s or C/c which are not ideal.

Using the example of the leopards, **B** stands for the dominant allele for a spotted coat and letter **b** stands for the recessive allele for the black coat. Two spotted parents who have a black cub must each be carrying an **B** and an **b**. The genetic diagram below shows the possible genotypes and phenotypes of the offspring.

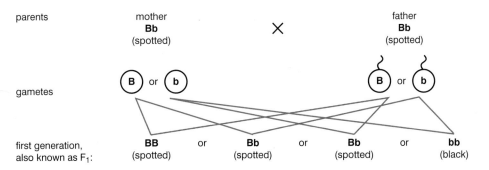

$\triangle$ Fig. 3.33 Three different genotypes are possible from the cross in this diagram. The probability of each genotype is 1BB: 2Bb: 1bb.

Because we are looking at a characteristic (fur colour) controlled by one gene, this is an example of a **monohybrid cross**. 'Mono' means one, and a 'hybrid' is produced when two different types breed or cross.

Another type of genetic diagram is known as a **Punnett square**. The example above can be shown in a Punnett square. The four boxes at the bottom right show the possible combinations in the offspring.

			male Bb spotted	
			gametes	
			B	b
female Bb spotted	gametes	B	BB spotted	Bb spotted
		b	Bb spotted	bb black

PROBABILITIES AND PREDICTIONS

In a monohybrid cross, when two heterozygous parents are crossed, the phenotype of the offspring with the dominant allele and the offspring with the recessive allele appears in the ratio of 3:1. The 3:1 ratio refers to the probabilities of particular combinations of alleles, so the probability of having an offspring with the phenotype of the dominant allele is three times the probability of having an offspring with the phenotype of the recessive allele.

In the example of the leopards, there is a 1 in 4 or 0.25 or 25% probability of a leopard cub being black. This is because there must be two recessive alleles (homozygous) in order for the phenotype to be expressed (visible).

With a large number of offspring in an *actual* cross of two heterozygous leopard parents, you would expect something near the 3:1 ratio of spotted to black cubs. However, because it is a matter of chance which sperm cell fertilises which egg cell, you should not be too surprised if a small litter – for example, of four cubs – contained two black cubs, or none.

Using the example of leopard coat colour and a Punnett square, we can also look at what happens if we cross homozygous and heterozygous individuals:

			male Bb spotted	
			gametes	
			B	b
female bb black	gametes	b	Bb spotted	bb black
		b	Bb spotted	bb black

The predicted outcome from this cross is a 1:1 ratio of spotted to black colouring. This gives a 1 in 2 or 0.5 or 50% probability of a cub from these parents being spotted, or being black.

REMEMBER

In order to gain higher marks, you will need to be able to predict probabilities of outcomes from any monohybrid cross. Practise drawing genetic diagrams to make sure you are confident with them. Although there are different types of genetic diagrams, if you are given the choice in an examination, it is usually best to choose to draw a Punnett square, as you are far less likely to make any mistakes.

Can you predict what will happen if a homozygous spotted coat leopard is crossed with a heterozygous spotted coat leopard?

			male Bb spotted	
			gametes	
			B	b
female BB spotted	gametes	B	BB spotted	Bb spotted
		B	BB spotted	Bb spotted

In this case, although some of the cubs born are likely to be homozygous and some heterozygous, they will all have spotted coats. They will have the same phenotype but not the same genotype.

QUESTIONS

1. Explain *monohybrid inheritance* in your own words.

2. Rabbits have a gene for coat colour – the allele for brown coat is dominant over the allele for black coat colour. Using the letter B for the dominant allele, and b for the recessive allele, write down all the possible genotypes and phenotypes for this gene. Explain your answers.

3. Using your answers from Question 2:

   a) Construct a genetic diagram to show the possible offspring from a cross between a male rabbit that is heterozygous and a female rabbit that is heterozygous.

   b) What is the probability of producing a black baby rabbit from this cross?

Developing investigative skills

In an investigation into the inheritance of a characteristic, students used red beads to represent dominant alleles and blue beads to represent recessive alleles.

As a homozygous dominant individual produces gametes that only contain the dominant allele, all the red beads were placed into a beaker to represent the gametes for this individual. As a homozygous recessive individual produces gametes that only contain the recessive allele, all the blue beads were placed into another beaker to represent the gametes for this individual.

To model what would happen in a cross between these two individuals, they took one gamete (bead) from one pot and paired it with one gamete (bead) from the other pot and wrote down the genotype and phenotype for that 'offspring'. This showed that all the offspring from these parents would be heterozygous (one red, one blue bead).

Devise and plan investigations

❶ Describe and explain how you would adapt this method to represent a cross between two heterozygotes. (Hint: Make sure you use enough beads to get a reasonable approximation of the actual result to the expected result.)

Analyse and interpret data

Some students carried out an investigation like this that started with two 'parents' heterozygous for a characteristic.

❷ Each pot started with 40 beads. How many red beads and how many blue beads were in each of the two pots? Explain your answer.

Only 20 selections were made from the two beakers, to produce the 'offspring'. The results are shown in this table.

Number of red/red pairs in 'offspring'	5
Number of red/blue pairs in 'offspring'	12
Number of blue/blue pairs in 'offspring'	3

❸ Draw a genetic diagram for this cross, to show the predicted probabilities of genotypes and phenotypes. (Hint: remember to choose letters for the alleles and explain which allele is modelled by the red beads and which by the blue beads.)

❹ Describe how the actual results differ from the expected results.

Evaluate data and methods

❺ Comment on the difference between the expected and actual results.

❻ Explain how you would adjust the method to help improve the results.

TEST CROSSES

Breeders can use a cross with a homozygous recessive individual to test whether an individual with the dominant phenotype is homozygous or heterozygous for the dominant allele. For example, a plant breeder may have a tall pea plant. The allele for tallness (T) is dominant over the allele for dwarfness (t). How can the plant breeder find out if the plant is TT or Tt?

- A cross between a homozygous dominant (TT) and homozygous recessive (tt) will produce all heterozygous (Tt) offspring, and so all offspring will have the phenotype of the dominant allele.

- A cross between a heterozygous dominant (Tt) and a homozygous (tt) recessive will have a 50% chance of producing offspring that have the phenotype of the dominant allele (tall) and a 50% chance of producing offspring with the phenotype of the recessive allele (dwarf).

(Try drawing genetic diagrams to confirm this for yourself.)

This kind of cross is known as a test cross.

CODOMINANCE

In the leopard example above, one allele of the gene pair for coat colour was dominant and the other was recessive. When both alleles of a gene pair in a heterozygote are expressed in the phenotype, with neither being dominant or recessive to the other, this is called **codominance**.

Human blood types are determined by three different alleles of the same gene: I^A, I^B and I^O (note that I represents the gene and the superscript letter shows the allele). I^A results in the production of the A antigen in blood, I^B results in the production of B antigen, I^O produces no antigen. The I^O allele is recessive but the I^A and I^B alleles are codominant. These three possible alleles can give us the following allele pairs:

- $I^A I^A$
- $I^B I^B$
- $I^O I^O$
- $I^A I^B$
- $I^A I^O$
- $I^B I^O$

These six different genotypes give us four different phenotypes: the four human blood groups A, B, AB and O.

The phenotype of blood group A can have the genotype of I^AI^A or I^AI^O because I^A is dominant over recessive I^O.

The phenotype of blood group B can have the genotype of I^BI^B or I^BI^O allele pairs because I^B is dominant over recessive I^O.

The phenotype of blood group AB has the genotype of the two codominant alleles, I^AI^B.

The phenotype of blood group O can only have the genotype I^OI^O, the recessive allele pair.

Alleles present (genotype)	Blood group (phenotype)
I^AI^A	A
I^AI^O	A
I^BI^B	B
I^BI^O	B
I^AI^B	AB
I^OI^O	O

△ Table 3.3 Genotypes and phenotypes for blood groups.

We can use genetic diagrams to predict the outcomes of crosses that involve codominant alleles.

	Mother	Father
Parental phenotype:	Blood group A	Blood group B
Parental genotype:	I^AI^O	I^BI^O
Possible gametes:	I^A $\qquad$ I^O	I^B $\qquad$ I^O
Possible offspring genotypes:	I^AI^B $\qquad$ I^AI^O	I^BI^O $\qquad$ I^OI^O
Possible offspring phenotypes:	Blood group AB $\qquad$ Blood group A	Blood group B $\qquad$ Blood group O

△ Fig. 3.34 A genetic diagram predicting the offspring of heterozygous parents of blood groups A and B.

A cross between a parent who is heterozygous for blood group A and a parent who is heterozygous for blood group B produces a predicted ratio of 1:1:1:1 for children with each of the four blood groups, giving a 25% or 0.25 or 1 in 4 probability that a child will inherit any one of the four blood groups.

Gregor Mendel (1822–1884) was the first person to study genetic inheritance in a thorough and scientific manner. He chose characteristics in peas to study because he could see clear differences in characteristics and patterns in their inheritance. He started by crossing plants with the same characteristics many times, until he was certain that they were pure-breeding (that is they would only produce offspring with that characteristic).

He then made hundreds of crosses of the same kind. He started by removing the anthers of each flower. Then he brushed pollen from a plant he had chosen for one parent on to the stigma of the other 'parent' and covered the flower to prevent other pollen getting in.

From his results, Mendel was able to show that alleles generally do not mix effects in the phenotype, but that a dominant allele in a heterozygote prevents the recessive allele being expressed.

1. Why was it important that the parent plants were pure-breeding?

2. Why did Mendel need to carry out hundreds of crosses before drawing a conclusion?

3. Pea flowers are pollinated by insects. How could Mendel be certain that no chance fertilisations took place?

4. At the time Mendel carried out his work, people could not understand how a characteristic could be present in one generation, 'disappear' in the next generation and then reappear in the next. Using genetic diagrams, and a characteristic of your choice, show how this happens when starting with pure-breeding parents for the dominant and recessive characteristics.

5. Explain the importance of a thorough and scientific method for drawing reliable conclusions.

POLYGENIC INHERITANCE

Although some characteristics are controlled by the alleles of a single gene, in the majority of cases more than one gene influences a particular phenotypic feature. For example, human hair colour, skin colour and height are each affected by more than one gene (and each of these genes may have several alleles). This is known as **polygenic inheritance** (*poly* means 'many'). This is part of the reason why there is so much variation in these features between different people.

QUESTIONS

1. Explain why the differences in some features, such as skin colour, can not be explained using a monohybrid cross genetic diagram.

2. Explain what is meant by *codominance*.

3. a) A man has blood group AB. Which alleles must he have in order to show this blood group?

b) A woman has blood group O. Which alleles must she have in order to show this blood group?

c) Draw a Punnett square to show the possible blood groups in the children from these two people.

d) State the probability of each possible blood group in the children.

FAMILY PEDIGREES

A genetic diagram predicts the possible outcomes of a cross for alleles of a particular gene. This is useful in species that produce large numbers of offspring, because chance is averaged out by the large number of crosses, so that the predicted and actual outcomes of a cross are similar. But in species like humans, where parents produce only a few offspring, it can be difficult to work out whether an allele is dominant or recessive from actual results.

Instead we use a **family pedigree** to investigate the inheritance of a particular characteristic. This shows the relationship between individuals in a large family and which of them has which version of that characteristic.

One example of a characteristic we can study like this is earwax type. Some people have wet earwax, which is an orangey colour and sticky. Other people have dry earwax which is paler and flaky.

Here is a family pedigree that shows the inheritance of earwax type. Note the key, which explains whether each individual in the diagram is male or female and which form of earwax they have.

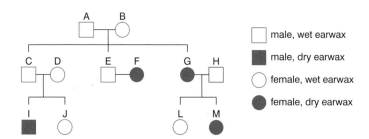

△ Fig. 3.35 A family pedigree for earwax.

This pedigree shows the following information:

- There are three generations shown for this family – A and B are the eldest generation, I, J, L and M are the youngest generation.
- Couple A and B had three children, couples C/D and G/H each have two children.
- This family includes four people with dry earwax; one is male and three are female.

This pedigree also shows that, although A and B do have wet earwax, they had a child (G) who has dry earwax. Similarly, couple C/D have a child with dry earwax. Using genetic diagrams we can show that:

- since G has dry earwax, at least one of the two parents must have the allele for dry earwax
- if the dry earwax allele was dominant, then either A or B should have dry earwax
- since neither A or B have dry earwax, the allele for dry earwax must be recessive
- since G has dry earwax, A and B must both be heterozygous for the dry earwax allele.

If you aren't sure about any of these statements, draw a genetic diagram to help you confirm what they are saying.

Using this kind of argument, you can work out the genotypes of all the individuals in the pedigree except E and J (because they have not had any children and may be homozygous dominant or heterozygous).

SCIENCE IN CONTEXT

INHERITED DISEASES

Some diseases are inherited because they are caused by a faulty allele. Doctors can use genetic diagrams and family pedigrees to help parents work out their chances of having a child with the disease and make choices about the future.

QUESTIONS

One human gene controls the presence of freckles on skin. Use this family pedigree to answer the following questions.

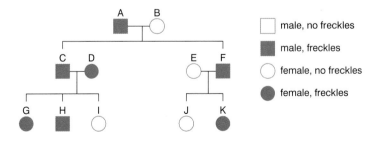

1. How many generations are shown?

2. How many children did couple A and B have?

3. How many daughters did couple C and D have?

4. Which allele is dominant? Explain your answer.

SEX DETERMINATION AND INHERITANCE

In the nucleus of a human cell there are 46 chromosomes that form 23 pairs. In all but one of these pairs, the two chromosomes of the pair always look identical. We call these 22 pairs the autosomal chromosomes. The chromosomes of the other pair are identical in women, but differ in men. These are called the **sex chromosomes**. The sex chromosomes of women are called X and X and in men are called X and Y.

When the gametes are produced, they each receive one of the sex chromosomes. So egg cells all contain an X chromosome, but sperm cells may contain an X chromosome or a Y chromosome. As a result of the way the sperm cells are produced, about 50% are X and 50% are Y.

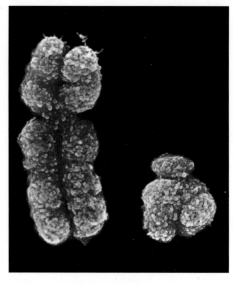

△ Fig. 3.36 The human X (left) and Y (right) chromosomes are different sizes.

During fertilisation, one sperm cell fuses with one egg cell. We can use a Punnett square to show the possible combinations outcomes of sex chromosomes in the offspring.

		father's gametes	
		X	Y
mother's gametes	X	XX (female)	XY (male)
	X	XX (female)	XY (male)

This shows that there is a 0.5 or 50% or 1 in 2 probability of any child being a boy or a girl. The ratio of boys to girls born in a family is often not 1:1, but over the whole human population about equal numbers of baby boys and baby girls are born.

QUESTIONS

1. Which sex chromosomes would be found in the cells of an adult woman?

2. Which sex chromosomes would be found in the cells of a baby boy?

3. A couple have 3 boys. What is the chance of their next child being a girl? Explain your answer.

CELL DIVISION

Mitosis

Organisms grow by the division of cells, when the body cells split in two. This kind of division is used in normal growth and repair. It is also the way that single-celled organisms reproduce and is the only type of cell division involved during asexual reproduction (reproduction that does not involve sex cells – see pages 191–192).

Before a cell divides, its chromosomes duplicate themselves. The new cells formed, sometimes called the **daughter cells**, contain chromosomes identical with the original cell. This type of cell division, in which the new cells are genetically identical to the original, is known as **mitosis**. Cells or organisms that are genetically identical to each other are known as **clones**.

Mitosis takes place in all normal body cells.

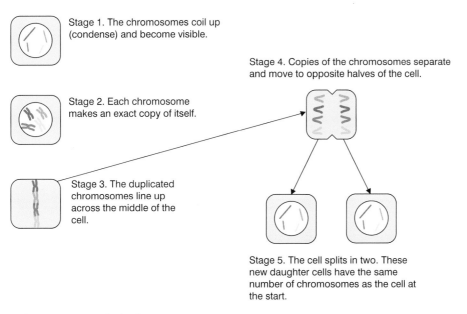

Stage 1. The chromosomes coil up (condense) and become visible.

Stage 4. Copies of the chromosomes separate and move to opposite halves of the cell.

Stage 2. Each chromosome makes an exact copy of itself.

Stage 3. The duplicated chromosomes line up across the middle of the cell.

Stage 5. The cell splits in two. These new daughter cells have the same number of chromosomes as the cell at the start.

Δ Fig. 3.37 Stages of mitosis.

Meiosis

During sexual reproduction, a male gamete fuses with a female gamete. If each gamete had the same number of chromosomes as a normal body cell, the zygote would end up with twice as many chromosomes as normal. Instead, gametes are produced by a different form of cell division called **meiosis**. Cells produced by meiosis have one chromosome of each pair – half the normal number of chromosomes. These cells are called **haploid** cells. When the gametes fuse during fertilisation, they restore the normal number of chromosomes, creating a **diploid** cell with pairs of chromosomes again. As human body cells contain 46 chromosomes we say that for humans the diploid number is 46. The haploid number for humans, as found in gametes, is 23.

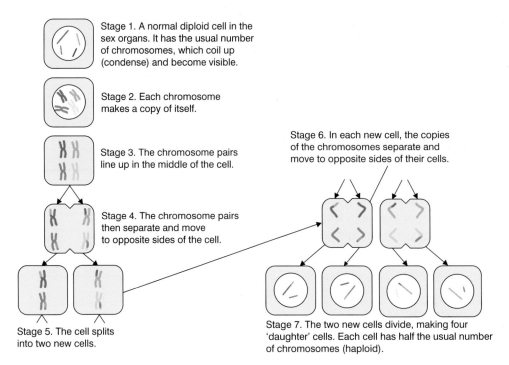

Stage 1. A normal diploid cell in the sex organs. It has the usual number of chromosomes, which coil up (condense) and become visible.

Stage 2. Each chromosome makes a copy of itself.

Stage 3. The chromosome pairs line up in the middle of the cell.

Stage 4. The chromosome pairs then separate and move to opposite sides of the cell.

Stage 5. The cell splits into two new cells.

Stage 6. In each new cell, the copies of the chromosomes separate and move to opposite sides of their cells.

Stage 7. The two new cells divide, making four 'daughter' cells. Each cell has half the usual number of chromosomes (haploid).

Δ Fig. 3.38 Stages of meiosis.

Cells produced by meiosis are not genetically identical. This means that, during fertilisation where there is a random chance that any one male gamete will fuse with the female gamete, the offspring produced will be different from each other. We say they show **variation**.

The table summarises the key similarities and differences between mitosis and meiosis.

	Mitosis	**Meiosis**
Starts with	Diploid body cell	Diploid cell in sex organ
Ends with	Diploid body cells	Haploid gametes
Number of daughter cells produced	Two	Four (although in women, three of these do not become egg cells)
Number sets of chromosomes in daughter cells produced	Two	One
Similarity in chromosomes of daughter cells	Chromosomes identical	Chromosomes show variability

Δ Table 3.4 Differences between mitosis and meiosis.

Practise drawing the stages of mitosis and meiosis, so that you become familiar with the differences between them. You will be expected to explain the effect of those differences in relation to asexual and sexual reproduction.

Variation from random fertilisation

Every gamete that forms has one of each pair of chromosomes. However, which one of each pair ends up in a particular gamete is purely random. An individual human can produce gametes with millions of different combinations of chromosomes.

The random nature of fertilisation, i.e. which combination of chromosomes in a sperm cell will join together with which combination of chromosomes in an egg cell, means that there is huge potential genetic variation between offspring. It is therefore not surprising that, even with the same parents, we can look very different from our brothers and sisters.

QUESTIONS

1. The diagram shows the life cycle of a human.

Copy the diagram and annotate it to show:

- when meiosis and mitosis occur
- which cells are diploid and which are haploid.

2. Draw a flowchart to summarise the stages of mitosis.

3. Draw a flowchart to summarise the stages of meiosis.

4. Explain the importance of meiosis in producing variation in offspring.

VARIATION

No two people are the same. Similarly, no two trees (even if they are of the same species) are exactly the same in every way. They have different heights, different trunk widths and different numbers of leaves. These differences are known as **variation**.

There are two types of variation. Discontinuous variation is where a characteristic can have one of a certain number of specific alternatives, for example gender, where you are either male or female, and blood groups, where you are either A, B, AB or O. Continuous variation is where a characteristic can have any value in a range, for example body weight and length of hair.

Causes of variation

The variation in appearance of characteristics in an individual may have various causes:

- **Environmental** causes – such as your diet, the climate you live in, accidents, your surroundings, the way you have been brought up and your lifestyle can all influence your characteristics.
- **Genetic** causes – in characteristics that are controlled by your genes, such as eye colour and gender.
- Environmental and genetic causes – Many characteristics are influenced by both *environment* and *genes*. For example, people in your family might tend to be tall, but unless you eat correctly when you are growing you will not become tall, even though genetically you have the tendency to be tall. Other examples are more controversial, such as human intelligence, where it is unclear how far environment or genes contribute to variation.

MUTATION

During the copying of DNA, which happens during cell division, sometimes there is a mistake in the copying so that the new DNA is not exactly the same as the original DNA. This is rather like mis-spelling a word when you are copying text, when you use one wrong letter and end up with a different word (such as *bold* instead of *bond*). The different word can completely change the meaning of the sentence. Similarly, the error in the DNA can produce a different form of the gene, that is, a new allele.

This change in a gene is called a **mutation**. Mutations are random changes, so they can occur anywhere in the DNA. They are also rare, happening, on average, only once in every 10 to 100 million DNA bases that are copied. If the mutation occurs in a body cell, then only the cells produced by mitosis from that body cell will also carry the mutation. However, if the mutation occurs in a cell that produces gametes during meiosis, there is a possibility that the mutation may be passed on to offspring, and so inherited.

Mutations may be:

- beneficial for the organism, giving it an advantage over other individuals of the species; for example, fair skin is a mutation that has happened in human populations several times as they moved into areas of northern Europe and Asia. It allows the individual to make more vitamin D from sunlight than those with darker skins.
- neutral, which means it has no obvious effect; many mutations fall into this category and can only be identified by looking in detail at the genetic code.
- harmful to the organism, either causing the early death of the embryo, or making the individual less able to survive than other individuals; many inherited diseases are caused by mutations of genes that produce proteins that are important in key processes.

Most mutations have no effect on the phenotype, some have a small effect and only rarely do they have a significant effect, whether that is beneficial or harmful.

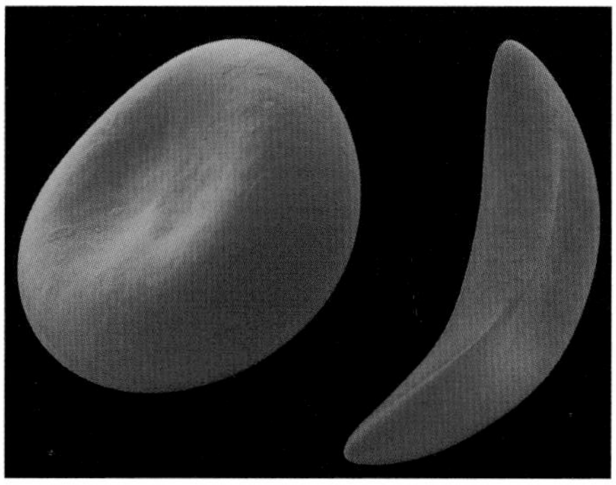

The reason that changes in DNA can sometimes affect an organism's phenotype is that if the sequence of bases in the DNA is altered then this may alter the sequence of amino acids in the protein that the DNA codes for. (See protein synthesis on pages 216–218.) The reason that some DNA changes may not affect the phenotype are that sometimes, even though the DNA may change slightly, this may not change the amino acid sequence of a protein, or sometimes, even if the protein does change slightly, this may not affect its functions.

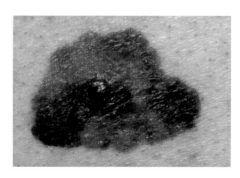

△ Fig. 3.39 A normal red blood cell (left) and a sickle cell (right). A mutation in the gene that produces haemoglobin can result in sickle-haemoglobin. This form of haemoglobin changes the shape of red blood cells when oxygen concentration is low, causing many health problems.

Causes of mutation

The incidence of mutation can be increased beyond the natural rate as the result of:

- exposure to **ionising radiation**, such as gamma rays, x-rays and ultraviolet radiation
- chemical **mutagens**, such as some of the chemicals in tobacco.

These factors may change the genetic code directly or cause it to be mis-copied more frequently during cell division.

△ Fig. 3.40 This skin cancer was caused by too much ultraviolet radiation from sunlight.

One of the most obvious effects of radiation or chemical mutagens is to damage the control mechanisms that instruct a cell to stop dividing at the right time. The continuing division of cells produces a lump of unspecialised tissue called a **cancer**. Cancers that take over the space of other tissues can eventually cause death. In Europe and North America, the different forms of cancer are one of the main causes of death from disease.

1. **a)** Name two causes of variation of characteristics in a species and give an example of each.

   b) Give one example of variation caused by a combination of these causes. Explain your answer.

2. Describe the term *mutation* in your own words.

3. Name two different causes of mutation and give an example of each.

EVOLUTION

Evolution means change over time – in terms of organisms, it means how species change in their characteristics over time. Evolution happens by the gradual accumulation of small changes. Some changes can be detected after a relatively short time, but other changes may take hundreds, thousands or even millions of years.

△ Fig. 3.41 Reconstruction, from fossil evidence, of a dinosaur with feathers.

Some evolution we can see in action, for example the increase in antibiotic resistance among bacteria (see pages 238–239). To show other examples of evolution that have taken place over longer periods of time, we have to rely on other forms of evidence, for example, looking at similarities between closely related species that have evolved from a common ancestor. For species that died out long ago we can use evidence from the fossil record. For example, over 100 million years ago some species of dinosaur evolved feathers, possibly to provide insulation of the skin over the whole body. Some longer feathers also evolved, possibly for display to attract a mate. Over time, some species of feathered dinosaur evolved larger forelimbs, with longer feathers, possibly to help them glide to escape predators. From these, the first birds evolved strong wing muscles that allowed them to fly.

Natural selection

Evolution happens by a process called **natural selection**. This was first explained by Charles Darwin in his book *The Origin of Species*, published in 1859.

Darwin realised that most organisms produce many more offspring than survive to adulthood. Many individuals die before they are old enough to reproduce. This may be, for example, because of predators or a lack of food. The reason why some die and others survive is sometimes due to chance, but is more often to do with how well adapted they are to the environment. Survival of the best adapted individuals is sometimes called *survival of the fittest*. If there is variation in how well individuals are adapted, and this variation has genetic causes, then those individuals who survive and reproduce pass on the genes (or alleles) responsible for the 'successful' adaptations, and these become more common in the population.

We will look at how this works using, as an example, the evolution of height in a species of plant.

△ Fig. 3.42 Taller plants are better adapted to receive more sunlight on their leaves and so are more likely to survive and reproduce than shorter plants.

- Some individual plants of this species will grow taller than others, as a result of their genes (genetic variation).
- Taller plants capture more sunlight for photosynthesis and shade shorter plants so that those get less sunlight.
- So taller plants are able to make more food, and therefore make more seeds than shorter plants. Some of the shorter plants may even die because of the lack of light.
- Embryos in the seeds from the taller plants will have inherited the 'tallness' genes from the parent plants.
- There will be more seeds with 'tallness' genes, so when they germinate there are likely to be a greater proportion of taller plants in the next generation.

△ Fig. 3.43 Over successive generations the average height of the plants increases.

- Over generations the height of this species of plant in this area will increase because in each generation taller plants are more likely to survive and reproduce compared with shorter plants. (In reality, this only happens up to a point where other factors limit the change – for example, a tall plant also needs more energy to grow as well as make more seeds, so there may come a point when growing even taller means making fewer seeds.)
- It is a factor of the environment (light intensity) that has caused this change in the population. We call this natural selection, because a natural factor appears to select individuals with some characteristics more than others, making it possible for them to pass on their genes to the next generation more successfully.

If the environment does not change, selection does not change. This will favour individuals with the same characteristics as the parents. If the environment changes, or a mutation produces a new allele, selection might now favour individuals with different characteristics or with the new allele. So the individuals that survive and reproduce will have a different set of alleles that they pass on to their offspring. This will bring about a change in characteristics of the species, that is it will produce evolution.

REMEMBER

When answering examination questions about natural selection, although the examples may change, the steps are always the same: variation, survival of the fittest, and inheritance of the successful characteristics which then become more common in the population.

Evolution of resistance to antibiotics

Antibiotics are chemicals that are used to kill bacteria when they cause infection. They were first used widely to treat injured soldiers during the Second World War and since then have saved millions of lives. However, more recently, the evolution of bacteria that are resistant to antibiotics has become a major problem, causing many human deaths from infections each year.

The evolution of antibiotic resistance is a good example of evolution through natural selection. It happens like this:

- A patient suffering from a bacterial infection is treated with an antibiotic, for example penicillin.
- The bacterial infection is caused by millions of one type of bacteria, and the individual bacteria within that population will show variation.
- Some bacteria, as a result of random mutation, may be resistant to the antibiotic which means that the penicillin does not kill them as quickly as the other bacteria, or even kill them at all.

- The numbers of resistant bacteria in the patient will increase, making it more likely that the resistant bacteria will be passed to another person.
- Newly infected people, if they become ill as a result of the infection, cannot now be successfully treated with penicillin because it will not kill all the resistant bacteria. So the doctor will have to use a different antibiotic to control the infection. We say that the bacteria have developed **antibiotic resistance**, in this case to penicillin.

Over time, bacteria have become resistant to a larger range of antibiotics, and some show multiple resistance – resistance to many kinds of antibiotics. Now there are few new antibiotics to use on multiple-resistant types of bacteria, and doctors are concerned that there will be an uncontrollable increase in the numbers of deaths from bacterial infections that used to be treatable.

EXTENSION

The rate of evolution of a new variation of a characteristic is related to how well it is favoured by natural selection. Antibiotics provide what is called a *strong selection pressure* – only those bacteria which are most resistant to them will survive and the rest die. So evolution of antibiotic resistance happens quickly. The other reason bacteria can evolve so quickly is that they have a very short generation time, with some reproducing as quickly as every 10 minutes.

QUESTIONS

1. Describe in your own words a) *evolution* and b) *natural selection.*

2. Explain the importance of the following factors in natural selection:

    a) variation in the population

    b) survival advantage.

3. Draw diagrams, such as a cartoon strip, to explain how antibiotic resistance can increase in a bacterial population.

End of topic checklist

An **allele** is one form of a gene, producing one form of the characteristic that the gene controls.

Antibiotic resistance is resistance to the effect of an antibiotic by bacteria, which normally kills them.

An **anticodon** is a section of three bases on a tRNA molecule that joins with the codon on the mRNA during protein synthesis.

A **base** is one of four molecules (adenine (A), thymine (T), cytosine (C) and guanine (G)) that join in pairs (A with T, C with G) that link the two strands within DNA.

A **chromosome** is a long DNA molecule that is found in a cell nucleus.

A **clone** is a cell or individual that is genetically identical to other cells or individuals.

Codominance is when both alleles for a gene are expressed in the phenotype.

A **codon** is a sequence of 3 bases on DNA or mRNA that codes for a specific amino acid.

A **daughter cell** is a cell produced by division of a parent cell.

Diploid describes a cell that contains two sets of chromosomes.

DNA (deoxyribonucleic acid) is the chemical that forms chromosomes and carries the genetic code.

A **dominant** allele is one that, if present, is always expressed in the phenotype.

Double helix describes the shape of the DNA molecule, rather like a twisted ladder.

Evolution is the change in the characteristics of a species over time.

A **family pedigree** is a diagram that shows the inheritance of different forms of a characteristic through the generations within a family.

A **gene** is a section of DNA that codes for a specific protein to produce a particular characteristic.

The **genetic code** is the code formed by the order of the bases in DNA that instructs cells how characteristics should be produced; it does this by telling cells to produce particular proteins.

A **genetic diagram** is a diagram that displays how a characteristic may be inherited by offspring from their parents' alleles.

An organism's **genome** means all its DNA or genes.

An individual's **genotype** describes its genetic makeup.

Haploid describes a cell that contains only one set of chromosomes, such as gametes.

Heterozygous is where the two alleles for a gene are different in the genotype.

Homozygous is where the two alleles for a gene are the same in the genotype.

Inherited characteristics are passed on from parent to offspring via the genes.

Ionising radiation is radiation, such as gamma rays, X-rays and ultraviolet radiation, that can damage cells and produce mutations in genes.

Meiosis is the form of cell division that produces four haploid, and genetically different, cells from a diploid parent cell, producing gametes.

Mitosis is the form of cell division that produces two identical diploid daughter cells from a diploid body cell, used for growth and repair in the body and in cloning and asexual reproduction.

A **monohybrid cross** refers to the inheritance of a characteristic produced by one gene.

mRNA (messenger RNA) molecules are copies of the DNA code for a particular protein, that travel to the ribosomes for protein synthesis.

A **mutagen** is a chemical that produces mutations in genes.

A **mutation** is a random change in a gene, sometimes producing a new allele.

Natural selection is the process by which evolution happens, as first described by Charles Darwin; it is the influence of the environment on survival and/or reproduction, so that organisms with some characteristics are more successful at producing offspring than others.

An organism's **phenotype** describes its physical characteristics; it is affected by both an organism's genes and its environment.

Polygenic characteristics are controlled by the alleles of more than one gene.

Protein synthesis is the process by which proteins are made at ribosomes.

A **Punnett square** is a form of genetic diagram.

A **recessive** allele is one that is only expressed if no corresponding dominant alleles are present.

Ribosomes are small organelles in the cytoplasm that are the site of protein synthesis.

RNA (ribonucleic acid) is the chemical that mRNA, tRNA and ribosomes are made from; it is similar to DNA but is usually single stranded and contains the base uracil (U) instead of thymine (T).

Sex chromosomes are chromosomes that control the sex of the individual, such as, in humans, XX in women and girls and XY in men and boys.

Transcription is the first part of protein synthesis in which an mRNA copy is made of a section of DNA coding for a gene; it happens in the nucleus.

Translation is the second part of protein synthesis in which amino acids are joined together in the order coded for by mRNA; it happens on the ribosomes in the cytoplasm.

tRNA (transfer RNA) molecules are small RNA molecules that transport amino acid molecules to the ribosomes for protein synthesis.

Variation refers to the differences in organisms.

The facts and ideas that you should know and understand by studying this topic:

◯ A cell contains a nucleus, which contains chromosomes that are long molecules of DNA. Small sections of the DNA are genes that control the form of characteristics by coding for specific proteins. All the DNA of an organism is called its genome.

◯ The DNA molecule is a double helix, like a twisted ladder, with two strands joined together with 'rungs' made of paired bases.

◯ The bases in DNA are adenine (A), thymine (T), cytosine (C) and guanine (G). In the pairs, A always joins with T and C always joins with G.

◯ RNA molecules are similar to DNA except they are usually single stranded and contain the base uracil (U) instead of thymine (T).

◯ Protein synthesis consists of the stages of transcription, where mRNA copies are made of DNA genes, and translation, which happens at the ribosomes, where anticodons on the tRNA join with codons on the mRNA to ensure amino acids are joined in the correct sequence to form a protein.

◯ Genes have alternative forms called alleles that produce variations in the characteristic.

◯ Dominant alleles in the genotype are always expressed in the phenotype, but recessive alleles are only seen in the phenotype when there is no dominant allele (there are two copies of the recessive allele).

End of topic checklist continued

○ A homozygous individual has two copies of the same allele, a heterozygous individual has different alleles for the gene.

○ An organism's phenotype refers to its physical characteristics, its genotype describes its genetic makeup.

○ Codominant alleles are expressed equally in the phenotype.

○ Most phenotypic features are controlled by more than one gene (polygenic inheritance) rather than by single genes.

○ Monohybrid inheritance is the inheritance of a characteristic controlled by one gene. The predicted outcomes of any cross can be shown in a genetic diagram, for example a Punnett square.

○ A family pedigree shows the inheritance of variations of a characteristic through several generations of a family.

○ In humans, women have two X chromosomes (XX) and men have an X and a Y chromosome (XY).

○ The sex of offspring is determined by the sex chromosomes inherited at fertilisation.

○ Mitosis is the division of diploid body cell to produce two diploid cells that are genetically identical to each other and the parent cell. Mitosis produces new cells during growth and repair of body cells and in cloning and asexual reproduction.

○ Meiosis is the division of a diploid cell to produce four haploid cells that are not genetically identical. Meiosis produces gametes before sexual reproduction.

○ Random fertilisation means that offspring from sexual reproduction show variation.

○ The diploid number of human cells is 46 and the haploid number is 23.

○ Variation in characteristics in a species may be caused by genes, environment or a combination of the two.

○ Mutation is a rare random event that produces a change in a gene (creating a new allele). Mutations in cells that produce gametes can be inherited.

○ A change in DNA (mutation) can affect the phenotype by altering the sequence of amino acids in the protein it codes for.

○ Most mutations have no effect on the phenotype, some have a small effect and rarely do they have a significant effect.

○ The rate of mutation can be increased by ionising radiation (such as gamma rays, X-rays and ultraviolet radiation), and by chemical mutagens (such as the chemicals in tobacco smoke).

○ Charles Darwin's theory of natural selection explains how evolution takes place. Natural selection is the effect of the environment on a population of organisms, where some individuals have variations in their characteristics that enable them to survive and reproduce more successfully than others.

○ Evolution is change in a species over time and can occur by natural selection when the environment changes or if a new beneficial mutation occurs.

○ The increase in antibiotic resistance in bacteria is an example of evolution by natural selection. The increase in antibiotic resistance can lead to some infections being difficult to control.

End of topic questions

1. Write two sentences that correctly link all the following words to explain how they are related: cell, chromosome, DNA, gene, nucleus. **(2 marks)**

2. The form of earwax in humans is controlled by one gene. The dominant allele produces wet earwax, and the recessive allele produces dry earwax.

   a) Using appropriate symbols, draw a genetic diagram to show the inheritance of earwax between a man with dry earwax and woman who is heterozygous for the characteristic. **(5 marks)**

   b) Describe the predicted probability of genotypes and phenotypes in their children. **(2 marks)**

   c) Explain why it is possible that their three children all have dry earwax. **(2 marks)**

3. A plant breeder had two plants of the same species that she knew were homozygous. One had pure white flowers and one had pure red flowers. She transferred pollen from one plant to the stigmas of the other plant.

   All the seeds produced grew into plants with flowers that were red with white splashes. Explain why the gene in this example shows codominance. **(3 marks)**

4. Skin cancer is the most common form of cancer in light-skinned people. The majority of skin cancers are not life-threatening if treated early.

   a) Cancer may be caused by a mutation to a gene in a cell. What does *mutation* mean? **(1 mark)**

   b) Name two types of causes of mutation. **(2 marks)**

   c) What is the most likely cause of skin cancer? Explain your answer. **(2 marks)**

   d) The graph shows the number of new cases of skin cancer in Sweden between 1970 and 2005.

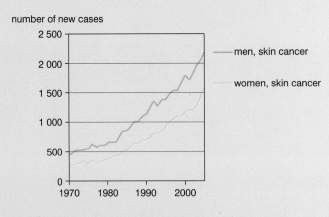

   i) Describe the curves shown on the graph. **(2 marks)**

   ii) Suggest one possible reason for the trend shown in both curves. Explain your answer. **(2 marks)**

5. Explain why a life cycle needs a stage in which meiosis occurs before fertilisation. **(2 marks)**

6. With reference to cell division, explain why asexual reproduction produces identical offspring but sexual reproduction produces variation in the offspring. **(4 marks)**

7. Explain the importance of family pedigrees for identifying whether or not a variation of an inherited characteristic is dominant or recessive. **(4 marks)**

8. Describe in words or pictures how bacteria that are resistant to many types of antibiotics have evolved. **(6 marks)**

9. **a)** How does RNA differ from DNA? **(2 marks)**

   b) Describe the roles of mRNA, tRNA and the ribosomes in protein synthesis. **(6 marks)**

Exam-style questions
Sample student answer

Question 1

People are either able to taste a chemical called PTC, or not. Being able to taste PTC is controlled by a single gene which has two alleles, T and t.

a) The diagram shows a pair of chromosomes:

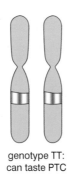

genotype TT:
can taste PTC

i) Is the allele for tasting PTC dominant or recessive? Explain your answer. **(2)**

Dominant ✓ ①

The letters are capitals ✗

ii) Write down the other possible genotypes related to tasting PTC, along with their phenotypes. **(2)**

Tt - the phenotype is can taste PTC ✓ ①

tt - the phenotype is cannot taste

PTC ✓ ①

iii) The diagram illustrates the cross correctly, but lacks detail.

The best way of illustrating the cross is to use a Punnett square, showing each stage of the cross:

the genotypes of the parents

the different alleles that could be passed on to the offspring from the mother and father (the alleles in the egg cells and sperm cells)

the possible combinations of alleles in the offspring (genotypes)

the possible phenotypes produced.

		Mother possible alleles in eggs	
		T	t
Father possible alleles in sperm	T	TT can taste PTC	Tt can taste PTC
	t	Tt can taste PTC	tt cannot taste PTC

A further point is the way in which the student has written the third possible genotype in their answer, as tT. Although not incorrect, the convention is to write the dominant allele first, so it should be written Tt.

Exam-style questions continued

b) A couple who can both taste PTC have children.

i) Give the possible genotypes of the man and the woman. (2)

The couple could be TT ✓ ①

or Tt ✓ ①

ii) The couple have children. Their first child cannot taste PTC; the second one can.

What does this tell you about the genotypes of the couple? Explain your answer fully. (4)

The genotype of both the man and

women must be Tt ✓ ①

Because otherwise, all the children

would be able to taste PTC ✓ ①

iii) Show the genetic cross involved. (4)

Tt   x   Tt   ✓ ①

↓

TT   Tt   tT   tt   ✓ ①

(Total 14 marks)

⑨/14

Question 2

A disease called cystic fibrosis is caused by a faulty allele of a gene that is recessive.

The family tree below shows the occurrence of cystic fibrosis in a family.

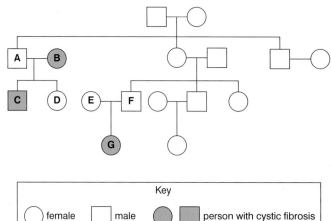

Key

○ female   □ male   ⬤ ⬛ person with cystic fibrosis

a) Using the symbols **f** for the allele causing cystic fibrosis and **F** for the 'healthy' allele, give the genotypes of the following members of the family, explaining how you come to your decision:

   i) father A (2)

   ii) daughter D (2)

   iii) parents E and F. (2)

b) Carriers are people who do not show a condition but may pass it on to their children. In the part of the family tree with the people A – G, identify the carriers of cystic fibrosis. (1)

(Total 7 marks)

Question 3

Codominance is seen in the genetic traits of a number of organisms, including humans.

a) The coat colour of some cattle shows codominance.

Some cattle have alleles that produce red hairs; others have alleles that produce white hairs. Cattle that have a combination of both alleles have red and white hairs. Their coat colour is described as *roan*.

   i) Explain the term codominance. (2)

   ii) Draw genetic diagrams to show the following crosses and predict the ratios of offspring produced.

   Use C^R to represent the allele for red hair and C^W to represent the allele for white hair.

   A red bull and a white cow. (7)

   A roan bull and a white cow. (10)

Exam-style questions continued

b) In the inheritance of human blood groups, the alleles for blood group A and blood group B are dominant to the allele for blood group O. The alleles for blood group A and B are codominant.

Use I^A to represent the allele for blood group A, I^B to represent the allele for blood group B, and I^o to represent the allele for blood group O.

Explain, using a genetic diagram, how it is possible for a father with blood group A and a mother with blood group B to have a child with blood group O. What is the probability of these parents having a child with blood group A? **(7)**

(Total 26 marks)

Question 4

a) The passage below describes the process of sexual reproduction.

Use suitable words to complete the sentences. **(4)**

To produce a new organism by sexual reproduction, two .... fuse. This process is known as ....

Usually, sexual reproduction involves .... parent organisms of the same species. The .... formed is genetically different from each of the parents.

b) Give one advantage and one disadvantage of:

i) asexual reproduction **(2)**

ii) sexual reproduction. **(2)**

c) **i)** Label the diagram of the human female reproductive system below. **(6)**

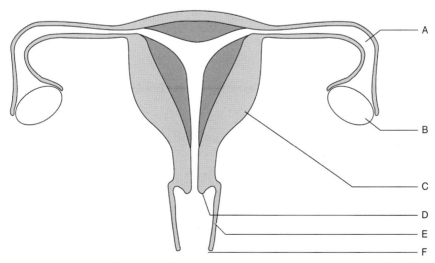

ii) Name the hormones involved in preparing the uterus for a fertilised egg, and explain their roles. **(4)**

(Total 18 marks)

Question 5

Flowers are adapted to be pollinated by insects or by the wind.

a) Name the structures of an insect-pollinated flower shown in the diagram below. **(10)**

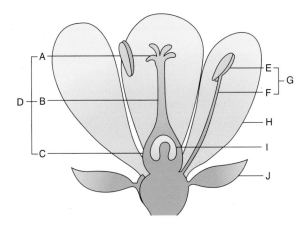

b) Explain how each of the following structures is adapted for pollination in wind-pollinated flowers:

  i) petals   **(2)**

  ii) stigma   **(2)**

  iii) stamens   **(2)**

  iv) pollen grains.   **(2)**

(Total 18 marks)

Exam-style questions continued

Question 6

A horse has a chromosome number of 64.

a) Put ticks in the table to show whether each statement refers to mitosis, meiosis, both or neither. (6)

	Mitosis	Meiosis
The chromosome number in each daughter cell is 64		
The daughter cells are haploid		
Two identical cells are produced		
The nuclear membrane disappears at the beginning, and is reformed at the end of the process		
Some variability occurs in the alleles of parent and daughter chromosomes		
Occurs when new red blood cells are produced in the blood of the horse		

b) Explain why meiosis is the type of cell division used to produce the horse's gametes. (2)

(Total 8 marks)

It is the result of Earth's unique position in the Solar System, and distance from the Sun, that there is such a vast abundance of living species on the Earth. We are sufficiently far from the Sun that water has not all evaporated from the Earth's surface, and not too far that it is solid ice. Liquid water provides a solution for many organisms to live in, and the solution inside their cells in which many cell reactions take place. As a result, millions of species share the surface of Earth, dependent on each other and on the air, land and water around them. Only the human species has developed advanced technology, and this technology now risks damaging the environment for humans as well as for all other living organisms.

STARTING POINTS

1. In order to study environments we need to know how abundant organisms are and how they are distributed. How can we do this?

2. There are many species in an ecosystem, but how can we show they are interdependent?

3. What are pyramids of number, biomass and energy and what do they show?

4. The Earth will not run out of carbon and nitrogen, so how are these recycled through living organisms and the environment?

5. Humans are having an ever-increasing effect on the environment, often causing pollution. What kinds of effects are we having and how can we reduce the damage we are causing?

SECTION CONTENTS

a) The organism in the environment

b) Feeding relationships

c) Cycles within ecosystems

d) Human influences on the environment

e) Exam-style questions

4

Ecology and the environment

△ Factories such as this one affect the environment.

△ Fig. 4.1 In the future, wild laws may help to protect tropical forests like this from the damage caused by logging, and maintain the ecosystem to properly support the organisms and people that live there.

The organism in the environment

INTRODUCTION

The idea of 'wild law' is fairly new. It acknowledges the rights of ecosystems and is intended to control human activity so that it does not damage the environment and species in an ecosystem. This changes the view that humans have the right to exploit an ecosystem and take whatever they want from it without thinking about the organisms that live there. The aim of wild laws is to protect ecosystems for the future.

KNOWLEDGE CHECK

✓ Explain the meaning of the terms *habitat* and *ecosystem*.
✓ Describe the abundance of organisms in a habitat.

LEARNING OBJECTIVES

✓ Understand the terms population, community, habitat and ecosystem.
✓ Practical: Investigate the population size of an organism in two different areas using quadrats.
✓ Understand the term biodiversity.
✓ Practical: Investigate the distribution of organisms in their habitats and measure biodiversity using quadrats.
✓ Understand how abiotic and biotic factors affect the population size and distribution of organisms.

DEFINITIONS IN ECOLOGY

In order to understand the effect of human activity on ecosystems, we need to study the environment. This study is called **ecology**. We will have to measure how the distribution and abundance of species is changing. Before we do that, we need to define the words that we use in ecology.

- An **ecosystem** is the interaction of all the organisms living within an area and the environment around them. It includes all the **abiotic factors** (physical factors) that affect organisms, such as rainfall, temperature and light intensity, plus all the **biotic factors** (biological factors) caused by the presence of other organisms (such as predation or competition for food or space). Examples of ecosystems are a lake, a desert, a tropical rainforest or a coral reef. An ecosystem is a fairly self-contained unit that is easy to distinguish from other ecosystems either by the physical factors (a lake is an area of water) or by biological factors (only particular plant and animal species live in a desert).

- All the organisms within an ecosystem are called a **community**. For example, the community living within a lake ecosystem might include many species of pond weed, many species of fish and many species of insects.
- Within an ecosystem are many **habitats** where individual species live. Each habitat has specific physical characteristics, such as a limited amount of light (within shade) or a high level of moisture (under a rotting log). Within a lake there are many habitats, such as out in the open water, in the mud at the bottom, within weed near the sides of the lake.
- The habitat will support different populations of organisms. A **population** is a group of individuals all of one species. For example, a population of perch (a species of fish) and a population of midge larvae may live in the open-water habitat of the lake. All the populations of organisms in the ecosystem make up the community.
- We often talk about the **environment** in which species live. This means all the biotic and abiotic factors that affect the species.

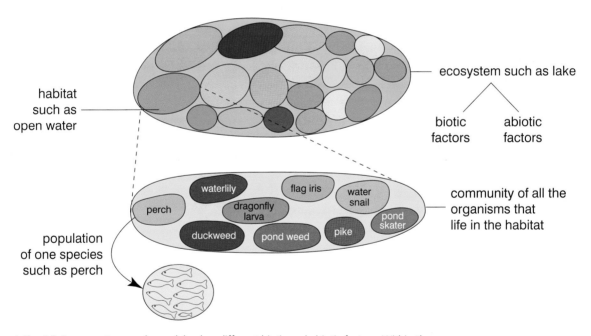

△ Fig. 4.2 An ecosystem, such as a lake, has different biotic and abiotic factors. Within the ecosystem are many different habitats. Within the habitats are communities made up of populations of different species.

▷ Fig. 4.3 A coral reef is a diverse habitat, home to many species and visited by many more.

BIOTIC AND ABIOTIC FACTORS

Biotic and abiotic factors affect both the population sizes of organisms, as well as where they live (their **distribution**).

Biotic factors include the following.

- **Predation** of animals, and **grazing** of plants, by other animals will often act to reduce population sizes. Often we only realise this when the predation or grazing stops, and the animal or plant species that is no longer being eaten greatly increases in numbers. On the other hand, some species only survive in areas where there are few or no **predators**.
- **Competition** for limited resources, such as food or shelter for animals, or light or water for plants, keeps populations in check.

Abiotic factors include the following.

△ Fig. 4.4 Dodos survived on the island of Mauritius because they had no known predators. Hunting by European sailors and the animals they brought to the island made the dodo extinct.

- Light, which is needed by plants to photosynthesise and grow. This is why plants do not grow in places such as deep caves where there is no light, and may struggle to grow in places with reduced light such as on a forest floor.
- Temperature affects organisms because extreme temperatures may kill them by stopping their enzymes working effectively. However, some species are especially adapted to live in very cold or very hot places.
- Water is needed by all living things so a lack of water will limit the population size of organisms, or determine which species can survive.

△ Fig. 4.5 The waxy needles of pine trees reduce water loss so they can survive even when there is no available water in the ground because it is all frozen.

QUESTIONS

1. In your own words, describe the term *ecosystem* and give two examples.

2. In your own words, describe the term *habitat* and give two examples.

3. Describe how an *ecosystem*, *habitat*, *community* and *population* are related.

4. What is the difference between *biotic* and *abiotic* factors? Give two examples of each.

ESTIMATING POPULATION SIZE

In order to measure population size, or **abundance**, of an organism, you need to calculate the number of individuals of that species in a given area, such as a habitat or ecosystem. It is rarely possible to count all the organisms of a species in an area, so the population size may be estimated. This can be done by counting the number of organisms in a small part of that area and then using that number to calculate the population size in the whole habitat.

For organisms that do not move much, you can use a **quadrat** to measure abundance in a given area. A quadrat is a square, often made from wire, plastic or wood. It is of a specific size, such as 10 × 10 cm or 50 × 50 cm. The quadrat is placed on the ground and the number of individuals of the species that lie within the quadrat are counted.

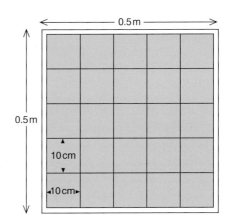

△ Fig. 4.6 This quadrat is 0.5 m × 0.5 m.

To get a reasonable estimate, you have to make sure to get enough samples that are representative of the whole area.

- You need to use a quadrat large enough to contain a fair number of individuals of the species. However, too large a quadrat will take too long to count all the individuals – there has to be balance between these needs.

- You need to make sure not to bias the results by choosing areas, such as ones that look more interesting or do not contain stinging weeds. To avoid this, quadrats are often placed randomly. This is sometimes done by carefully throwing the quadrat over your shoulder without looking. However, it is much better to use measuring tapes to indicate the axes of a grid over the area, with squares the same size as the quadrat, and then use random numbers generated by a calculator to decide the coordinates for placing each quadrat.

△ Fig. 4.7 Sampling an ecosystem using a quadrat.

- You need to take enough repeat measurements to make sure you average out natural variation between the numbers counted in each quadrat. Again, the number of repeats will probably be limited by the time available.

The results from all the quadrats are then averaged, to give an average number per quadrat. The population size in the whole habitat can then be calculated using the number of quadrats that fit within the area of the habitat. For example:

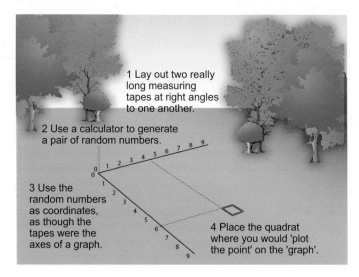

1 Lay out two really long measuring tapes at right angles to one another.

2 Use a calculator to generate a pair of random numbers.

3 Use the random numbers as coordinates, as though the tapes were the axes of a graph.

4 Place the quadrat where you would 'plot the point' on the 'graph'.

△ Fig. 4.8 How to lay out a quadrat.

- if there is an average of 5 plants of a particular species in a quadrat of size 1 × 1 m,
- the whole area of 10 × 10 m will contain 100 quadrats,
- so the estimate of population size over the whole area will be 5 × 100 = 500.

This technique is useful for comparing two different habitats, with different abiotic or biotic factors, to see if they affect the distribution of a species. For example, measurements from trampled and untrampled areas of a school playing field may show how different plant species respond to trampling.

If the number of individuals is too great to count, time is limited, or you just want a general idea of abundance, then you can use an abundance scale instead of counting. The letters ACFOR stands for the groups used in the estimating: Abundant, Common, Frequent, Occasional, Rare. Different people using the scale must first decide how to define the different categories, otherwise their data will not be comparable.

Alternatively, you can estimate the percentage cover of each species within a quadrat, which is the proportion of the quadrat covered by that species. You can then use these values to estimate percentage cover over the whole area.

Developing investigative skills

Students were asked to investigate the effect of light intensity on plant species, by estimating the abundance of two plant species on the north-facing and south-facing slopes of a hill, each slope being approximately 50 m by 20 m in area.

Devise and plan investigation

❶ Write a plan for this investigation, bearing in mind that the students were given two hours to carry out their practical work. Remember to include all the abiotic factors they should measure, and the equipment they will need.

Analyse and interpret data

The table shows the results from one group.

	Number of plants in quadrat 100 × 100 cm				
	Quadrat 1	Quadrat 2	Quadrat 3	Quadrat 4	Quadrat 5
Species A (south)	18	15	7	12	10
Species B (south)	4	10	12	3	6
Species A (north)	7	3	6	12	1
Species B (north)	15	16	7	9	13

❷ Calculate averages for each species in each area, and from those estimate the abundance of each species in each area of the hillside.

❸ Describe the results for each species.

❹ The light intensity measured on the two sides of the hill showed that it was greater on the south side than on the north side of the hill. Draw conclusions about these results.

Evaluate data and methods

❺ In evaluating their conclusions, different students made the following comments. Explain how you would respond to each of these comments if you were going to carry out this investigation, so that you could get more reliable results.

a) 'We moved two of our randomly placed quadrats because they were in the middle of thorny plants.'

b) 'There was a stream at the bottom of the north-facing slope but not on the south-facing slope. This might have affected our results.'

c) 'It was cloudy when we were sampling the north-facing slope but not when we were sampling the south-facing slope. This means the plants on the north-facing slope were getting less light.'

(a) counting individuals within a quadrat

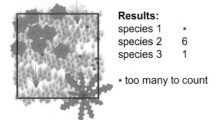

Results:
species 1   *
species 2   6
species 3   1

* too many to count

(b) estimating percentage cover in a quadrat

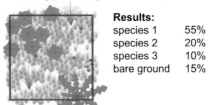

Results:
species 1   55%
species 2   20%
species 3   10%
bare ground 15%

species 1
species 2
species 3
bare ground

Δ Fig. 4.9 Counting or estimating populations.

QUESTIONS

1. What is a quadrat?

2. What do we mean by the abundance of an organism?

3. Why is the random sampling method more likely to give reliable results than sampling where you choose to place a quadrat?

4. Why is the ACFOR scale useful?

SCIENCE IN CONTEXT

THE IUCN 'RED LIST OF THREATENED SPECIES'

The IUCN (International Union for Conservation of Nature and Natural Resources) gathers data about the distribution and abundance of species of plants and animals all over the world, from scientists studying the organisms in the wild. With these data the IUCN creates a 'Red List of Threatened Species', indicating how global distribution and abundance are changing over time as a result of factors such as human activity and climate change. Countries do this on a national level, and local councils and wildlife trusts may do this on a regional level.

Δ Fig. 4.10 A Royal Bengal tiger, member of an endangered species.

The data show which species are most at risk of **extinction,** and so those that need the greatest protection and **conservation** efforts. This information not only decides where money for conservation should be targeted, but also helps councils and governments plan where best to build so that the vulnerable species are protected.

BIODIVERSITY

Some ecosystems, such as a rainforest, contain a large number of different organisms, made up of many different species, and we say therefore that there is a high **biodiversity**. Other ecosystems, such as a sandy desert, contain a much smaller number of different species, as well as fewer organisms in total, so we would therefore say that there is a low biodiversity. Biodiversity is a measure of the variety of living things in an area.

Δ Fig. 4.11 The rainforest has a much higher biodiversity than the desert.

INVESTIGATING THE DISTRIBUTION AND BIODIVERSITY OF ORGANISMS

You can use quadrats also to sample how organisms are spread in an area, their distribution, as well as get a measure of biodiversity. Distribution is often measured in relation to a changing factor. For example, you might want to look at how the distribution of different species of plants changes between a field and the edge of a nearby pond. For this, you use a **line transect**, which is a line between the two areas often marked by a long measuring tape. Quadrats are placed at regular intervals along the line, and the number of each species of plant is counted. The number of quadrats counted will probably be limited by the time available.

The results from the line transect shown in the diagram would probably tell you which plants need to grow in dry soil and which plants grow better in damp or waterlogged soil. You might investigate how shade from a large tree affects the distribution of plants by using a transect from the base of the tree out into full sunshine. You could also measure the effect of waves on the distribution of seaweeds or species of seashells between high tide and low tide marks.

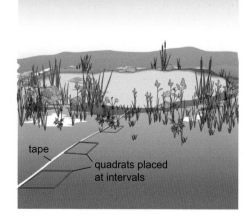

Δ Fig. 4.12 A line transect.

By counting the number of different species in each quadrat along a line transect, you would also have a measure of how biodiversity changes along the transect. Or you could compare the biodiversity of two completely different areas by using results from quadrats to compare the number and types of species between the two places.

QUESTIONS

1. What do we mean by the distribution of an organism?

2. What is a line transect?

3. Explain why a line transect is useful for measuring the distribution of an organism.

4. What is meant by the term *biodiversity*?

5. a) Name an ecosystem that has a high biodiversity.

   b) Name an ecosystem that has a low biodiversity.

REMEMBER

When evaluating studies that estimate population size or distribution of organisms from counting individuals in quadrats, make sure you comment on the reliability of the results based on the methods used. You should also describe what could have been done within the time available that would have increased the reliability of the results.

End of topic checklist

Abiotic factors are non-living or physical factors, such as temperature or rainfall, that affect the population sizes and distributions of organisms.

Abundance means the number of organisms of a particular species in an area, related to the population size.

Biodiversity is a measure of the variety of organisms in an area.

Biotic factors are factors affecting the abundance and distribution of organisms that are caused by other organisms, such as through competition for food.

Community refers to all the populations of organisms that live in an area or ecosystem.

Distribution means how organisms are spread out in an area, where they are found.

Ecology is the study of living organisms and their environment.

An **ecosystem** consists of all the organisms and physical factors in a fairly self-contained area, such as a lake or desert.

The **environment** consists of organisms and the factors that affect them.

A **habitat** is a small part of an ecosystem where a species lives.

A **line transect** is a line along which quadrats are placed to sample organisms.

A **population** is all the organisms of one species living in the same habitat.

A **quadrat** is a square frame used for sampling the abundance and distribution of organisms.

The facts and ideas that you should know and understand by studying this topic:

○ Ecosystems may contain many habitats, each with its own community of different populations.

○ Quadrats can be used to estimate the population size of an organism in two different areas.

○ Biodiversity is a measure of the variety of organisms in an area.

○ Both biotic and abiotic factors affect the population size and distribution of organisms.

○ Quadrats can be used for sampling the distribution of organisms in their habitats and for measuring biodiversity

End of topic questions

1. Using your own words, define the following terms:

   a) ecosystem (1 mark)

   b) habitat (1 mark)

   c) population (1 mark)

   d) community (1 mark)

   e) abiotic factor (1 mark)

   f) biotic factor. (1 mark)

2. Using an ecosystem of your own choice, give an example of each of the terms in Question **1**. (6 marks)

3. Describe how you would use a quadrat to measure the population size of a species of snail living in a garden that is 100 m². Identify what you would do to make sure your results are as reliable as possible. (3 marks)

4. A pupil wanted to estimate the population size of one species of flowering plant in a field. She used a quadrat that was 1 m² and the field was 200 m². She used the quadrat 10 times and counted a total of 25 plants of that species. Estimate the population size for that species in the field. (2 marks)

5. A transect survey of plants and algae was carried out in a tidal saltmarsh. The transect started in the middle of a creek (at point 0 on the chart in Fig. 4.13 on the next page) and was measured going from the middle of the creek onto the land to a maximum distance of 8 metres. At high tide the creek was filled with sea water, but at low tide it was drained of water.

Fig. 4.13 shows the results. Each species is displayed separately, and the abundance at any particular distance is shown by the width of the kite shape.

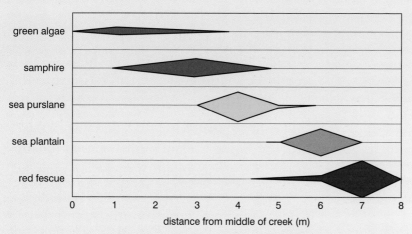

△ Fig. 4.13

a) How many species were sampled on this transect? (1 mark)

b) Which of the following abiotic factors might vary along this transect? Explain your answers.

    i) light intensity (2 marks)

    ii) time submerged in sea water (2 marks)

    iii) temperature. (2 marks)

c) The creek was 6 metres wide *in total*. Which species were able to grow within the creek? Explain your answer. (2 marks)

d) **i)** Suggest one explanation for the distribution of the algae. (1 mark)

    ii) How would you test your answer to part **i)**? Explain your answer. (2 marks)

e) **i)** What distances from the middle of the creek show the highest biodiversity? (1 mark)

    ii) Explain your answer to part **i)**. (1 mark)

△ Fig. 4.14 Alligators can survive for months without food.

Feeding relationships

INTRODUCTION

All animals need to eat to provide the fuel for respiration. Some animals like the shrew need to consume two or three times their body weight of insects, slugs and worms every day in order to survive. They live life quickly, being on the hunt for food for most of the time, especially at night. In contrast, alligators only need to feed about once a week and can live for months without food. They live life much more slowly than shrews, waiting in ambush for **prey** to get close before attacking. Most animals eat somewhere between these extremes, although adult mayflies have no mouthparts and never eat. They live just a few days, using energy stored from earlier stages in their life cycle, as their only purpose is to reproduce, after which they die.

KNOWLEDGE CHECK

✓ Give examples of food chains.
✓ Use a combination of food chains within a habitat to produce a food web.
✓ Describe how organisms in a food webs are interdependent.

LEARNING OBJECTIVES

✓ Understand the names given to different trophic levels, including producers, primary, secondary and tertiary consumers and decomposers.
✓ Understand the concepts of food chains, food webs, pyramids of number, pyramids of biomass and pyramids of energy transfer.
✓ Understand the transfer of substances and energy along a food chain.
✓ Understand why only about 10% of energy is tranferred from one trophic level to the next.

TROPHIC LEVELS

Food chains

You should be familiar with food chains from your earlier work. A **food chain** shows 'who eats what' in a habitat. For example, in Fig. 4.15, owls eat shrews, shrews eat grasshoppers, grasshoppers eat grass. (Remember, the arrows in a food chain show the direction in which the food passes.)

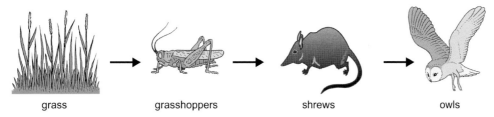

△ Fig. 4.15 An example of a food chain.

Each level in a food chain shows a separate **trophic level**, or level at which that species is feeding. So the diagram shows:

- grass – The **producer** level, because grass is a plant and produces its own food using light energy during photosynthesis.
- grasshoppers – These are the **primary consumers**, 'consumer' because they eat the grass and 'primary' because they are the first eaters of other organisms in the food chain. This level may also be called **herbivores** because they eat plant material.
- shrews – These are the **secondary consumers** because they eat the primary consumers. Also called **carnivores**, because they eat meat.
- owls – These are the **tertiary consumers** because they eat the secondary consumers. They are also carnivores.

If anything ate owls, they would be quaternary consumers, but food chains often do not reach that level. Animals at the highest trophic level in a food chain may also be called the top consumers.

All animals are consumers, because they eat other organisms to get their food, in contrast to plants that are producers. Predators are any animals at secondary consumer level or above, because they kill other animals in order to eat them. The animals they kill are called their prey. Scavengers eat dead plants or animals, and **omnivores** eat both plants and animals.

△ Fig. 4.16 The hyphae of this fungus are growing through the dead tree and secreting enzymes that cause the wood to break down into simpler chemicals.

What is not shown in a food chain is what happens to all the dead plant and animal material that is not scavenged. This material decays as a result of the action of **decomposers**, i.e. fungi and bacteria. Fungi digest their food by secreting enzymes outside their hyphae; they then absorb the dissolved food materials. Many bacteria also do this. However, only some of the digested food materials are absorbed – the rest are released into the environment. Decomposers play an essential role in ecosystems (see pages 284–286).

OTHER PRODUCERS

Not all producers are plants, and not all producers use light energy. There are species of bacteria that produce their own food without the presence of light energy from the Sun. Instead they get the energy they need for the formation of sugars from chemical reactions.

These bacteria are the source of food for food chains and webs that exist where there is no sunlight, such as deep in oceans and in deep underground caves. Be careful to avoid the statement that 'all life on Earth depends on the Sun' as this is an oversimplification and not totally accurate.

QUESTIONS

1. Use your own words to describe the following terms:
   a) producer, b) primary consumer, c) secondary consumer,
   d) decomposer.

2. At which trophic levels are the following groups of organisms placed? Explain your answers.

   a) herbivores, b) carnivores, c) omnivores.

3. Explain why decomposers are not shown in food chains.

FOOD WEBS

If we look more closely at the trophic levels in a food chain, it is rare to find an organism that is eaten by just one other species, or a predator that feeds on just one prey. It may also be the case that a predator may feed at more than one trophic level – an omnivore, for example, is a primary consumer when feeding on plants, but a secondary or tertiary consumer when eating other animals. So food chains within a habitat are linked together to form a **food web**. A food web is a better description of the feeding relationships in a habitat and shows how living organisms are interdependent.

Food webs still usually try to group the organisms according to their trophic level. For example, in Fig. 4.17, the rabbit, squirrel, rat, seed-eating bird and herbivorous insect are all primary consumers and are placed just above the producer level.

There are usually many more species in a habitat than shown in the diagram, and linking them all in one food web can get confusing. So food web diagrams may focus on the relationships between key

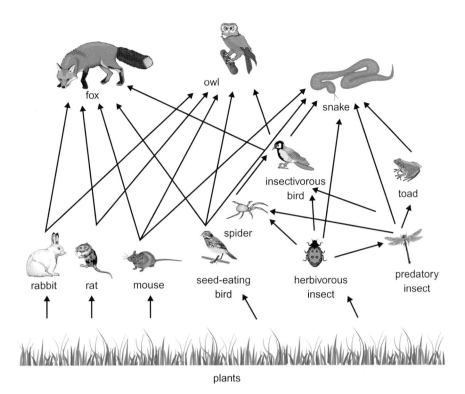

△ Fig. 4.17 A simplified food web.

organisms rather than all of them. For example, they may only include the most numerous species, or focus on the most vulnerable species. This can be helpful if you want to use the food web to predict what would happen to the ecosystem if the food web were changed in some way, such as by human activity.

You could use the food web shown to predict what would happen if the plants were sprayed with an insecticide. This would kill the herbivorous insects and so reduce the amount of food available to all the animals that feed on them.

QUESTIONS

1. In the diagram of a food web, name one tertiary consumer.

2. In the food web, at which trophic levels is the fox feeding? Explain your answer.

ENERGY TRANSFERS IN FOOD WEBS

Food chains and food webs not only show who eats what: because 'food' is a store of chemical energy, they also show the flow of energy through the trophic levels, from producers to top consumers. You can investigate this transfer of energy using simple diagrams.

Pyramids of number

In many food chains, if you look at the number of organisms in each trophic level you will find far more producers than primary consumers, and far more primary consumers than secondary consumers. For example, a few tigers eat a larger number of antelopes, which eat a far larger number of grass plants. We can use these data to create a **pyramid of number** as shown in the diagram. Each bar in the pyramid represents a different trophic level, arranged in order starting with producers at the bottom and ending with the top consumer at the top. The width of each bar is drawn to scale, representing the numbers of individuals in the trophic level.

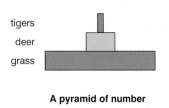

tigers
deer
grass

A pyramid of number

△ Fig. 4.18 A typical pyramid of number.

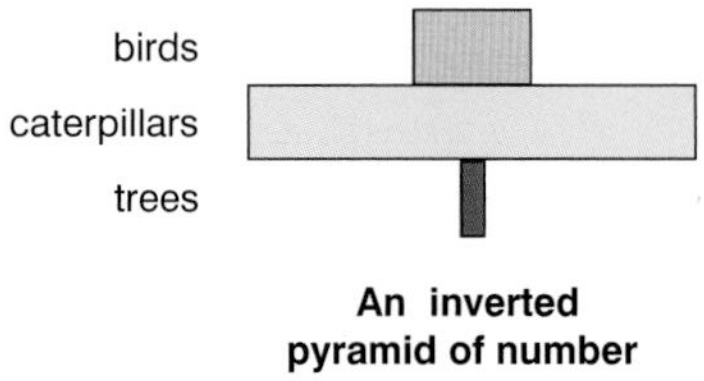

birds
caterpillars
trees

An inverted pyramid of number

△ Fig. 4.19 Inverted pyramid of number.

These diagrams are called 'pyramids' because their shape is often wider at the base and narrower towards the top. However, this is not always the case. Imagine one large tree, on which hundreds of caterpillars feed, on which many birds feed. This pyramid of number produces a shape with a very narrow base, sometimes called an *inverted* pyramid of number.

Gathering data for a pyramid of number is relatively simple, because you just have to count the number of organisms in each trophic level within the area of observation.

Pyramids of biomass

It is not surprising that, in the example of tree/caterpillars/birds, you get an inverted pyramid of numbers because one tree is huge in comparison with the tiny caterpillars that feed on it. If you measure **biomass** (the mass of living material) in the organisms instead of number, you can avoid this problem.

The biomass of each trophic level is usually calculated as the average dry mass of one individual multiplied by the total number of individuals. This is usually within a given area, so the values will be in mass per unit area. If you draw these values to scale, you can produce a **pyramid of biomass**. Fig. 4.20 shows what happens if you produce a pyramid of biomass for the tree/caterpillars/birds example from before. Using the biomass of the organisms produces the pyramid shape as expected.

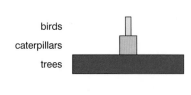

birds
caterpillars
trees

A pyramid of biomass

△ Fig. 4.20 A pyramid of biomass.

Note that dry mass is used when constructing pyramids of biomass. This is the mass without any the water in the body, because water is relatively heavy and its content in a body can vary a lot depending on the state of hydration. Measuring dry mass may involve killing at least one organism of each type, so that the tissues can be dried fully in an oven. If more individuals are used the masses can be averaged to give an average biomass per individual. Tables of average biomass can be used in order to avoid killing any more organisms, but this adds another level of estimation to the process, and so may decrease the reliability of the data in the pyramid even further.

Pyramids of biomass produce problems of their own. The mass of the tree is actually a measure of the biomass accumulated over many years. This is often called the standing crop. The insects living off it may have been produced and consumed in a matter of days and so do not accumulate biomass in the same way. This produces the correct shape for the pyramid but can create a different problem.

The growth of plankton in an area of sea at a certain time of year was measured over a few days and a pyramid of biomass created (see Fig. 4.21). Again the pyramid is inverted. This is due to 'under-sampling' of the algae in the food web, which is caused by the relatively short life span of the algae compared with the longer-lived herbivorous organisms. Unlike the tree, algae do not produce a significant standing crop so do not provide a significant base layer for the habitat.

To help you understand this, think of a shop that sells fruit. In the morning, the shopkeeper will fill the shelves with fruit, and during the day customers will take fruit from the shelves to buy. The 'standing crop' of fruit on the shelves will look smaller in the afternoon than in the morning. Only if the shopkeeper has someone to continually restock the shelves will the standing crop be the same throughout the day.

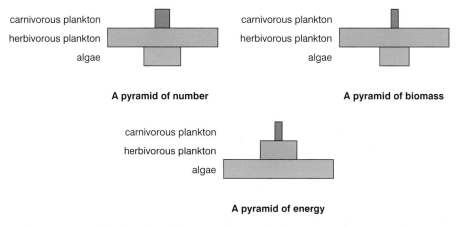

Δ Fig. 4.21 A pyramid of number, biomass and energy for the same marine community. (Herbivorous and carnivorous plankton are tiny animals that float in the top layers of the open oceans.)

Developing investigative skills

Some students were collecting data on the abundance of plants and snails on some rough wasteland so that they could construct ecological pyramids.

Devise and plan investigations

❶ Draw a food chain for these organisms, and identify each trophic level.

❷ Write a plan for this investigation, to explain how they could gather reliable data for a pyramid of number.

Demonstrate and describe techniques

❸ Explain fully how they could convert the data for a pyramid of number into a pyramid of biomass.

Analyse and interpret data

The students collected the following data for their pyramid of number.

Sample site	1	2	3	4	5
Number of plants	46	75	39	28	22
Number of snails	4	8	5	1	2

❹ Calculate a mean number for each trophic level.

❺ Use your mean values to draw a pyramid of number.

❺ The students found out the following mean values of dry mass for one organism in each trophic level: plant 38 g, snail 6 g.

❻ Use these values to construct a pyramid of biomass for these organisms.

❼ Describe and explain the shape of both pyramids.

Pyramids of energy

Remember that each trophic level represents the amount of chemical energy in the organisms that are available to the organisms that feed on them when they digest their food. So you can create a **pyramid of energy** in units of energy per area per time such as kJ/ha/yr to show the amount of energy in each trophic level. This avoids the problem we saw with pyramids of biomass, because it includes a value for time as well as area. If you look at pyramids of number, biomass and energy for the same food chain, you can see the effect of this.

Although pyramids of energy always produce a pyramid shape each time, they are more difficult to calculate because you have to measure the energy content of each population within the area. This involves burning the plant or animal tissue in a calorimeter to measure the energy it contains and then multiplying by the number of individuals. Alternatively, using estimates of the average fat and muscle content of an individual animal, you can estimate the average energy content in an individual animal. All these phases of estimation reduce the reliability of the calculated values.

ENERGY TRANSFER IN DINOSAURS

In 1976 a scientist named J.O. Farlow attempted to construct a pyramid of energy for a community of dinosaurs that lived in part of North America about 85 million years ago. This community included several herbivorous dinosaur species as well as carnivorous tyrannosaurs. At the time that Farlow carried out his calculations, there was a major debate about whether dinosaurs were 'warm-blooded' (like modern mammals and birds) or 'cold-blooded' (like modern reptiles).

△ Fig. 4.22 Dinosaurs: cold-blooded or warm-blooded?

He calculated the energy requirements of the herbivorous and carnivorous dinosaurs using estimates from modern mammals, and also from modern reptiles. Using energy values from mammals, he calculated that the carnivores in the community needed far more energy to support their needs than could be provided by the herbivores. However, using lizard energy values, he was able to produce a pyramid of energy of the right shape. He used this as evidence to suggest that large dinosaurs were cold-blooded like modern lizards, although he admitted that the number of assumptions he had made in his calculations did not prove that they were not warm-blooded like mammals. Since Farlow's time, other evidence has suggested more strongly that many dinosaur species were 'warm-blooded'.

The reason that a pyramid of energy is always a pyramidal shape is that each trophic level can only use a proportion of the energy in the food that they eat (or, in the case of plants, the food they make using the light energy they absorb). At each stage in the food chain, some of the energy within the trophic level is lost to the environment, as heat energy from respiration, or as chemical energy in waste substances that pass to the environment and the decomposers.

QUESTIONS

1. Describe the terms **a)** pyramid of number, **b)** pyramid of biomass, **c)** pyramid of energy.

2. Explain, with an example, how you would produce a pyramid of number for a community with three trophic levels.

3. Explain why a pyramid of biomass may have an inverted shape.

EFFICIENCY OF ENERGY TRANSFER

If you compare the amount of energy that enters a trophic level with the amount of energy available to the next trophic level, you can see that there are several sources of energy loss at each stage. From this you can calculate the efficiency of energy transfer, which is the proportion of energy passed to the next trophic level compared with the energy that entered the trophic level – for example, the amount of energy in the plants that a herbivore converts to animal tissue compared with the amount of energy in all the plants that it could eat.

Energy losses in plants

The amount of light energy that falls on the Earth's surface varies at different times of day and year, and varies in different parts of the world (places near the Equator receiving more light energy than places nearer the poles). On average, tropical areas receive between 3 and 5 kWh/m^2 per day (which is about the same energy as a one-bar electric heater left on for 3–5 hours).

Plants use only a tiny proportion of this for many reasons, as shown in Fig. 4.23. It has been estimated that most plants only convert about 1% to 2% of the light energy that falls on them into chemical energy in biomass. This is the energy available to a herbivore that eats the plant.

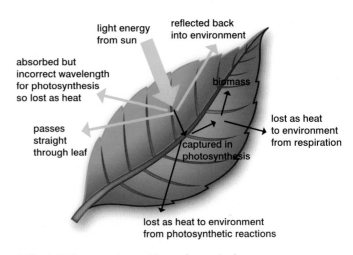

△ Fig. 4.23 Energy gains and losses from a leaf.

Energy losses in animals

When an animal eats, the food is digested in the alimentary canal and the soluble food molecules are absorbed into the body. The undigested food in the alimentary canal is egested as faeces. Absorbed food molecules may be used for different purposes in the body:

- to produce new animal tissue or gametes for reproduction
- used as a source of energy for respiration
- converted to waste products in chemical reactions.

The chemical energy in the food molecules remains as chemical energy in body tissue, or in the waste products, such as urea, which are excreted to the environment.

When food molecules are broken down during respiration chemical energy in the molecules is converted to heat energy. This heat energy is lost to the environment by conduction and radiation. So only a proportion of the chemical energy in the animal's food is converted into chemical energy in its body tissues. This is what increases the animal's biomass.

Different groups of animals have different efficiencies of energy transfer, depending on their food and on their body chemistry. For example, plant material is more difficult to digest than animal material, so carnivores generally have greater energy transfer efficiencies than herbivores. In addition, mammals and birds maintain their internal body temperature at a constant level, often above the temperature of the environment, and this requires the release of additional heat energy from respiration. So mammals and birds generally have a lower energy transfer efficiency than reptiles and amphibians, which do not maintain a constant internal body temperature. On average only about 10% of the energy in one trophic level is transferred to the next.

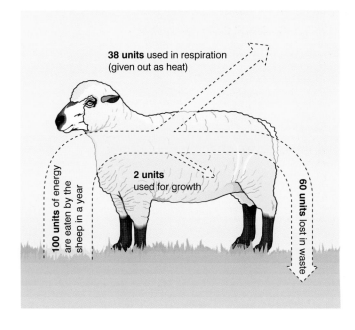

38 units used in respiration (given out as heat)

100 units of energy are eaten by the sheep in a year

2 units used for growth

60 units lost in waste

◁ Fig. 4.24 The energy flow in a young sheep.

Calculating energy transfer efficiencies involves the estimation of many values. This means that the values you may find in textbooks and on the internet are only best estimates and must not be taken as exact. In addition, many sources quote a value of 10% as the efficiency of energy transfer between any trophic level and the one above. Calculations of efficiency vary from about 0.2% to around 20% for different organisms in different ecosystems. This gives an *average* of 10%, but over such a large range this is not very reliable.

The fact that there are energy losses to the environment at each trophic level explains the shape of pyramids of energy. It also explains the fact that food chains rarely contain more than 4 or 5 trophic levels. Any organism expends energy when looking for food, and the more scattered the food, the more energy is lost in moving about to find it. Top consumers usually have to hunt over large distances to find enough food. If an organism fed exclusively on them, it would expend more energy hunting for its food than it would gain from eating it.

EXTENSION QUESTIONS

Humans are omnivores and grow a range of crop plants and animals for food. For example we may grow wheat grain and grind it into flour for making bread, pasta and other foods. Or we might grow wheat grain and feed it to chickens that we then eat.

1. Explain what is meant by 'Humans are omnivores'.

2. Draw diagrams for the two food chains described above.

3. Sketch pyramids of energy for each of the food chains.

4. Using your pyramids of energy, comment on the efficiency of growing wheat that is then fed to chickens which we eat.

5. Some people say that, as the human population continues to grow, we should all become vegetarian.

    a) Describe the argument for this.

    b) Suggest at least one argument for not doing this based on ecological principles.

QUESTIONS

1. Draw a flowchart to show the energy gains and losses from a plant leaf.

2. Draw a flowchart to show the energy gains and losses from a herbivore, such as a cow.

3. Explain fully why a pyramid of energy has that particular shape.

TRANSFER OF SUBSTANCES ALONG FOOD CHAINS

Food contains chemicals as well as energy, so food chains and food webs can be considered as diagrams of the transfer of substances as well as energy through the organisms in a community. Different substances, such as carbon and nitrogen, are taken into the body in different forms by plants and animals. These will be discussed further in the next chapter.

What you will see is that substances can continually cycle between the living and physical parts of the environment. This is unlike energy, which enters communities as light energy and is ultimately released to the environment as heat energy that cannot be converted to a useful form in organisms again.

QUESTIONS

1. Give one example of a substance that is transferred along a food chain.

2. Compare the transfer of energy and substances along food chains.

End of topic checklist

Biomass is the mass of living material, such as the mass of a living organism.

A **decomposer** is an organism that causes decay of dead material, such as many fungi and bacteria.

A **food chain** shows the sequence of a producer, a primary consumer that eats it, a secondary consumer that eats the primary consumer, and so on.

A **food web** shows many interdependent food chains.

A **primary consumer** is an animal that eats plants (also a herbivore).

A **producer** is an organism that produces its own food, such as plants using light energy in photosynthesis to produce glucose.

A **pyramid of biomass** diagram shows the biomass in different trophic levels of a food chain, often a pyramid shape.

A **pyramid of energy** diagram shows the energy content of different trophic levels of a food chain, always a pyramid shape.

A **pyramid of number** diagram shows the number of individual organisms in different trophic levels of a food chain, often a pyramid shape.

A **secondary consumer** is an animal that eats primary consumers.

A **tertiary consumer** is an animal that eats secondary consumers.

A **trophic level** is a feeding level in a food chain or food web, such as producer, primary consumer.

The facts and ideas that you should know and understand by studying this topic:

◯ Different trophic levels include producers, primary consumers, secondary consumers, tertiary consumers and decomposers.

◯ Food webs are made up of many interdependent food chains.

◯ Differences between trophic levels can be shown as pyramids of number, pyramids of biomass and pyramids of energy.

◯ Pyramids of energy are always a pyramid shape, but the other two kinds of pyramid diagrams may not always be so.

◯ Both substances and energy are passed along food chains.

◯ Energy transfer to the next trophic level is always a small proportion of the energy gained, about 10% on average.

End of topic questions

1. The photograph shows lions feeding on the carcass of a zebra. When a lion catches a zebra, it will share the meat with other lions. Before the lion started chasing the zebra, the zebra had been feeding on grass.

   a) Is the lion a carnivore or herbivore? Explain your answer. **(2 marks)**

   b) At which trophic level does the zebra feed? **(1 mark)**

   c) Draw a food chain for the organisms shown in the photograph. **(2 marks)**

   d) Lions also feed on the herbivores gazelle and wildebeest. Use all these organisms to draw a food web for the African grassland. **(3 marks)**

2. In a community of organisms in a garden there are 5 lettuces. There are 40 caterpillars feeding on the lettuces until 2 thrushes (insectivorous birds) eat all the caterpillars.

   a) Draw a pyramid of number for this community. **(3 marks)**

   b) Describe the limitations of this pyramid. **(2 marks)**

   c) Describe the difficulty of preparing the data for a pyramid of biomass for these organisms. **(2 marks)**

3. Use the food web on page 271 to predict what would happen to the numbers of the following species if all the herbivorous insects were killed by insecticide. Explain your answers.

   a) predatory insects **(2 marks)**

   b) insectivorous birds **(2 marks)**

   c) mice **(2 marks)**

   d) snakes. **(2 marks)**

4. Fig. 4.25 shows a pyramid of number for a food chain early in the year. Later in the year the caterpillars change into butterflies and fly away to feed on flowers.

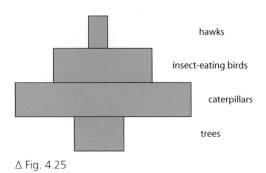

△ Fig. 4.25

a) Which organisms are the primary consumers in this food chain? Explain your answer.

(2 marks)

b) Explain why this pyramid is not the usual pyramid shape. **(1 mark)**

c) Suggest what this pyramid would look like for the same food chain in late summer. Explain your answer. **(2 marks)**

d) The data in this pyramid were used to draw a pyramid of energy. Suggest the shape of the pyramid of energy. Explain your answer. **(2 marks)**

5. **a)** Explain as fully as you can why only a small proportion of the energy in the producer trophic level is passed to the primary consumer trophic level. **(2 marks)**

b) Explain as fully as you can why only a small proportion of the energy in the primary consumer trophic level is passed to the secondary consumer trophic level. **(3 marks)**

6. Compare the difficulty of producing a pyramid of number, pyramid of biomass and pyramid of energy. **(6 marks)**

Cycles within ecosystems

INTRODUCTION

The light energy from the Sun, on which most of life on Earth depends, seems to be an inexhaustible supply. (In fact, in 4 to 5 billion years the Sun will fade and stop supplying that energy.) Light energy from the Sun is converted to chemical energy in the tissues of living organisms and then transferred to the environment as heat energy, which organisms can no longer use. However, all the chemicals that are found in living tissue, such as carbon and nitrogen, are limited to what is currently on Earth. So it is just as well that these substances are continually cycled between the living and non-living parts of ecosystems.

△ Fig. 4.26 Sunlight provides the energy for all plant growth.

KNOWLEDGE CHECK

✓ Carbohydrates, proteins and lipids all contain carbon.
✓ Proteins also contain nitrogen.
✓ Decomposers digest dead organic material releasing some of the products of digestion into the environment.

LEARNING OBJECTIVES

✓ Describe the stages in the carbon cycle, including respiration, photosynthesis, decomposition and combustion.
✓ Describe the stages in the nitrogen cycle, including the roles of nitrogen-fixing bacteria, decomposers, nitrifying bacteria and denitrifying bacteria.

THE CARBON CYCLE

Carbon is cycled through the living and non-living parts of ecosystems, in different forms at different stages of the **carbon cycle**. Carbon dioxide from the atmosphere is converted to complex carbon compounds in plants during photosynthesis. This is often called the 'fixing' of carbon by plants. Respiration in plants returns some of this fixed carbon back to the atmosphere as carbon dioxide. Carbon in the form of complex carbon compounds passes along the food chain.

At each stage, some of this carbon is released as carbon dioxide to the atmosphere as the result of respiration.

When organisms die, their bodies decay as they are digested by decomposers (bacteria and fungi). This is also known as **decomposition**. Carbon dioxide is released when the decomposers respire using the carbon compounds from the dead organisms.

If dead organic material is buried too quickly by sediment or water for decomposers to cause decay, and remains buried, then it may be converted to other complex carbon compounds. Peat is formed when mosses and other plants are buried in swampy ground for hundreds of years. Over many millions of years, where there were once huge forests growing in swampy regions, heat and pressure have turned the organic material to coal. Heat and pressure over many millions of years also produce oil and natural gas from the decaying bodies of tiny marine organisms that were buried in sediment at the bottom of oceans. Peat, coal and oil are **fossil fuels**. We can release the carbon from the complex carbon compounds in fossil fuels, as well as from other sources such as wood, into the air as carbon dioxide during **combustion**, when we burn them.

△ Fig. 4.27 Water, like sediment, excludes air, which prevents decay organisms from respiring. So dead plant material builds up over time, forming peat. Peat can be burnt as a fuel, although this is now discouraged so that peat bog habitats can be protected.

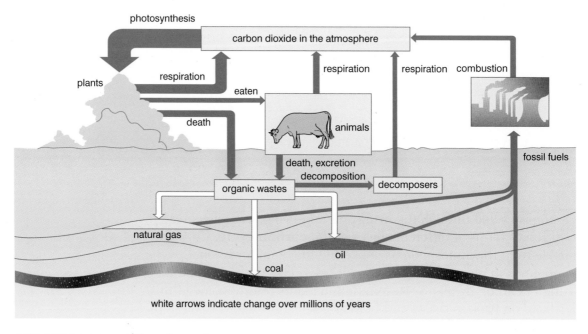

△ Fig. 4.28 A summary of the carbon cycle.

CARBON DIOXIDE IN THE ATMOSPHERE

Over the past 10 000 years or so, as a result of photosynthesis and respiration and other physical processes, the exchange of carbon between organisms and the atmosphere resulted in little change in the amount of carbon dioxide in the atmosphere. On average, over one year about 120 billion tonnes of carbon dioxide are removed from the atmosphere by photosynthesis, and a similar amount returned by respiration.

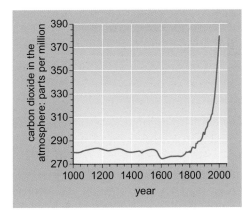

△ Fig. 4.29 Atmospheric carbon dioxide concentration from 1000 AD to recent times.

During the past 250 years, however, human activity has added increasing amounts of carbon dioxide to the atmosphere. Today about 5.5 billion tonnes of carbon dioxide are added to the atmosphere every year through human activity, particularly through combustion of fossil fuels.

Compared with 120 billion tonnes through natural processes, this may not seem a lot, but there is no process that balances this addition. So the concentration of carbon dioxide in the atmosphere is increasing. It is this additional carbon dioxide in the atmosphere that most people believe is causing global warming and climate change.

QUESTIONS

1. Describe the role of the following in the carbon cycle:

   a) respiration

   b) photosynthesis

   c) decomposition.

2. In what form is carbon when it is in the following stages of the carbon cycle?

   a) Earth's atmosphere

   b) plant tissue

   c) fossil fuels.

3. Explain the significance of combustion in the carbon cycle.

THE NITROGEN CYCLE

Living things need nitrogen to make proteins, which are used, for example, to make new cells during growth. Air is 79% nitrogen gas (N_2), but nitrogen gas is very unreactive and cannot be used by plants or animals.

The **nitrogen cycle** describes the way in which nitrogen passes between the living and non-living parts of an ecosystem. Animals take in nitrogen in the form of proteins when they eat plant tissue or animal tissue. They break down the proteins to amino acids in digestion and convert them into new proteins in their own body tissue. Since plants do not eat, they can't take their nitrogen in this form. Instead they absorb nitrogen in the form of nitrate ions (NO_3^-) using active transport from the soil water around their roots. They then convert these ions into the proteins they need.

Nitrate ions are present in the soil as a result of several processes, one non-living and the others as the result of living organisms. The non-living process is lightning, which generates large amounts of energy that cause atmospheric nitrogen to react with oxygen. This forms nitrate ions, which dissolve in rain and fall to the ground where they remain in soil water.

The biological processes that produce nitrate ions are important. The first begins with organic material cont nitrogen compounds, such as proteins in the form of 1 urine from animals, and as dead plant and animal tiss

This material decays as a result of decomposers, relea the form of ammonium ions into the soil.

Some bacteria in the soil can take in ammonium ions nitrite ions (NO_2^-). Other soil bacteria can take in nitr produce nitrate ions. Both kinds of bacteria are called **bacteria**.

The other biological process that makes nitrogen avail nitrates also involves bacteria. These bacteria are unu: can convert nitrogen gas from the atmosphere directly plants can use. They are called **nitrogen-fixing bact** these bacteria live free in the soil and some grow in ro small bumps on the roots, of some kinds of plants esp (plants of the pea and bean family).

Nitrifying bacteria can only grow in aerobic conditions, when there is plenty of oxygen in the soil. In waterlogged conditions, air in the soil is pushed out, so these bacteria cannot grow. These conditions favour another type of bacteria, which can respire anaerobically. As they grow, they convert nitrates back to nitrogen gas, which escapes to the atmosphere. They are called **denitrifying bacteria**, because they remove nitrates from the soil and make it less fertile for growing plants.

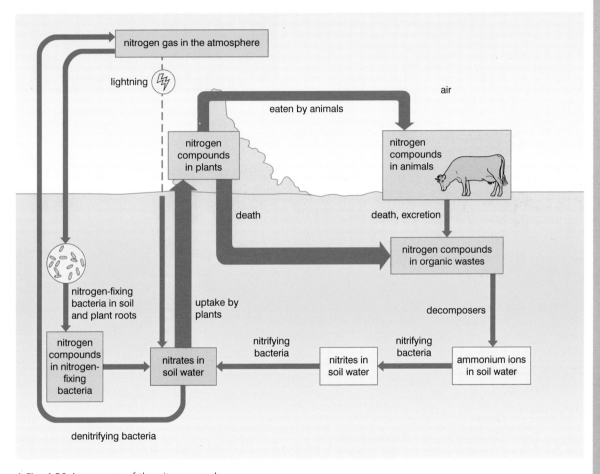

△ Fig. 4.30 A summary of the nitrogen cycle.

RETURNING NITRATES TO THE SOIL

As plants grow, they remove nitrates from the soil. In natural communities, those nitrates are returned to the soil when the plants die and decay, or when the animals that ate them die and decay. However, when we grow crops, we usually remove the plants from the ground, taking the nitrates (in the form of nitrogen compounds in plant tissue) away. This reduces the amount of nitrates in the soil each year.

Farmers can return some nitrates to the soil by spreading animal manure (faeces and urine mixed with straw) or by growing legume crops in some years.

Farmers sometimes plant legume crops in order to add nitrogen to the soil. The following year another crop, such as potatoes or cabbages, will grow better without additional fertiliser.

Millions of tonnes of nitrogen-containing fertiliser are made each year by chemical processes. These are used to fertilise fields to provide food for the growing human population.

△ Fig. 4.31 Some legume plants produce special nodules on their roots in which nitrogen-fixing bacteria live.

These processes often use energy from fossil fuels, which are non-renewable resources. There is concern that, if we do not find other ways of making these fertilisers, global food production may decrease as fossil fuels reserves are used up.

QUESTIONS

1. Describe the roles in the nitrogen cycle of:

   a) nitrifying bacteria

   b) nitrogen-fixing bacteria

   c) denitrifying bacteria.

2. Explain the importance of nitrifying bacteria in the nitrogen cycle for the fertility of soils.

3. Explain the importance of decomposers in the cycling of nitrogen.

Developing investigative skills

Students were given some young wheat plants and some young legume plants growing in separate pots. The plants had been grown from seed. The seeds for the legume plant had been inoculated with nitrogen-fixing bacteria so that they developed root nodules containing the bacteria. The students were also given two nutrient solutions for watering the plants: one that contained all nutrients and one that contained all nutrients except nitrogen.

Devise and plan investigation

❶ Write a plan, using this equipment, to investigate the effect of nitrogen on the growth of legume and non-legume plants. Make clear how your investigation is designed to produce reliable results.

❷ Write a suitable prediction for your investigation.

Analyse and interpret data

The table shows the results of the investigation carried out by the students.

Plants	Wheat		Legume	
Nutrient solution used for watering	All nutrients	Without nitrogen	All nutrients	Without nitrogen
Height at start (cm)	5.2	4.8	3.6	4.1
Height at end (cm)	20.6	13.6	18.1	19.3

❸ For each plant calculate the percentage increase in height from the start to the end of the experiment.

❹ Describe the differences between the growth of each plant.

❺ Explain the results for the wheat plants.

❻ Explain the results for the legume plants.

❼ Draw a conclusion for this experiment.

Evaluate data and methods

❽ Identify any weaknesses in this method.

❾ Explain how the method should be improved so that a more reliable conclusion can be produced.

End of topic checklist

The **carbon cycle** can be represented as a diagram that shows how the element carbon cycles in different forms between living organisms and the environment.

Combustion means burning, such as the burning of fossil fuels.

Decomposition means the decay of dead plants and animals and their waste materials.

Denitrifying bacteria are soil bacteria that convert nitrates in the soil to nitrogen gas which is released to the atmosphere.

Fossil fuels are fuels formed from organic material, such as peat, coal, oil and natural gas.

Nitrifying bacteria are soil bacteria that convert ammonium ions to nitrite ions, and nitrite ions to nitrate ions, important in the nitrogen cycle.

The **nitrogen cycle** can be represented as a diagram that shows how the element nitrogen cycles in different forms between living organisms and the environment.

Nitrogen-fixing bacteria are bacteria found in the soil and in root nodules of some plants that can convert atmospheric nitrogen into a form of nitrogen that plants can use.

The facts and ideas that you should know and understand by studying this topic:

○ The processes of respiration, photosynthesis, decomposition and combustion are all important in the carbon cycle.

○ Nitrogen-fixing bacteria, decomposers, nitrifying bacteria and denitrifying bacteria are all important in the nitrogen cycle.

End of topic questions

1. The graph in Fig. 4.32 shows the change in carbon dioxide concentration above a forest over two days, and the light intensity just above the top of the trees.

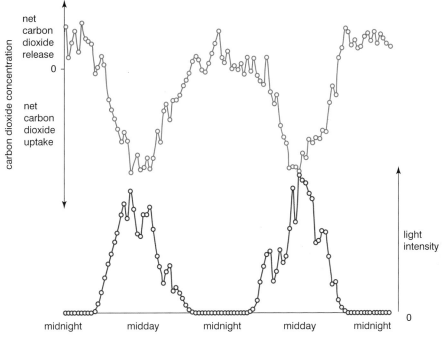

△ Fig. 4.32

a) Explain the changes in light intensity shown in the graph. **(2 marks)**

b) Explain the changes in carbon dioxide concentration shown in the graph. (Remember there are more organisms than just the trees in the forest.) **(4 marks)**

2. a) Explain why waterlogged soils, such as swamps and bogs, usually have very low concentrations of nitrates. **(3 marks)**

b) Sundew plants live in waterlogged soils of swamps and bogs. They have sticky hairs on some leaves that are special adaptations for catching insects. Once the insect is trapped, the leaf rolls up and enzymes are secreted to digest the animal. The digested liquid is absorbed by the plant. Suggest why these adaptations are important to the sundew. **(3 marks)**

3. In a tropical forest, the layer of dead leaves (called the leaf litter) on the forest floor is usually very thin at all times of the year. In temperate woodlands (as in the photograph) where there are seasons of summer and winter, many trees drop their leaves in the autumn and grow new ones in the spring.

a) Tropical trees drop a few leaves at a time at any time of year. What happens to the leaves on the ground? Explain your answer as fully as possible. **(2 marks)**

b) The leaf litter in a temperate woodland is deep all through winter when it may be cold enough for snow, until it gets warm again in spring. Then the leaf litter disappears. Explain these observations as fully as you can. **(3 marks)**

4. 'Without bacteria in the soil, there would be no plants and no animals.' Explain this statement. **(5 marks)**

Human influences on the environment

△ Fig. 4.33 Before humans arrived on St Helena this landscape would have been covered with dense tropical rainforest.

INTRODUCTION

St Helena is an isolated island in the Atlantic Ocean. It was uninhabited until it was discovered in 1502. The human population of the island slowly increased to over 1000 people in the 1700s, and around 4500 people live there now.

The first people to arrive on the island found many plants and animals that were unique to the island. As people cleared the dense tropical forest, to make space for building and for growing crops and keeping herd animals, many species became extinct.

The introduction of animals that did not naturally live there, such as cats, goats and rats, had a devastating effect on wildlife. Cats catch and kill small animals and birds, and rats steal and eat eggs from bird nests. Many of the lower areas near the sea are now completely bare of vegetation as a result of grazing by goats.

KNOWLEDGE CHECK

✓ The Earth's atmosphere is affected by human activity (such as deforestation and combustion of fuels) and by natural processes (such as volcanoes).
✓ Development of the environment can be sustainable or non-sustainable.
✓ Organisms that cannot adapt to changing conditions fast enough may go extinct.

LEARNING OBJECTIVES

✓ Understand the biological consequences of pollution of air by sulfur dioxide and carbon monoxide.
✓ Understand that water vapour, carbon dioxide, nitrous oxide, methane and CFCs are greenhouse gases.
✓ Understand how human activities contribute to greenhouse gases.
✓ Understand how an increase in greenhouse gases results in an enhanced greenhouse effect and that this may lead to global warming and its consequences.
✓ Understand the biological consequences of pollution of water by sewage.
✓ Understand the biological consequences of eutrophication caused by leached minerals from fertiliser.
✓ Understand the effects of deforestation, including leaching, soil erosion, disturbance of evapotranspiration and the carbon cycle, and the balance of atmospheric gases.

AIR POLLUTION

Acid rain

Since the Industrial Revolution began in northern Europe in the 1700s, humans have burned increasing quantities of fossil fuels (such as coal, oil and natural gas) to provide energy for industrial processes. The effects of this can now be seen all over the world.

Burning fossil fuels gives off many gases, including sulfur dioxide and various nitrogen oxides (NO_x gases).

Sulfur dioxide combines with water droplets in clouds to form sulfuric acid. Nitrogen oxides combine with water to form nitric acid. When the rain falls, it is more acidic than usual, so we call it **acid rain**.

Acid damages cells and delicate tissues directly. If breathed in, the acid can damage the delicate tissues lining the lungs. Plants may have their leaves damaged, so they can no longer photosynthesise and grow well.

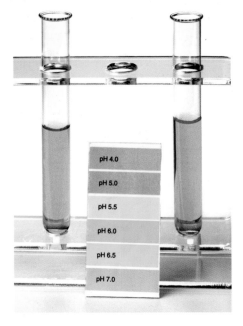

△ Fig. 4.34 The pH of normal rain water (tube on the left) is about 5.6, but acid rain (tube on the right) can have a pH of less than 3.0.

Animals with soft skins, such as fish and amphibians (frogs and toads), may have their skins damaged by the acid rain falling in the ponds, lakes and rivers where they live. Single-celled organisms, such as protoctists, are even more likely to be damaged.

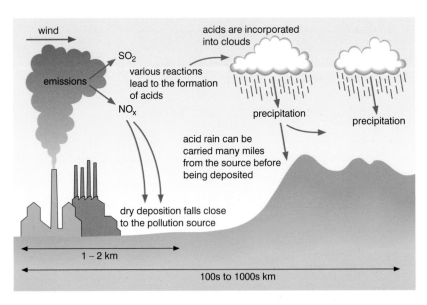

△ Fig. 4.35 Acid rain can be transported many kilometres from where the acidic gases were released into the air.

Acid rain can also cause damage indirectly. In soil, it can cause some mineral ions to dissolve into the soil water. Some of these ions, such as aluminium ions, are poisonous to some species. Others may be more easily washed out of the soil, away from plants that need them, so that plant growth is reduced.

Damaging some organisms in a food web has an impact on other organisms because organisms in an ecosystem are interdependent. Species that depend on those damaged by acid rain will either fail to thrive or move away from the area. However, some species can tolerate acidity better than others, so they will benefit by having more space to live in.

▷ Fig. 4.36 This species of lichen can only grow where there is no pollution in the air.

 SCIENCE IN CONTEXT

REDUCING SULFUR DIOXIDE EMISSIONS

Most of the fossil fuels that we burn are used either in industry, particularly to generate electricity, or in vehicle engines. Over the past few decades, particularly in Europe and the US, efforts have been made to reduce emissions of sulfur dioxide.

- Sulfur dioxide is removed ('scrubbed') from the gases given off from combustion as they pass up the chimneys of factories and power stations, so that it is not released into the atmosphere.
- Sulfur compounds are removed from petrol and diesel fuels before they are burned in vehicle engines.

These efforts have resulted in a major decrease in sulfur dioxide concentration in the air. However, more needs to be done to solve this problem completely.

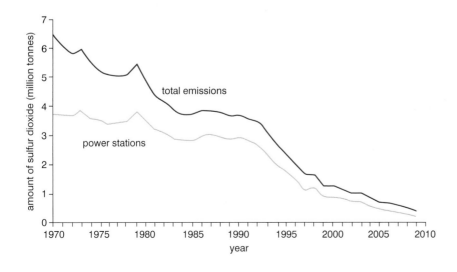

◁ Fig. 4.37 Sulfur dioxide emissions from the UK.

Developing investigative skills

You can investigate the effect of acid on the germination of seeds by adding dilute acid to the water used to water growing seedlings.

Devise and plan investigations

❶ Write a plan for an investigation on the effect of different acidic pHs on the germination of seeds.

Demonstrate and describe techniques

❷ Describe any hazards with your plan and how you should protect against them.

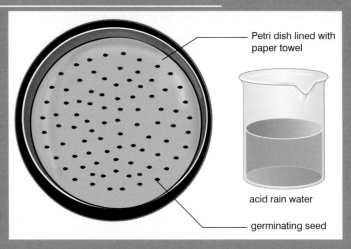

△ Fig. 4.38 The effect of acid rain on seedlings.

- Petri dish lined with paper towel
- acid rain water
- germinating seed

Analyse and interpret data

The graph in Fig. 4.39 shows the results from an investigation into the effect of different pHs on the germination of wheat seeds.

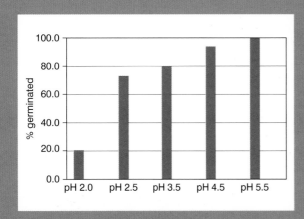

◁ Fig. 4.39 Results of investigation.

❸ Use the graph to draw a conclusion about the effect of acid rain on the germination of wheat.

Carbon monoxide

Carbon monoxide is a colourless, odourless gas. It is formed when substances containing carbon are burned in a limited supply of air, such as when a Bunsen burner hole is partially closed.

Fossil fuels such as petrol, natural gas, coal and wood can release carbon monoxide when they are burned. Exhaust from motor vehicle engines are the main source of carbon monoxide pollution in urban areas.

Carbon monoxide can have serious health effects on humans and animals. When inhaled, it reduces the supply of oxygen to tissues and organs. This is particularly damaging for the heart and brain. An increase in carbon monoxide can be a serious problem for people with cardiovascular disease and for young children who are still growing rapidly.

△ Fig. 4.40 A blue Bunsen flame occurs when there is complete combustion because there is a good supply of air through the air hole. A yellow Bunsen flame shows that the gas is burning incompletely because there is a limited supply of oxygen (air hole closed) – the gases given off will include carbon monoxide.

SCIENCE IN CONTEXT

CARBON MONOXIDE POISONING

Carbon monoxide poisoning occurs when a person has breathed in too much carbon monoxide. The carbon monoxide binds to the haemoglobin and prevents the haemoglobin binding with oxygen instead. So the effects of carbon monoxide poisoning are mainly caused by the reduction of oxygen reaching the body tissues.

At low concentrations in the air, carbon monoxide may produce a headache, dizziness and confusion, as a result of not enough oxygen getting to the brain. At higher concentrations this can also produce a feeling of sickness, and convulsions. If a person is exposed to carbon monoxide for too long, it will cause death.

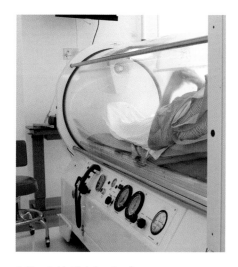

△ Fig. 4.41 High levels of oxygen can remove carbon monoxide from the blood.

Treatment for carbon monoxide poisoning may be with air that contains a high proportion of oxygen, which is known as hyperbaric treatment. The high concentration of oxygen causes the carbon monoxide to be released from the haemoglobin, making it possible for it to carry oxygen again.

1. Use Fig. 4.35 showing how acid rain is formed to:

   a) explain how human activity contributes to acid rain

   b) explain why acid rain is not just a problem for places where there is a lot of industry.

2. Describe the different ways in which acid rain can damage organisms and ecosystems.

3. Explain the dangers of carbon monoxide to human health.

THE EFFECTS OF GREENHOUSE GASES

The greenhouse effect

There are many gases in the Earth's atmosphere, but one group plays an important role in keeping the Earth's surface warm. These gases are called the **greenhouse gases** and they include:

- water vapour – water that has evaporated from the Earth's surface
- carbon dioxide – produced naturally from respiration of living organisms (see the carbon cycle on pages 283–285)
- nitrous oxide (N_2O) – released naturally from soil and water due to soil bacteria
- methane – produced during the decay of organic material, such as in swamps, and in the digestion of food in the alimentary canal
- CFCs (chlorofluorocarbons) – artificial chemicals produced for use in cooling systems in refrigerators and in aerosols. These are no longer widely used because they damage the ozone layer in the atmosphere.

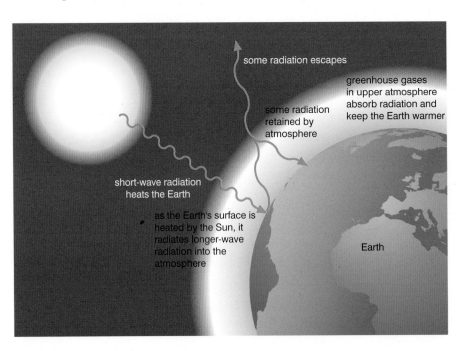

some radiation escapes

greenhouse gases in upper atmosphere absorb radiation and keep the Earth warmer

some radiation retained by atmosphere

short-wave radiation heats the Earth

as the Earth's surface is heated by the Sun, it radiates longer-wave radiation into the atmosphere

Earth

△ Fig. 4.42 The greenhouse effect on Earth.

Short-wave radiation from the Sun warms the ground, and the warm Earth gives off heat as longer-wave radiation. Much of this radiation is stopped from escaping from the Earth by the greenhouse gases and returns to warm the Earth's surface. This is known as the **greenhouse effect**.

The greenhouse effect is normal, and it is important for life on Earth. It keeps the Earth warmer than it otherwise would be. Without it, it is estimated that the surface of the Earth would be about 33 °C cooler than it is now. All water on the surface of the Earth would be frozen, and very little life could exist in these conditions.

CONDITIONS ON VENUS AND MARS

The importance of the greenhouse effect can be seen by comparing the conditions on the surfaces of Earth, Venus and Mars. Mars has a relatively thin atmosphere, and although this is composed mainly of carbon dioxide, its effect as a greenhouse gas is limited. The average surface temperature on Mars is about –55 °C, varying from about 27 °C during the day at the equator to –143 °C at the poles at night. In contrast, Venus has a much denser atmosphere than Earth, consisting mainly of carbon dioxide with clouds of sulfur dioxide. These gases create a very strong greenhouse effect, heating the surface of Venus to an average of around 460 °C. It is not surprising, therefore, that Earth is the only planet in our Solar System where we know life has evolved.

△ Fig. 4.43 People once thought that Martians lived on Mars, but scientific exploration suggests there is probably no life on the planet.

The enhanced greenhouse effect

Human activities have increased, and are continuing to increase, the proportion of some greenhouse gases in the Earth's atmosphere.

- Carbon dioxide is increasing as a result of the combustion of fossil fuels (see pages 284–285).
- Nitrous oxide is increasing partly from the combustion of fossil fuels, but mainly from the breakdown of artificial nitrogen fertilisers, which are added to fields to encourage crop growth (see page 288).
- Methane is increasing as a result of the increasing numbers of people and farm animals that we keep, and from the release of the gas from artificial wetlands such as rice paddy fields.
- CFCs are still in the atmosphere due to their use in refrigerators and aerosols.

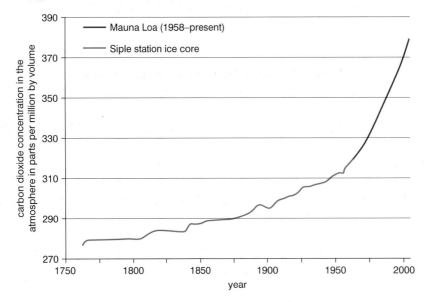

△ Fig. 4.44 Change in the concentration of carbon dioxide in the Earth's atmosphere over the last two and a half centuries.

Average global surface temperature has also been rising over the past few centuries. Global temperatures can vary over a wide range due to many natural factors, such as the amount of radiation received from the Sun which varies due to a predictable but complex cycle. However, many scientists are certain that the recent increases in temperature are the result of increased emissions of greenhouse gases from human activity, leading to an **enhanced greenhouse effect** resulting in what is commonly called **global warming**.

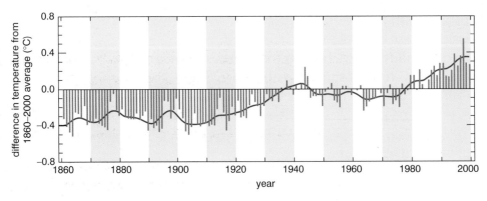

△ Fig. 4.45 Variation of the Earth's surface temperature, calculated as the difference from the average temperature between 1860 and 2000.

Be very careful not to confuse the natural greenhouse effect, which is essential for life on Earth, with the enhanced greenhouse effect and global warming as a result of the release of additional greenhouse gases from human activity.

Many features of our climate are the result of differences in local surface temperatures in different places, such as the speed of winds or amount of precipitation. Predictions from computer modelling of the effects of global warming suggest that different parts of the world, at different times, may experience an increase in:

- the number and strength of storms
- drought
- flooding as a result of increased rainfall and rising sea levels
- hotter summers and warmer winters
- cooler, wetter summers and colder winters.

These changes will not only affect humans, but whole ecosystems, potentially increasing the rate at which other species become extinct. This will have even more of an impact as a result of the interdependency of organisms through food webs in communities.

QUESTIONS

1. Distinguish between the greenhouse effect and the enhanced greenhouse effect.

2. Give examples of natural causes and human causes of emissions of the following gases:

   a) carbon dioxide

   b) nitrous oxide

   c) methane.

3. Give three examples of possible consequences of global warming.

WATER POLLUTION

Sewage pollution

Sewage is human waste, faeces and urine, which we all produce and need to dispose of. Faeces and urine contain high concentrations of many nitrogen-containing substances, and so are good sources of nutrients for plants and microorganisms. In fact, farmers often use animal waste as manure to spread on their fields to increase crop production.

In areas where many people live, sewage disposal is a big problem. Many cities have sewage management systems to carry the sewage in pipes to treatment centres where it can be broken down, or away from the city to prevent drinking water being contaminated. However, adding large amounts of sewage to water can cause water **pollution**, where the environment is damaged and living organisms are harmed.

△ Fig. 4.46 There are cities in the world where there are no proper systems for removing and treating sewage.

When untreated sewage is added to water, the nutrients in it dissolve into the water.

- Bacteria and other microorganisms living in the water increase in numbers rapidly as a result of the increased concentrations of nutrients.
- The more microorganisms there are, the more respiration is taking place in order to make new materials for growth and reproduction.
- Respiration takes dissolved oxygen from the water, reducing the oxygen concentration of the water.
- Other aquatic (water-living) organisms find it increasingly difficult to get the oxygen they need from the water for respiration, and may die.
- The decay of dead organisms in the water provides more nutrients, so more bacteria grow, respire and take more oxygen from the water.
- Eventually, most of the large aquatic plants and animals in the water may die.
- Low oxygen concentration in the water can also encourage the growth of anaerobic bacteria, some of which are pathogenic. This makes the water dangerous for drinking.

The situation can be made worse by excessive plant growth when the nutrients are first added. The leaves of the plants can cover the surface of the water, making it more difficult for oxygen from the air to dissolve into the water and replace the oxygen used by respiration by aquatic organisms. The plants on the surface also block light from the aquatic plants below the surface, which is another reason why they may die.

Eutrophication

The adding of nutrients to water is called **eutrophication**. Eutrophication can be caused by sewage pollution, and can lead to the death of aquatic organisms, particularly the larger plants and animals. It can be caused by other factors that lead to an increase in nutrient minerals in water. For example, if a farmer adds more minerals in the form of fertiliser to a field than the crop plants can absorb, the remaining minerals will soak away in ground water (this is called **leaching**) into nearby streams and rivers. Also, if there is heavy rainfall soon after the fertiliser has been spread on a field, the minerals will dissolve in the rain water and run off the surface of the field into streams and rivers. This will cause eutrophication, which may lead to pollution and the death of aquatic organisms.

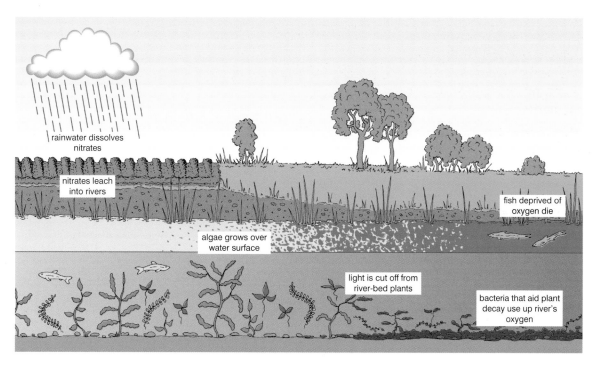

Δ Fig. 4.47 Eutrophication caused by fertilisers.

▷ Fig. 4.48 A mass of dead fish is often a sign of eutrophication or water pollution.

Eutrophication is often wrongly defined as the pollution of water and death of aquatic organisms. This is incorrect – eutrophication is simply the adding of nutrients. It comes from the Greek word *eutrophia,* meaning healthy or adequate nutrition. Adding nutrients that the ecosystem can use normally may be an advantage, but adding them in excess may lead to the death of aquatic organisms as a result of the depletion of dissolved oxygen in the water. So excess nutrients can cause pollution.

QUESTIONS

1. Explain what we mean by *eutrophication.*

2. Give two reasons why the use of artificial fertiliser on a field could cause eutrophication of a nearby stream.

3. Draw a flow diagram to show how sewage can cause eutrophication leading to water pollution.

EXTENSION

Water that we have used in our houses needs to be recycled in a way that makes it safe to return to water systems. This includes the waste water from cooking and washing as well as water used to flush away human sewage. Treatment is partly through physical processes and partly through biological processes.

- The sewage is first allowed to settle and then screened to remove heavy solids and floating material.

- The liquid is then passed into treatment beds that contain large numbers of bacteria and other microorganisms. These organisms break down any organic materials in the liquid. The beds are continually aerated by bubbling air through them.

- Sludge drained from the settling and treatment beds may be digested by other microorganisms in an enclosed space. This produces methane which can be burnt. Alternatively the drained sludge may be dried and used as a natural fertiliser on fields.

- Water drained from the treatment beds is treated with chlorine before passing it back into the water system.

△ Fig. 4.49 Sewage treatment.

1. What role do the bacteria and microorganisms have in the treatment beds?

2. Explain fully why the treatment beds are continually aerated.

3. Explain the effect of the bacteria and microorganisms in the treatment beds on the liquid from the sewage.

4. Suggest why the solid sludge is digested in enclosed conditions.

5. Explain the use of dried sludge as a natural fertiliser.

6. Chlorine is a disinfectant. Explain why the last stage of water treatment is the use of chlorine.

7. Biochemical oxygen demand (BOD) is the amount of dissolved oxygen consumed by organisms in water over 5 days at 20 °C. The BOD of sewage can be more than 600 mg/dm^3, while the water from the treatment plant that is passed to the water system may be 20 mg/dm^3 or lower. Explain this change.

DEFORESTATION

Deforestation is the permanent destruction of large areas of forests and woodlands. It usually happens in areas that provide quality wood for furniture and to create farming or grazing land.

Forests act as a major carbon store because carbon dioxide is taken up from the atmosphere during photosynthesis and used to produce the chemical compounds that make up trees. When forests are cleared, and the trees are either burned or left to rot, this carbon is released quickly as carbon dioxide. This disturbs the carbon cycle and the balance of gases in the atmosphere by rapidly increasing the proportion of carbon dioxide compared with oxygen in the air surrounding the forest. On the scale of deforestation in the Amazon Basin, the amount of carbon dioxide released is so great that it cannot be brought back into balance as a result of photosynthesis. This additional carbon dioxide contributes further to the enhanced greenhouse effect.

Deforestation also has an effect on the natural cycling of water. Trees draw ground water up through their roots and release it into the atmosphere by transpiration. Trees may also shade

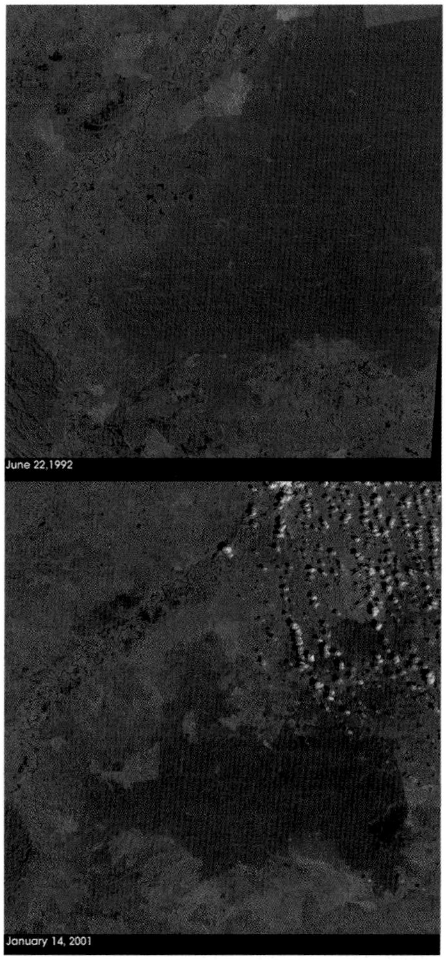

June 22, 1992
January 14, 2001

▷ Fig. 4.50 Satellite images of the island of Sumatra show the extent of deforestation between 1992 (top photo, land mostly green) and 2001.

the ground and reduce the amount of water that evaporates directly from the soil and other surfaces. The transfer of water from the ground to the atmosphere by both evaporation and transpiration is called **evapotranspiration**. As forest trees are removed, the amount of water that can be held in an area decreases, and evapotranspiration is also disturbed, which in turn, depending on whether ground water evaporates or drains away, can cause either increasing or decreasing rainfall in the area.

Removing the protective cover of vegetation from the soil can also result in **soil erosion**. This is where the soil is washed away by rain. This in turn can lead to blocked rivers and flooding. The top layers of soil are the ones that contain the most minerals, from the decay of dead vegetation, so soil erosion removes essential minerals from the land. Soil minerals are also lost by leaching, which is the soaking away of soluble minerals in soil water because there are few plant roots in the soil to absorb the minerals and lock them away in plant tissue. This loss of minerals from the soil is permanent, and makes it very difficult for forest trees to regrow in the area, even if the land is not cultivated.

All these effects change the ecosystem of an area and may result in the loss of species.

▷ Fig. 4.51 This satellite image of a river estuary in Madagascar shows large amounts of soil in the water (orange). This is a result of deforestation in the surrounding area.

QUESTIONS

1. Explain these terms:

   a) deforestation

   b) soil erosion

   c) leaching.

2. Explain how deforestation may affect:

   a) evapotranspiration

   b) soil fertility

   c) the carbon cycle and atmospheric carbon dioxide.

End of topic checklist

Acid rain is rain that contains higher than normal concentrations of dissolved acidic gases, such as sulfur dioxide and nitrogen oxides, which causes the rain to have a lower pH than normal.

Deforestation is the destruction of large areas of forest or woodland.

The **enhanced greenhouse effect** is an increase in the greenhouse effect most likely caused by the release of additional greenhouse gases from human activity.

Eutrophication is the addition of nutrients to water, which may lead to water pollution.

Evapotranspiration is the transfer of water from the ground to the atmosphere by the combined effect of both evaporation from the soil and transpiration from plants.

Global warming is a warming of the Earth's surface and atmosphere, possibly as a result of an enhanced greenhouse effect.

The **greenhouse effect** is the warming effect caused by greenhouse gases in the atmosphere that prevent some of the heat energy radiated by the Earth's surface escaping into space.

Greenhouse gases are atmospheric gases, such as carbon dioxide and water vapour, that trap some of the heat radiated from the Earth's surface and prevent it escaping into space.

Leaching is the loss of dissolved mineral nutrients in soil water as it soaks deep into the ground beyond the reach of plant roots.

Pollution is damage to the environment, people and other organisms, often as a result of adding chemicals to the air, water or land.

Sewage is human waste, faeces and urine, which needs to be disposed of.

Soil erosion is the washing away of soil as a result of rainfall when there is little vegetation to hold on to the soil.

The facts and ideas that you should know and understand by studying this topic:

○ Sulfur dioxide and carbon monoxide can cause air pollution and damage living organisms.

○ Water vapour, carbon dioxide, nitrous oxide, methane and CFCs are all examples of greenhouse gases.

○ Human activities, such as burning fossil fuels, contribute to the production of greenhouse gases.

○ An increase in greenhouse gas emissions causes an enhanced greenhouse effect, which may lead to global warming.

○ Possible consequences of global warming include rising sea levels and changing weather patterns.

○ Untreated sewage can lead to eutrophication and water pollution through depletion of dissolved oxygen.

○ Leached minerals from fertilisers may cause eutrophication of water.

○ Effects of deforestation include leaching, soil erosion, disturbance of evapotranspiration and the carbon cycle and the balance of gases in the atmosphere.

End of topic questions

1. **a)** Explain the meaning of the term *pollution*. (1 mark)

   b) Name two human activities that are major sources of sulfur dioxide in the atmosphere. (2 marks)

   c) Explain how sulfur dioxide can lead to acid rain. (2 marks)

   d) Sketch a diagram to show how sources of sulfur dioxide in one region could result in acid rain in another region. (3 marks)

   e) Explain why sulfur dioxide causes pollution. (2 marks)

2. In parts of Europe, farmers now use satellite technology to help them see which parts of a field need additional fertiliser, and how much fertiliser they need.

   a) Why do farmers add fertiliser to their fields? (1 mark)

   b) What might happen if a farmer adds too much fertiliser to a field? (1 mark)

   c) Explain as fully as you can how this could result in water pollution. (4 marks)

   d) Explain the advantage of this use of technology. (2 marks)

3. Rivers and lakes that are used for water supplies may be monitored to make sure the water in them is safe for use. This is done by taking samples of water and testing them.

   One way of monitoring is to measure the amount of oxygen that is used by the water (the oxygen demand) over a period of 5 days.

   a) Why does the concentration of oxygen decrease in the water? Explain your answer. (2 marks)

   b) In this test would polluted water use more oxygen than unpolluted water? Explain your answer. (2 marks)

Another way of monitoring the water is to sample the small organisms that live in it. Some species, such as worms, are better adapted for living in water that has a low oxygen concentration. Other species, such as mayfly larvae, need a high concentration of oxygen in the water.

    c) Which of the two species above would be more common in polluted water? Explain your answer.   **(2 marks)**

    d) What does *adapted* mean?   **(1 mark)**

    e) Why might sampling the organisms be a better measure of the long-term health of the water than measuring the oxygen demand of the water?   **(2 marks)**

4. a) Give two reasons for large-scale deforestation.   **(2 marks)**

    b) Describe the effects of large-scale deforestation on the organisms in the region.   **(3 marks)**

    c) Explain why after deforestation, even if the land is left undisturbed so the vegetation regenerates naturally, plant growth will be much slower than before.   **(3 marks)**

5. Explain as fully as you can why scientists think that human activities are leading to global warming.   **(3 marks)**

EXAMINER'S COMMENTS

a) **i)** Incorrect. The student has not realised that the figures for methane are on the right-hand axis. The correct answer is carbon dioxide.

ii) The answers given for carbon dioxide and nitrous oxide are detailed and correct.

Again the student has not appreciated that the graph for methane has been drawn to the scale on the right-hand axis, which ranges from 0.6 to 2.0 ppm. This has meant that although the trends have been described, the values of methane concentration are incorrect.

It is important to check scales carefully when reading data from graphs. The answer should therefore be:

The concentration of methane has shown a very slow, slight upward trend from 0 to around 1700, ranging from 0.65 ppm to around 0.75 ppm.

Then after a slight dip, a steep increase to around 1.925 ppm in 2000.

Exam-style questions
Sample student answer

Question 1

This question is about the enhanced greenhouse effect and global warming.

a) The graph shows the concentration of some greenhouse gases in the air from the year 0 to 2000.

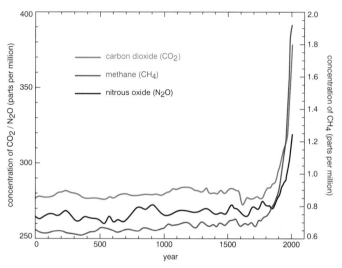

i) According to the graph, which greenhouse gas was present in the highest concentration in the air in the year 2000? **(1)**

methane ✗

ii) Describe the trends shown on the graph in the changes of concentration of each greenhouse gas from the year 0 to 2000. **(6)**

The concentration of carbon dioxide has been fairly stable at around 280 parts per million from 0 to 1600 ✓ ①

Then after a dip, shows a steep increase to 380 ppm in 2000 ✓ ①

The concentration of nitrous oxide has shown a little fluctuation from 0 to around 1800, ranging from 260-275 ppm ✓ ①

Exam-style questions continued

But there has been a steep increase to

around 320 ppm in 2000 ✓ ①

The concentration of methane has

shown a very slow, slight upward

trend from 0 to around 1750, ranging

from 255 ppm to around 260 ppm ✗

But then a steep increase to around

390 ppm in 2000 ✗

iii) The candidate has written a good answer for the contribution of the burning of fossil fuels to the increase in carbon dioxide. These all refer to the burning of fossil fuels, however, and the candidate could have picked up the second mark by referring to deforestation.

iii) How has human activity contributed to the change in the concentration of carbon dioxide in the air?

(2)

Carbon dioxide production has increased from the burning of fossil

fuels in transport, heating and cooling, and in manufacture ✓ ①

b) The table gives information on several greenhouse gases.

Gas	Chemical formula	Lifetime (years)	Global Warming Potential*
Carbon dioxide	CO_2	variable	1
Methane	CH_4	12	21
Nitrous oxide	N_2O	114	310
CFC-11	CCl_3F	45	3 800
CFC-12	CCl_2F_2	100	8 100
Sulfur hexafluoride	SF_6	3 200	23 900

*The **Global Warming Potential (GWP)** is a measure of how much heat a greenhouse gas traps in the atmosphere relative to that trapped by the same mass of carbon dioxide. A GWP is calculated over a time interval. The values in the table are for a 100-year time scale.

b) i) Correct.
Water vapour is the most abundant and important greenhouse gas in the atmosphere, but human activity has only a small effect on its concentration in the atmosphere.

ii) The candidate has picked up two marks, but for the third mark, has not mentioned the fact that sulfur hexafluoride has the longest lifetime – a greenhouse gas that is around for a shorter time will make less of a contribution to the greenhouse effect.

iii) This is a good answer, but the student has not mentioned the 'enhanced greenhouse effect'. The final marking point could have been extended:

But increases in greenhouse gases as a result of human activity are leading to the enhanced greenhouse effect ✓

This is leading to a significant warming of the Earth called global warming ✓

Exam-style questions continued

i) Name one other greenhouse gas not listed in the table. **(1)**

Water vapour ✓ ①

ii) Which greenhouse gas in the table contributes most to global warming? Explain your answer. **(3)**

Sulfur hexafluoride ✓ ①

It has the highest GWP. ✓ ①

iii) Explain how greenhouse gases result in the greenhouse effect and global warming. **(6)**

Shortwave radiation from the Sun passes through the Earth's atmosphere and warms the ground ✓ ①

The warmed Earth gives off longer wave radiation that is prevented from leaving the Earth by greenhouse gases in the atmosphere ✓ ①

The trapping of the radiation leads to the Earth warming up, which is called the greenhouse effect ✓ ①

The greenhouse effect is important, because without it, the temperature on the Earth would be 33° C lower - the Earth would be uninhabitable ✓ ①

But increases in greenhouse gases as a result in human activity is leading to a significant warming of the Earth called global warming ✓ ①

(Total 19 marks)

Question 2

The table below gives information on how the land area covered by forest in seven countries has changed from 1990 to 2005.

Country	Area covered by forest, millions of hectares		Area of forest lost from 1990 to 2005, %
	1990	**2005**	
Bolivia	109.9	58.7	46.6
Brazil	851.5	477.7	
Colombia	113.9	60.7	
French Guiana	9.0	8.1	
Peru	125.5	68.7	
Suriname	16.3	14.8	
Venezuela	91.2	47.7	

a) i) Calculate the area of forest lost for each country, as a percentage of the area in 1990. The first one has been done for you. (6)

ii) During the time period 1990 to 2005, in which country is there:

the greatest percentage deforestation? (1)

the least percentage deforestation? (1)

iii) Suggest two reasons for deforestation. (2)

b) Explain the effects of deforestation, referring to:

- soil erosion (1)

- leaching (2)

- evapotranspiration (3)

- atmospheric oxygen and carbon dioxide. (4)

(**Total 20 marks**)

Exam-style questions continued

Question 3

The food web below shows the feeding relationships of some of the organisms on a rocky shore.

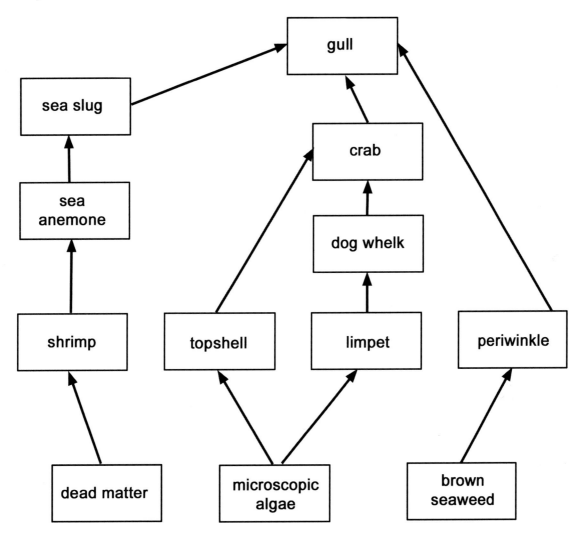

a) In the food web, state which organisms are:

   i) producers (2)

   ii) primary consumers (3)

   iii) secondary consumers. (3)

b) Within the food web, the food chain that includes microscopic algae, limpets, dog whelks, crabs and gulls is five organisms long. Explain fully why food chains as long as this are rare. (5)

Exam-style questions continued

c) A student investigated the distribution of a species of brown seaweed and periwinkles down a rocky shore. Her results are shown below.

Distance below High Water on seashore in metres	Distribution of organisms	
	Observed distribution of brown seaweed	Density of periwinkles, mean number of periwinkles per m^2
0	absent	0
10	rare	16
20	occasional	52
30	abundant	156
40	abundant	128
50	occasional	44
60	rare	12

i) Suggest two reasons for the distribution of the periwinkles. **(2)**

ii) Describe a method the student could have used to estimate the density of periwinkles at different distances down the seashore. **(7)**

(Total 22 marks)

Exam-style questions continued

Question 4

Nutrients are cycled in nature.

a) **i)** The passage below describes the stages in the carbon cycle.

Use suitable words to complete the sentences in the passage. (9)

.... from the atmosphere is converted to complex carbon compounds in .... by the process of .... This is often called carbon ....

Plants are then often eaten by .... , which build up their own complex carbon compounds.

The process of .... , in both plants and animals, returns some of this carbon back to the atmosphere as ....

When organisms die, their bodies decay because of the action of .... Some of the complex carbon compounds are taken into the bodies of these organisms, where some may be converted to carbon dioxide during their ....

ii) By what other process does carbon dioxide enter the air? (1)

Exam-style questions continued

b) The diagram below shows the nitrogen cycle. Name the bacteria labelled A, B, C and D on the diagram. (4)

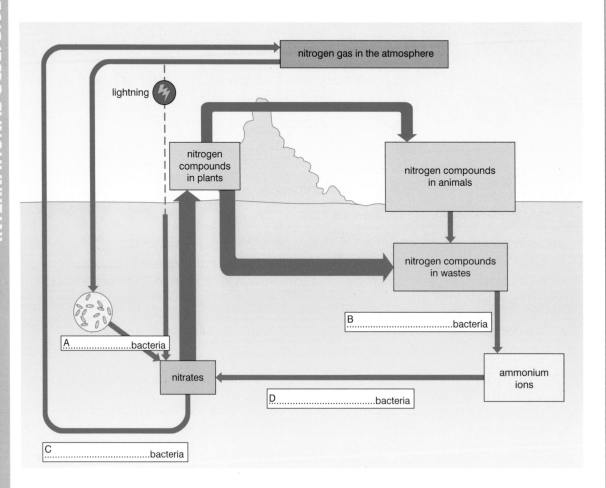

(Total 14 marks)

Question 5

The effect of sewage into a river was monitored over a number of years. The results are shown in the graphs below.

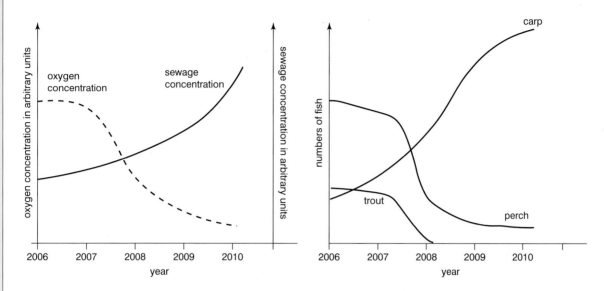

a) i) Describe the trends in sewage and oxygen concentration between 2006 and 2010. **(2)**

   ii) Explain why sewage had this effect on the oxygen concentration in the water. **(3)**

b) Describe and suggest an explanation of the effects of the changing oxygen concentration on the different fish populations. **(6)**

(Total 11 marks)

Before the development of farming around 10 000 years ago, humans were hunter-gatherers, taking food from the wide range of plants and animals that lived in their community. As they developed the skills of farming, humans had to choose plants and animals that grew well in the local environment and provided the most food for the rapidly growing human population.

Today we rely on a small number of plant and animal species to provide all our food. Over thousands of years, these species have been changed as they have become domesticated. Selective breeding has developed breeds and varieties that produce more of what we need, such as sheep that produce more wool, animals that have much larger muscles for the meat we eat, grain crops like wheat that produce much larger seeds. As the human population continues to grow, we need even more food. Many people hope that the new techniques of genetic engineering and cloning will help improve crop plants and farm animals so that we can continue to produce enough food for everyone.

STARTING POINTS

1. Plants grow best under certain conditions. What are the best conditions for plant growth and how can we manipulate the environment to create them?

2. Pests reduce the yield of crops. What methods can be used to control crop pests and what are the advantages and disadvantages of each method?

3. We use microorganisms to make many foods. What conditions do microorganisms need for growth, and what foods can we produce using them?

4. We are increasingly farming fish in order to provide the food we need. What conditions do fish farms need to provide to maximise the growth of the fish?

5. Most of our animal breeds and crop plant varieties have been developed through selective breeding. How is this done?

6. Genetic modification is a technique that we hear about frequently in the media. What is it and what are its advantages and problems?

7. Cloning is another technique that is being used to develop plants and animals with the characteristics we need. How is cloning done, and what could it be used for?

SECTION CONTENTS

a) Food production

b) Selective breeding

c) Genetic modification (genetic engineering)

d) Cloning

e) Exam-style questions

5
Use of biological resources

△ The wheat in this field has been bred selectively to produce a better yield.

Food production

△ Fig. 5.1 This large wild salmon needed more food to reach this size than a farmed salmon of the same size would need.

INTRODUCTION

The human population continues to increase. By 2050 it is expected to reach, or even exceed, 9 billion. To avoid starvation and to avoid long-term damage to the environment that will reduce the ability to grow food, we need to consider new ways to produce food. Growing food organisms in more tightly controlled conditions can improve the rate of growth. This can increase the yield, the amount of food that the crop plant or farm animal produces. For example, it takes about 2–4 kg of food to produce a farmed salmon and 10 kg of food for a wild salmon of the same size. This is because farmed salmon do not have to waste energy hunting their food.

KNOWLEDGE CHECK

✓ Plants need light, water and carbon dioxide for photosynthesis, and nutrients from the soil to make proteins and other compounds.
✓ Aerobic respiration requires oxygen; anaerobic respiration does not.
✓ Animals need oxygen, water and nutrients for healthy growth.

LEARNING OBJECTIVES

✓ Describe how glasshouses and polythene tunnels can be used to increase the yield of certain crops.
✓ Understand the effects on crop yield of increased carbon dioxide and increased temperature in glasshouses.
✓ Understand how the use of fertiliser can increase crop yield.
✓ Understand the reasons for pest control and the advantages and disadvantages of using pesticides and biological control with crop plants.
✓ Understand the role of yeast in the production of food including bread.
✓ Practical: Investigate the role of anaerobic respiration by yeast in different conditions.
✓ Understand the role of bacteria (*Lactobacillus*) in the production of yoghurt.
✓ Understand the use of an industrial fermenter, and explain the need to provide suitable conditions in the fermenter, for the growth of microorganisms.
✓ Understand the methods used to farm large numbers of fish to provide a source of protein.

CROP PLANTS

Protected cultivation

Plants may be grown in glasshouses or polythene tunnels (sometimes called polytunnels) to protect them from the environment. Temperature, levels of water, fertiliser and even carbon dioxide can be carefully controlled to ensure the maximum **yield**. Increasing the availability of all of these factors that are normally found in outdoor conditions tends to increase growth up to a point.

△ Fig. 5.2 Polytunnels are now supplying green vegetables that previously could not be grown in cold regions.

The level of carbon dioxide normally present in the outside air allows plants to grow well, but increasing the level of carbon dioxide increases the rate of photosynthesis and thus the rate of growth. Increasing the temperature tends to increase the rate of metabolic reactions, although temperatures that are too high are damaging. The increased growth due to additional carbon dioxide and warmth creates a greater demand for nutrients and water. Protection also allows crops to be produced in areas that would be too cold for them, or at times of year that are outside their normal growing season.

REMEMBER

For the best marks, be prepared to apply all your knowledge of the factors that limit photosynthesis to the growth of crop plants in glasshouses and polytunnels.

QUESTIONS

1. Describe four factors affecting plant growth which can be controlled in glasshouse or polytunnel cultivation.

2. Explain how the factors you gave in Question **1** can be controlled to maximise plant growth.

3. Explain how using glasshouses or polytunnels might help a farmer grow crops when they cannot be grown in open fields.

FERTILISERS

Plants need just water, carbon dioxide and light for photosynthesis, but need other substances to produce all the other chemicals in their tissues and grow well. You saw in Topic 2e that plants need several mineral ions which they absorb through their roots from the soil.

- Nitrogen is needed (as nitrate ions) to form amino acids (to make proteins), nucleic acids and other chemicals. If nitrate ions in the soil are limited, plant growth is stunted and crop yield is low.

- Magnesium ions are needed to make chlorophyll. If magnesium ions in the soil are limited, then the plant will not be able to make enough chlorophyll and photosynthesise as rapidly. So growth will be restricted and crop yield reduced.

The amount of nutrients in the soil may be limited for several reasons.

- If another crop has recently been grown in the field, it may have absorbed many of the mineral ions that were in the soil.
- Some soils naturally contain fewer mineral ions than others, because the rocks below them contain few of the important minerals. Or substances in the soil, particularly calcium carbonate, may hold on to the mineral ions and not release them into the soil water. Soils that cover chalk and limestone often suffer from this problem.

To maximise rate of growth for every crop, farmers need to add more minerals to the soil. This can be done in several ways.

△ Fig. 5.3 Fertiliser can increase the growth of plants. The ear of maize on the left has been grown with fertiliser, the ear on the right has not.

- **Crop rotation** is where a different crop is grown in the field each season and one of the crops grown is a leguminous crop (bean/pea family). Legumes have root nodules containing nitrogen-fixing bacteria, which help to add nitrate ions to the soil (see pages 286–287).
- Adding natural **fertiliser** such as compost or manure (a mixture of animal faeces and urine often mixed with straw to help the rotting process). Compost and manure contain high concentrations of nitrogen and other mineral ions. Decay of the compost or manure by microorganisms releases the mineral ions into the soil.
- Artificial fertilisers are produced by chemical manufacture, containing an ideal balance of the main mineral ions that plants need for growth, such as nitrogen, phosphorus and potassium (NPK fertilisers). These fertilisers are produced as powders or granules that can be spread dry on fields or dissolved in water before spraying the fields.

◁ Fig. 5.4 Spreading muck (manure) to fertilise a field before planting.

For the best marks, be prepared to apply all you have learned in earlier chapters on the digestive system and formation of urine, the nitrogen cycle, and decay by microorganisms, to explain why soil can be improved for plant growth by crop rotation or natural or artificial fertilisers.

SCIENCE IN CONTEXT

GUANO

In the 1800s, much of the fertiliser used on fields in the US and Europe was *guano*, the dry faeces of mainly seabirds, which contains high concentrations of nitrogen-containing ammonium ions and phosphate ions.

During the breeding season these birds nest on islands, many off the coast of Chile and Peru in South America, and their faeces built up over centuries.

Trade in guano during the 1800s made Chile a much richer country. However, in the early 1900s, methods had been developed to produce artificial fertilisers starting with ammonia (NH_3).

△ Fig. 5.5 Birds nesting on this island off the coast of Peru leave layers of guano.

Today over 100 million tonnes of fertiliser are used worldwide each year to improve crop growth and yield. Production methods require a lot of energy from the burning of fossil fuels, so a lot of research is being carried out into producing fertilisers in a way that can be continued without damaging the environment.

QUESTIONS

1. Explain why a farmer might add fertiliser to a field containing a crop.

2. What is meant by *crop rotation* and how can it help crops to grow better?

3. Give two examples of fertiliser.

PEST CONTROL

Plants provide food for many organisms as well as humans. A wide range of other animals and insects will eat crop plants if they are not protected. These animals damage the plants by eating parts of their leaves (as do caterpillars), or taking up sap (as do aphids). Damaging the plants reduces their ability to make food and produce new tissue, so they do not grow as well and do not produce as great a crop yield. We call these animals **pests**, because they are a problem to us.

△ Fig. 5.6 Aphids are pests of many plant species. They insert their mouthparts into the phloem in a plant vein and take up the sap, which has a high concentration of sugars.

Traditionally pests were controlled by hand, picking the pests off the plants, or by using domesticated animals, such as chickens, to eat them. In the huge crop fields of today, the main control is by using pesticides. **Pesticides** are chemical poisons that kill pests. They are sprayed onto the plants and eaten by the pests when they feed on the plants. This helps to protect the crop from damage, and so increase yield, but there are problems with using pesticides.

The problems with pesticides

Some pesticides kill not only the pest species but also other species in the community. This can have two drawbacks.

- Killing lots of different types of insect in an area will reduce the amount of food available for any animals that specialise on eating insects, such as insectivorous birds. This will affect other organisms in the food web, because of interdependence.
- If the other species killed include predators of the pest then, once the pesticide has been washed away by rain, it is possible for the pest species to return and increase in number even more rapidly, causing even more damage to the crop.

Some pesticides can cause other problems higher up the food chain, through **bioaccumulation** when the toxin is stored in tissue. A good example of this is DDT, a pesticide used widely in the 1950s and 1960s.

DDT is stored in fatty tissues in animals. A small amount in the body may have no noticeable effect on the animal, so predators that eat insects treated with DDT may not be obviously harmed. However, the more insects they eat, the more DDT is stored in their tissues – the DDT accumulates. Predators that feed on these animals end up containing much higher concentrations of DDT than are in the environment, and store the DDT in their tissues in turn.

At high concentrations DDT is toxic to larger organisms too. In birds it can cause eggs to be laid with thinner shells than normal, which break more easily, killing the developing chick inside. In the 1960s it became clear in the US and Europe that numbers of birds of prey, which are top consumers in food webs, were decreasing rapidly as a result of poisoning by DDT and eggshell thinning. DDT was then banned for use in agriculture in the US and Europe.

A growing problem with pesticides is that pest species are evolving resistance to the chemicals. This is because any individuals that survive the use of a pesticide are more resistant than those that are killed, so the individuals that reproduce have offspring that carry the genes for resistance. Farmers have responded by using greater amounts of pesticide. However, scientists are researching alternative ways of controlling pests, to avoid further damage to the environment.

Biological control

Another way of controlling pests uses other organisms, which is called **biological control**. This can be done in several ways.

- Introducing a predator of the pest to the crop can help to reduce pest numbers and so reduce damage. This can work well in glasshouses and in polytunnels, but is more difficult in open fields where the predators can move away from the area and so limit their effectiveness for the crop. It also works best when the introduced predator focuses mainly on the pest species, and not other species which could damage the food web and create greater problems. Introducing predators that are not native to the area can also have problems, as there will be no natural predator of the introduced species and therefore no control of their numbers.
- Introducing a parasite of the pest can also help to reduce the population size of the pest and so reduce damage to the crop.
- Companion planting of other plants next to the crop can work in several ways. They may help to attract higher numbers of natural predators to the area, by providing a food for a different stage of the predator's life cycle. For example, hoverflies need nectar, but their larvae eat large numbers of aphids. So planting the right kinds of flowers near to a crop can attract more hoverflies to lay their eggs on the crop plants. Companion planting can also help by masking the smell of a crop plant, so that flying pest species are not as easily attracted to the area.

△ Fig. 5.7 The plant at the bottom was treated with *Bacillus thuringiensis* but the top plant was not. Then caterpillars were added to both plants. The bacteria protect the plant from caterpillar damage.

- There are some natural biological pesticides. One example is made by the bacterium *Bacillus thuringiensis,* which makes a toxin. Farmers can spray crops that are particularly susceptible to caterpillar attack with a liquid containing the bacteria. If a caterpillar eats these bacteria, the toxin poisons and kills the caterpillar. These sprays are not as damaging to the environment because only organisms that eat the bacteria are affected. However, there is evidence that some species of caterpillars are evolving resistance to the toxin.

Sometimes a combination of biological control and pesticides is used to control pests, using predators or parasites to keep the pest population smaller than it would be and only using pesticides in limited amounts when pest numbers get out of control. This is known as integrated pest management.

THE CANE TOAD MISTAKE

The cane toad was introduced to Australia in 1935 to control cane beetle, which damaged the crops of sugarcane in Queensland.

The numbers of cane toads increased rapidly, eating mice, lizards, birds and many invertebrates. Unfortunately they did not have any effect on the cane beetles which lived too high on the sugarcane stalks for the toads to reach them.

△ Fig. 5.8 Cane toad.

The cane toads, and their tadpoles, are poisonous, and there is no natural predator in Australia. So the cane toad has now become a pest in its own right, reducing the population sizes of many native species by predation and outcompeting other native animals for food.

QUESTIONS

1. Explain why pests reduce crop yields.

2. Explain why biological control may work better in a glasshouse or polytunnel than in an open field.

3. Draw up a table to show the advantages and disadvantages of pest control using pesticides and using biological control.

Pakistan is a major producer of cotton. Most cotton farmers have only a few hectares of land and income from selling their cotton supports them and their family.

The most serious cotton pest in Pakistan is cotton leaf curl virus which is carried by whitefly. In the 1980s farmers used very few pesticides, but damage caused by the virus persuaded many farmers to spray with insecticides. By 1997 almost half the farmers were spraying at least 7 times a season with pesticides. The chemicals they were using were not only dangerous to people, they also kill a large range of insect species.

The graph shows the amount of pesticide used in Pakistan and the yield of cotton between 1980 and 1996.

1. Describe the change in pesticide use shown by the graph.

2. Describe the change in yield of cotton shown by the graph.

3. Suggest an explanation as to why the yield changed like this.

The natural predators of whitefly include four insect species that normally increase in number as whitefly numbers increase.

4. Explain why predator numbers normally increase as whitefly numbers increase.

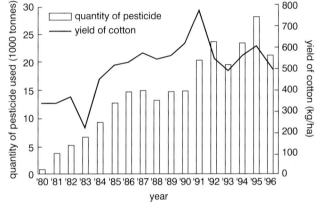

Δ Fig. 5.9 Pesticide use and yield of cotton.

5. Suggest what happened to predators as a result of the change in pesticide use.

6. Explain as fully as you can why the use of pesticide could result in greater damage to the cotton than without pesticide use.

7. Many farmers in Pakistan are now trained to apply pesticides specific for whitefly only once or twice at key points in the growing season. They may also release additional whitefly predators in fields. Explain how these practices can improve the yield of cotton.

MICROORGANISMS

Food from microorganisms

We use microorganisms to produce many types of food and drink. Here are some examples.

- Bread – A **yeast** (fungus) and sugar mixture is added to flour to form a dough. The respiration of the yeast produces carbon dioxide, which forms bubbles in the dough, making it light and fluffy. Other bakery products, such as pizza dough, are also made using yeast.

- Cheese – A culture of bacteria is added to warm milk and the resulting curds are separated from the liquid whey. The curds are dried and formed into cheeses. Sometimes other bacteria and fungi (moulds) are added to give flavour and colour to the cheese.
- Soy sauce – Cooked soya beans and roasted wheat are fermented using microorganisms including yeast (fermentation means anaerobic respiration by the microorganisms), filtered and then pasteurised (heated quickly to a high temperature to kill microorganisms) before bottling.
- Single-cell protein (SCP), also called mycoprotein – This is a fungus that is mixed with carbohydrate and kept in warm conditions so that it grows rapidly. The fungus is then separated from the mixture, treated to improve its flavour and dried before use.

◁ Fig. 5.10 Cheese is made from milk that has had bacteria added to it.

The production of yoghurt

If milk is left for several days at a warm temperature, it soon starts to break down to form solid curds and liquid whey, as in cheesemaking. If a culture of **Lactobacillus** bacteria is added instead to fresh milk and kept warm, the milk turns into yoghurt. The milk must first be pasteurised by heating quickly to a high temperature for a brief time. This kills the microorganisms that would turn it into curds and whey. The heat also breaks down the milk proteins a little, which stops them forming a solid product in the process.

Lactobacillus bacteria feed on lactose (the sugar in milk) and convert it to lactic acid. Lactic acid reacts with the proteins in milk, making them coagulate (stick together), which gives yoghurt its thicker texture. The acid also gives the yoghurt its characteristic tangy flavour.

△ Fig. 5.11 Culturing milk with *Lactobacillus* produces a food with a different taste and texture.

Acetaldehyde is also produced, which adds to the flavour. Once the yoghurt has formed, it is cooled to keep it in perfect condition for as long as possible. Yoghurt is considered to have many health benefits, because it contains all the nutrients in milk but is easier to digest.

QUESTIONS

1. Name four food products that are made using microorganisms.

2. Draw a flowchart to show the stages in yoghurt production. Name the bacterial culture used and explain its role.

INVESTIGATING ANAEROBIC RESPIRATION BY YEAST

Yeast is a living organism, and like all living organisms it grows better and respires more quickly in certain conditions. These conditions can be investigated by measuring the volume of carbon dioxide given off as a result of anaerobic respiration:

glucose $\rightarrow$ ethanol + carbon dioxide (+ ATP/energy)

- The effect of temperature. If you set up a series of closed tubes containing a solution of yeast and sugar, of the same concentrations, and connect each tube to a gas syringe to measure the volume of gas given off, you can investigate the effect of temperature on the rate of respiration by placing each tube in a water bath at a different temperature. (Note: This can be carried out more simply by collecting the gas in a rubber balloon fixed over the neck of the tube, but measuring the size of the balloon is more difficult to do accurately.) This will show that up to an *optimum* temperature, the rate of respiration will increase, but beyond that temperature it will decrease due to denaturation of enzymes and other proteins.
- The effect of glucose concentration. This can be investigated in a similar way to above, or using the method described in the investigation below.

Developing investigative skills

You can investigate the effect of glucose concentration on the rate of anaerobic respiration in yeast using the equipment shown.

Five sets of tubes are set up as shown in Fig. 5.12. 2 cm³ of yeast solution is added to each tube. Then the following amounts of glucose are added to each tube:

tube 1, none; tube 2, 0.5 g; tube 3, 1.0 g; tube 4, 1.5 g; tube 5, 2.0 g.

The solution in each boiling tube is mixed thoroughly. A thin layer of vegetable oil is floated on the top of the solution in each tube, and the tubes placed in a warm electric water bath for an hour.

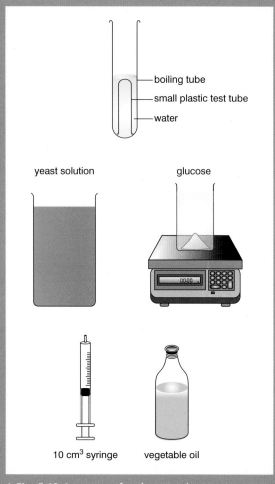

△ Fig. 5.12 Apparatus for the experiment.

Demonstrate and describe techniques

❶ Explain why the vegetable oil is used.

❷ Explain why the tubes were placed in a warm water bath rather than kept at room temperature.

Analyse and interpret data

After an hour the test tubes have risen by different amounts within their boiling tubes, as shown by the table.

Tube	Glucose added (g)	Height of small tube (cm)
1	0	0
2	0.5	1
3	1.0	2
4	1.5	4
5	2.0	6

❸ Explain what caused the test tubes to rise up.

❹ Describe the results shown in the table.

❺ Explain the results using your knowledge of the respiration of yeast.

❻ Draw a conclusion from this investigation.

Evaluate data and methods

❼ Explain how you could improve the method.

USING MICROORGANISMS ON A LARGE SCALE

Microorganisms can be used to make chemicals on a large scale. For example, by growing the fungus *Penicillium* in a large vessel called a **fermenter**, the antibiotic penicillin can be made in large quantities. Conditions inside the fermenter need to be controlled to help the microorganisms grow as quickly as possible.

- Nutrients needed for growth are added to replace those that are used up. These include an energy source, such as glucose, for respiration. They also include other nutrients needed for the synthesis of new cell materials, such as a nitrogen source (for example ammonia) for making proteins and nucleic acids.

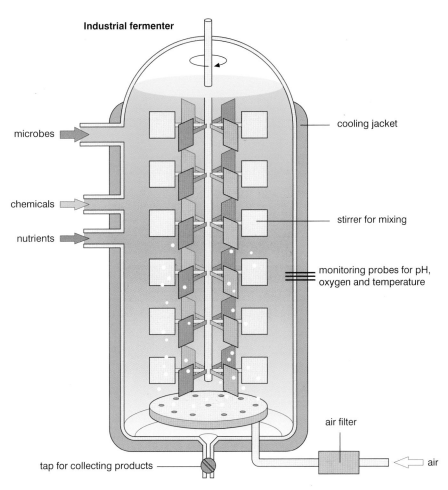

Δ Fig. 5.13 The structure of an industrial fermenter for the growth of microorganisms.

- The reactions of respiration release heat energy. If the temperature of the mixture gets too high, this could affect the rate of enzyme action and slow down the rate of growth. A cooling jacket is used to remove excess heat, and the temperature inside the fermenter is monitored continually to maintain the optimum temperature for growth.
- The pH is continually monitored because if it varies too far from the optimum pH for the microorganisms' enzymes, the rate of growth will slow. If needed, the pH of the solution is adjusted using buffer chemicals.
- *Penicillium* and many bacteria which are also cultured in fermenters, are aerobic. So they need a continuous supply of oxygen (**oxygenation**), in the form of air bubbled through the mixture in the fermenter.
- The mixture in the fermenter is continually **agitated** by stirring, to make sure the cells do not all settle to the bottom and so that all the materials in the fermenter are well mixed.
- Before a new batch of culture is added to the fermenter, it is sterilised by passing steam through it. Also all solutions added to the fermenter are sterilised. These **aseptic precautions** make sure that no other microorganisms are added to the fermenter, which could affect the growth of the microorganism in the culture.

Once the process has been going long enough for the mixture in the fermenter to contain enough product, some of it is drained through a tap at the bottom of the fermenter. The mixture is then processed to extract the product. More culture and nutrients can be added to the fermenter to replace what has been drained off, so that the process can continue without stopping.

REMEMBER

Make sure you can explain what each feature of an industrial fermenter does and why it is needed.

QUESTIONS

1. Explain what is meant by an *industrial fermenter*.

2. a) Name four conditions inside a fermenter that are controlled while the microorganisms are growing.

   b) For each condition named in **a)**, explain why it might change and why it needs to be controlled.

3. a) What is meant by *aseptic precautions*?

   b) Why are these needed when preparing the fermenter before adding the microorganisms?

FISH FARMING

Fish is an excellent low-fat, high-protein food and is becoming more popular every year. Unfortunately, this increasing demand is causing problems of overfishing of wild stocks, so to supply this growing market suppliers are building fish farms where they can make sure the fish are kept in optimum conditions for growth.

△ Fig. 5.14 An open-water fish farm in La Spezia, Italy.

Freshwater fish farms may be little more than a large natural pond or lake where the fish are kept. This is known as an *open* farming system. Increasingly, though, farmed fish are kept in artificial ponds built specially for this purpose to make care of the fish as easy as possible. These are *closed* farming systems. Farmed sea fish are grown in large tanks suspended in the sea so that sea water can flow through them.

The fish are harvested when they have reached a suitable size, although some may be kept until they are sexually mature and are used for breeding. To stock the farm, eggs and sperm are removed from adult fish and then mixed in the laboratory. The eggs are hatched in tanks of gently flowing water, and when the young fish are a suitable size they are added to the tanks. Choosing the best fish to breed is an example of **selective breeding**. (You can read more about this in the next chapter on pages 342–347.)

Water quality

In open systems, water quality is maintained by water moving through the pond, lake or sea tank. The location of the fish farm will have been chosen where the quality of the water can be relied on throughout the year to contain sufficient oxygen and to be clear of silt and chemicals that might affect the fish. However, there is evidence now to show that, in the areas around sea fish farms, the natural community of organisms is being damaged by the large amounts of waste from the farmed fish.

Some large closed systems remove waste products from the fish and left-over food on a continual basis. In smaller systems these build up, encouraging the growth of microorganisms and other organisms that might cause disease. So the ponds need to be emptied on a regular basis and cleaned out.

Control of pests and diseases

Keeping fish at much higher population densities than occur in nature makes it easier for pests and diseases to be passed from fish to fish. The higher density also stresses the fish, making them more susceptible to disease. In 1984, around 80% of the salmon in a large salmon hatchery in Norway died from one kind of virus. Pests and diseases can also damage natural sea fish stocks, as has been seen in wild salmon fish stocks in areas where salmon are being farmed.

Fish farmers treat the fish with pesticides and with antibiotics to control pests and diseases. In open systems, these chemicals can pass to the natural ecosystem and affect the organisms in the natural community too. There is also a concern that some of these chemicals remain in the fish and so pass into our bodies when we eat the fish.

Control of predation

Having large numbers of fish in an area is very attractive to predators of the fish – this is **interspecific predation**, predation *between* species. In freshwater fish farms the farmer must protect the fish from birds and mammals, such as otters, that would naturally feed on them. This is done by placing netted cages over the ponds. In sea fish farming, the cages are also protected from birds by covering with nets, and from sea predators such as seals and dolphins by the strong metal used to make the cages.

Many of the species of fish that are farmed are carnivorous and predators on smaller fish. So, in large concentrations, **intraspecific predation** (*within* species) or cannibalism can occur. This is managed by keeping a smaller number of fish in each tank or pond.

Feeding the fish

Several of the species kept in freshwater systems are herbivorous, such as carp and tilapia. The ponds in which they are kept are fertilised to encourage the growth of pondweed on which the fish feed. As long as the fertiliser is kept within the pond, it will not cause eutrophication and pollution in other natural water systems.

Most of the species kept in seawater tanks are carnivorous, for example salmon and seabass. They are fed on fish meal, because they do not digest plant-based food well. On a dry-weight basis, between 2 and 4 kg of wild-caught fish are needed to produce 1 kg of farmed salmon. Unfortunately, this means that wild fish still need to be caught in order to make the fish meal, so this still risks overfishing natural fish stocks.

The future of fish farming

Many people think that fish farming is the only solution to preventing wild fish stocks being completely destroyed. In addition, because fish are 'cold-blooded' (do not maintain their body temperature above that of the environment), they convert more of their food into body tissue than mammals. So, for each kilogram of protein produced, a cow needs to be fed the equivalent of about 61 kg of grain but a fish needs the equivalent of only 13 kg of grain. Fish farming also produces fewer greenhouse gases (such as carbon dioxide and methane) than farming mammals.

In 2005, on a global scale, the total amount of protein produced by fish exceeded that of beef cattle. About 34 per cent of the global fish protein was produced by farmed fish, up from just under 4 per cent in 1970.

Fish farmers maximise production by:

- stocking fish at high densities
- controlling the quality of the food and the frequency of feeding the fish
- choosing the best fish to breed from
- protecting against pests with chemical pesticides
- using antibiotics to minimise risk of disease, e.g. infectious salmon anaemia
- protection from predators (mammals, birds, reptiles, other species of fish)
- removing waste products.

QUESTIONS

1. Explain what we mean by *fish farming*.

2. List three factors that need controlling in fish farming.

3. For each factor listed for Question **2**, explain why it needs to be controlled.

△ Fig. 5.15 Fish are kept at high population densities in fish farms.

End of topic checklist

Agitation means mixing up, such as in a fermenter to keep the microorganisms in contact with the ideal amounts of nutrients and oxygen.

Aseptic precautions involve killing all unwanted microorganisms to prevent their growth, such as in a fermenter.

Biological control means using organisms, such as a predator or parasite, to control the numbers of pests.

An industrial **fermenter** provides the optimum conditions for the growth of microorganisms.

Farmers may add **fertiliser** to soil to provide mineral ions to improve plant growth, either natural fertiliser (e.g. compost or manure) or artificial fertiliser.

Interspecific predation is predation of one species by another, for example predator and prey.

Intraspecific predation is predation of individuals in a species by other individuals of the same species.

Lactobacillus bacteria are used to turn milk into yoghurt.

Oxygenation is the adding of oxygen to water, usually by bubbling air through it.

A **pest** is an organism that causes damage to crop plants or farm animals.

A **pesticide** is a chemical used to kill pests.

Yield refers to the amount of food produced from a crop or farm animal.

The facts and ideas that you should know and understand by studying this topic:

⭕ Using glasshouses and polythene tunnels to protect crop plants makes it possible to control growing conditions, such as temperature and carbon dioxide concentration, and protect crops from pests, so maximising the rate of growth of the crop and increasing the yield.

⭕ Fertilisers increase crop yields by providing mineral ions that the plants need in sufficient quantities for rapid growth.

⭕ Pests on crops can be controlled using chemical pesticides, but pests may develop resistance to the chemicals, and other organisms may be damaged by the chemicals.

○ Biological control, using predators, parasites or companion planting, can help to control pests on crops, but introduced organisms may cause more problems if they target species other than the pest and have no natural predator to control their numbers.

○ Microorganisms produce many foods, such as bread and yoghurt.

○ Yeast in bread making uses sugar to produce carbon dioxide when it respires anaerobically.

○ The effect of factors such as temperature on the rate of anaerobic respiration of yeast can be investigated by measuring the amount of carbon dioxide produced.

○ *Lactobacillus* bacteria break down lactose in milk to make lactic acid in the production of yoghurt.

○ Microorganisms can be grown on a large scale in fermenters, where suitable conditions for growth can be provided, such as the optimum temperature and pH, oxygenation, nutrients, agitation, and the exclusion of other unwanted microorganisms using aseptic precautions.

○ Fish to provide a source of protein can be farmed in large numbers in restricted areas such as ponds and tanks, where conditions for rapid growth can be controlled, such as water quality and removal of waste products, interspecific and intraspecific predation, pests and disease, and the quality and frequency of feeding. Selective breeding of farmed fish can improve the quality of fish grown.

End of topic questions

1. The table shows data given to farmers to encourage them to grow lettuces and tomatoes in a glasshouse.

Crop grown in glasshouse	No added carbon dioxide	Carbon dioxide added
Fresh mass of 10 lettuce plants (kg)	0.9	2.5
Fresh mass of saleable tomato fruits (kg)	4.4	6.4

   a) What factor that affects plant growth has been changed in the glasshouse? **(1 mark)**

   b) How has this factor affected the plants? **(1 mark)**

   c) Explain why this factor affects plant growth. **(2 marks)**

   d) What other factors should have been controlled in the experiment shown in the table, so that the comparison of the results is fair? Explain your answers. **(6 marks)**

   e) What would farmers need to consider before changing from growing crops in open fields to this method? **(3 marks)**

2. Large areas of crops attract a large number of pests. Different pests harm crops in different ways and reduce crop yield.

   Explain as fully as you can why each of these pests can reduce crop yield.

   a) Large herbivores, such as rabbit or antelope, that eat large amounts of plant leaves. **(2 marks)**

   b) Stalk borer caterpillars that eat the soft tissues in stems of crop plants. **(2 marks)**

3. a) Discuss the benefits of introducing a new predator to a crop to control a pest species compared with using pesticides. **(2 marks)**

   b) Discuss the risks of introducing a new predator to a crop to control a pest species compared with using pesticides. **(2 marks)**

4. Yeast is used in the making of bread. A solution of yeast is mixed with a little sugar to start the yeast growing. It is then mixed with wheat flour to make a dough. The yeast cells break down sugars in the wheat dough as they respire.

a) The respiration of yeast in bread-making is anaerobic even though there is oxygen available. Write down the reaction for the anaerobic respiration of yeast. **(1 mark)**

b) What effect does the yeast have on the dough? **(1 mark)**

c) Explain fully how it has this effect on the dough. **(2 marks)**

d) Name one other food that is produced using microorganisms. **(1 mark)**

5. Many people consider fish farming as a sustainable system of food production for the future.

a) Comparing fish farming with producing protein from mammals:

   i) describe two advantages of fish farming **(2 marks)**

   ii) describe two disadvantages of fish farming. **(2 marks)**

b) Farmed fish may be given antibiotics to improve growth.

   i) Explain why antibiotics can improve growth. **(2 marks)**

   ii) Describe and explain one disadvantage of giving antibiotics to animals that we grow for food. **(2 marks)**

c) Sustainability is the ability to continue something without damaging the environment. Evaluate the sustainability of fish farming. **(3 marks)**

Selective breeding

INTRODUCTION

All domesticated dogs belong to the same species, even though there are many different breeds of different shapes and sizes. The evolution of the domestic dog seems to have begun tens of thousands of years ago, as some wolves began to live and work alongside humans. This may have been useful when humans were hunting wild animals for food.

Since then humans have selected different characteristics in dogs to produce different breeds for different purposes. Wolfhounds are a large breed that can hunt alongside people on horseback. Jack Russell terriers are a small breed that chased rabbits out of their burrows when people hunted rabbits for food.

△ Fig. 5.16 The tiny chihuahua was bred to be a gentle pet. The large mastiff was selected for its aggression, to act as a guard dog.

KNOWLEDGE CHECK

✓ Genes control many characteristics in organisms and these characteristics can be passed on to offspring.
✓ Sexual reproduction produces variation in offspring.

LEARNING OBJECTIVES

✓ Understand how selective breeding can develop plants with desired characteristics.
✓ Understand how selective breeding can develop animals with desired characteristics.

BREEDING FOR CHARACTERISTICS

Selective breeding is when a breeder chooses which parent organisms to breed from. The choice is usually for a particular characteristic, such as size of farm animals, or colour of flowers in plants grown for horticulture. Since many characteristics are controlled by genes, by breeding together organisms that show the nearest form to the desired characteristic (such as a large body), the breeder is more likely to get offspring that also have these characteristics. If the breeder continues to select individuals with the largest size, over time the average size of the organisms will increase.

Selective breeding may also be done to combine particular combinations of characteristics. For example, by breeding a wheat plant that has a long stalk and a large seedhead with a wheat plant that has a short stalk and small seedhead, there may be some offspring with a short stalk and large seedhead. Selecting from the offspring that have the best combinations of these characteristics over many generations can produce a new variety with the perfect combination.

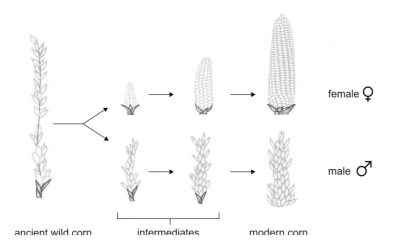

female ♀

male ♂

◁ Fig. 5.17 Over centuries of selective breeding, the ancient wild corn (maize) has developed into the modern breeds of corn that have the large seed cobs that we harvest.

ancient wild corn   intermediates   modern corn

Note that plants of the same species but with distinctively different characteristics are called **varieties**, while animals of the same species with distinct characteristics are called **breeds**. Since they are still of the same species, different plant varieties can interbreed and different animal breeds can interbreed.

REMEMBER

For the best marks, be prepared to explain the inheritance of characteristics through selective breeding in terms of the inheritance of genes and alleles you have already learnt.

SELECTIVE BREEDING IN PLANTS

Plants are selectively bred in farming to improve characteristics such as the **yield** or flavour of the food from a crop. Cross-breeding plants involves taking pollen from one parent plant and using it to pollinate another parent with the desired characteristics. This requires great care to prevent the flowers being pollinated with other pollen beforehand, and is usually done in the laboratory. Plants usually produce a lot of seed – the seeds from the cross are planted and grown, and the best individuals are selected and bred from in the following year.

△ Fig. 5.18 The variety of rice on the left produces more grain than the one on the right. It also has a longer stalk and so is more easily blown over, which makes it more difficult to harvest. Selectively breeding these two varieties could produce a high-yielding variety with a short stalk, which would be more useful.

Improved yield can be produced in many ways:

- increasing the size of the part of the plant we eat, such as seeds in wheat, maize and rice; tubers in potatoes and yams; leaves in cabbages
- decreasing the size of the parts of the plant we do not eat, such as stalks in wheat, because less energy is then 'wasted' by the plant growing parts that we do not want and it is easier to harvest
- improving pest and disease resistance, as less damage to the plant means it will grow faster
- improved growth in adverse conditions, such as drought or cold
- improving the taste or colour of the crop.

Other factors can also help, such as reducing stalk length so that rice and wheat plants aren't blown over as easily in strong winds and so are easier to harvest.

 SCIENCE IN CONTEXT

TULIP MANIA

Plants are also bred in horticulture, for gardens, for houseplants and cut flowers, to improve the colour, shape and form of the flowers and leaves. This is because people like new things.

For example, tulips were introduced to Europe in the 1500s from Turkey. They were so exotic that they became a luxury item that all wealthy people had to have. Plant breeders rapidly developed new varieties through selective breeding, such as flowers with different-coloured lines or specks on the petals.

At the peak of 'tulip mania' in the Netherlands in the 1630s, single tulip bulbs were being sold for more than 10 times the annual income of a skilled craftsman. Prices suddenly collapsed in 1637.

△ Fig. 5.19 A completely black flower is almost impossible to breed, but that does not stop people trying to produce it because many people would pay a lot of money for something so rare.

Selective plant breeding is not all a success story. For example, rice plants from around the world were crossed in breeding experiments to produce so-called 'miracle rices'. But the plants required extra fertiliser and plenty of water to produce the high yields. The modern seeds were also very expensive. If conditions were not perfect, the new varieties could sometimes do worse than the traditional varieties, and in some countries productivity actually went down. Scientists began to appreciate how important the environment was to the way the genes worked. The old-fashioned varieties had evolved over thousands of years to cope with local environmental conditions.

QUESTIONS

1. a) Explain why some characteristics can be bred for in selective breeding programmes.

   b) Explain why some characteristics cannot be bred for in selective breeding programmes.

2. Give three characteristics that have been selectively bred for in crop plants to improve crop yield.

3. For each of the characteristics you have given in Question **2**, explain how these improve crop yield.

4. Explain why plants are selectively bred in horticulture.

EXTENSION

One of the problems with selective breeding is that, when you breed from only a small number of individuals, you reduce not only the variation in the characteristics you are selecting for, but also the variation in other alleles. This means that you can lose other characteristics which might be useful in the future.

To protect against this, many wild varieties of rice, wheat, potatoes and other plants are collected and grown in case we need their characteristics in the future.

△ Fig. 5.20 There are many wild varieties of rice, but we eat only a few varieties selectively bred for particular characteristics such as larger grain size.

1. Using what you know about sexual reproduction, suggest why the amount of variation between selected individuals is smaller than in wild populations of a plant.

2. Why is it useful that selectively bred varieties have only limited genetic variation? Explain your answer as fully as you can.

3. Why could it be a problem in the future that selectively bred varieties have only limited genetic variation? Explain your answer as fully as you can.

4. Growing wild varieties of crop plants to keep them for the future takes a lot of space and time to look after them. This space and time could be used to grow varieties that produce more food. Do you think it is worth keeping wild varieties like this? Explain your answer as fully as you can.

SELECTIVE BREEDING IN ANIMALS

Humans have controlled the breeding of animals since they were first domesticated thousands of years ago. By selecting animals that have suitable characteristics, and breeding only from these, we have created animals that are dramatically different from their wild ancestors. Even within one species, breeds can be selected for particular purposes: for example, one breed of sheep may be good for meat production while another produces better wool, or more milk.

△ Fig. 5.21 A merino sheep from Australia.

In the past, sheep were more difficult to manage than modern breeds. They grew more slowly and produced less meat or wool. Over thousands of years farmers have bred for:

- increased growth and muscle formation
- increased milk production
- softer, longer wool (easier for making into yarn)
- docility (quiet/calm nature)
- resistance to disease.

Different breeds have different advantages, often linked to the conditions where they are farmed or the product (milk, wool or meat) required.

Sheep breed	Original home	Features
Merino	Spain	Fine wool in large amounts
East Friesian	northern Germany	Excellent milk producer. Cross bred with other breeds to produce milk sheep for hotter climates.
West African Dwarf	central Africa	Produced for meat. Little affected by local parasite (trypanosome) that reduces meat production in other breeds.

△ Table 5.1 Some different breeds of sheep.

Chickens have been bred for increased muscle production (for meat), earlier maturity (so they can be sold earlier) and for greater egg production. We also breed other domesticated animals, such as horses for the ability to run fast (for racing) or carry loads (for riding), dogs for different activities (such as hunting or being good pets), cats (for appearance) and fish (for display in fish tanks).

As with plant breeding, animal breeding can have drawbacks. For example, some dog breeds have been bred together for so long that they have developed a greater risk of problems such as cancer or mental health issues.

△ Fig. 5.22 Chickens are bred selectively for more meat, more eggs and earlier maturity.

QUESTIONS

1. Explain how animals can be selectively bred.

2. For each of the following characteristics in sheep, explain why different breeds with these characteristics have been bred:

   a) increased muscle production

   b) resistance to disease

   c) docility (quiet, calm nature).

3. Explain why we cannot breed one breed of sheep for all needs.

End of topic checklist

Selective breeding is when individuals are chosen for breeding because of their characteristics.

The facts and ideas that you should know and understand by studying this topic:

○ Plants can be selectively bred for characteristics such as increased yield, increased resistance to pests or diseases, greater rate of growth in difficult conditions, leaf and flower colour and flavour of crop.

○ Animals can be bred selectively to produce increased yields of meat or milk, greater resistance to diseases or pests, greater tolerance of environmental conditions, for their ability to hunt, and for the way they look.

End of topic questions

1. **a)** Describe one advantage of using selective breeding to produce a new variety of crop plant. **(1 mark)**

    b) Describe two disadvantages of using selective breeding to produce a new variety of crop plant. **(2 marks)**

2. A gardener has one variety of a plant with large white flowers, and another variety of the same plant species with small red flowers. Explain how she might be able to breed these plants to produce a plant with large red flowers. **(3 marks)**

3. Sheep are an important source of meat and wool in India. Different areas in the country have their own breeds of sheep that have lived in the areas for hundreds of years. Some breeds are particularly well adapted to the dry hot climate and drought.

    To help increase meat production, sheep breeds that produce more meat have been brought from Europe and Australia to cross with the local breeds.

    a) Explain why farmers in India want to improve their breeds of sheep. **(1 mark)**

    b) Suggest why Indian farmers have not just replaced their breeds with the European and Australian sheep breeds. **(2 marks)**

    c) Describe how a farmer might selectively breed from a European sheep breed and an Indian sheep to produce a heat-tolerant breed that develops a lot of muscle. **(3 marks)**

4. People keep fish as pets in ponds and fish tanks for display. These two fish are different varieties of one species, goldfish.

    a) How could we show that these two fish are of the same species? Explain your answer. **(1 mark)**

    b) Suggest how the black colour of goldfish first arose. **(1 mark)**

    c) Suggest why the different varieties of goldfish have been bred. Explain your answer. **(2 marks)**

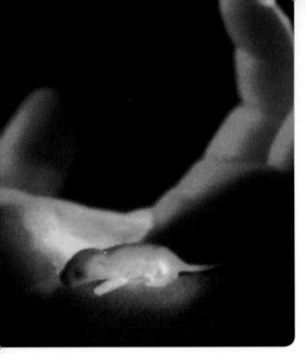

△ Fig. 5.23 This mouse has been modified with a jellyfish gene that makes it glow in the dark.

Genetic modification (genetic engineering)

INTRODUCTION

All organisms are constructed using information coded in their DNA. Since all living organisms on Earth have evolved from a common ancestor the information stored in their DNA is decoded in almost exactly the same way. This means that a piece of DNA taken from one organism, such as the gene that causes a jellyfish to glow in ultraviolet light, and placed in the nucleus of an organism of a completely different species, such as a mouse, will be decoded in the same way as it was originally: so that the mouse glows like a jellyfish in ultraviolet light.

This may seem not very useful, but it has important applications. Inserting the glow gene into cancer cells in humans would help researchers see how the cells move around inside the body and would help with treatment.

KNOWLEDGE CHECK

✓ A gene is a short section of DNA.
✓ Characteristics may be controlled by the genetic code in a cell.
✓ Different genes code for different characteristics.

LEARNING OBJECTIVES

✓ Understand how restriction enzymes are used to cut DNA at specific sites and ligase enzymes are used to join pieces of DNA together.
✓ Understand how plasmids and viruses can act as vectors, which take up pieces of DNA, and then insert this recombinant DNA into other cells.
✓ Understand how large amounts of human insulin can be manufactured from genetically modified bacteria that are grown in a fermenter.
✓ Understand how genetically modified plants can be used to improve food production.
✓ Understand that the term transgenic means the transfer of genetic material from one species to a different species.

THE TOOLS OF GENETIC ENGINEERING

Genetic modification (or **genetic engineering**) is taking a gene from one organism and inserting it into the DNA of another organism so that the gene is expressed (displays its characteristic). The process is sometimes called recombinant DNA technology because it involves the recombining of DNA. The DNA that is formed when the gene is inserted is sometimes called **recombinant DNA**, and the organism containing the recombinant DNA is called a genetically modified organism (GMO). If the DNA has been transferred from a different species the genetically modified organism is described as being **transgenic**.

Genetic engineering requires some special tools.

Restriction enzymes

To extract a gene from an organism, a method of separating it from the rest of the DNA is needed. **Restriction enzymes** are used to cut the DNA at particular sites. Different enzymes cut at different sites, so the right enzyme for the gene that is to be extracted must be chosen.

Vectors

Getting DNA into the nucleus of a different cell is tricky, because it must be taken through the cell membrane and into the cytoplasm or nucleus without damage. Viruses naturally get their own genetic material into cells when they infect them. Bacteria are able to pass short circles of DNA called plasmids from one to another. The technology of genetic engineering has used these natural processes to get recombinant DNA into cells.

- With viruses, the required gene is inserted into the protein coat of the virus to protect it. Then the virus is used to infect a cell. Viruses that will not cause disease are used for this.
- Remember that some bacteria have two kinds of chromosome – a single loop that contains the main instructions for building and controlling the cell, and separate circles of genetic material called **plasmids** (see page 20) which they can pass between one another.

Structures for inserting genes into cells in genetic engineering, such as viruses and plasmids, are called **vectors**.

SCIENCE IN CONTEXT

VECTORS FOR CYSTIC FIBROSIS

Cystic fibrosis (CF) is an inherited disease involving a gene that produces a particular carrier protein that is found in cell membranes. A person with cystic fibrosis has a faulty protein, and this causes thick sticky mucus to line the lungs and other organs instead of thin runny mucus.

The gene for the protein was discovered in 1989, and since then scientists have tried many ways of introducing the normal allele to the lungs of CF patients with vectors so that the cells make the right protein. One problem is that the cells that line the lungs are replaced every few days, so treatment is needed on a regular basis as new cells are made.

Viruses used as vectors have been tested, but usually they are only successful in getting the normal allele into a body cell the first time treatment is tried. After that the body attacks the viruses. Other vectors have been tried, but CF by gene replacement (gene therapy) still needs a lot of research to produce a successful method of treatment.

Ligase enzymes

Once the required gene is inside the cell, it must be joined into the cell's own DNA to form recombinant DNA. This allows it to be used by the cell to produce the new characteristic. Another kind of enzyme is used to do this, called a **ligase enzyme** which joins pieces of DNA together.

An example of genetic modification

One of the best and oldest examples of genetic modification is the insertion of the gene that codes for human insulin into bacteria, causing the bacteria to produce human insulin.

- The human insulin gene is identified and many copies of the gene are made.
- Bacterial plasmids are extracted and cut open using a restriction enzyme.
- The plasmids and genes are mixed together, and a ligase enzyme added to join them.
- The plasmids are put back into bacterial cells which are then tested to check if they contain recombinant DNA. Any bacteria that do not have the inserted gene are discarded.
- These bacteria with recombinant DNA can now produce human insulin. When they reproduce asexually by cell division the daughter cells also all carry the human insulin gene.

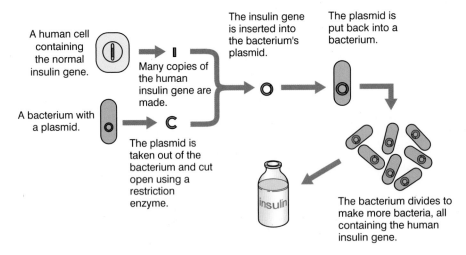

A human cell containing the normal insulin gene.

Many copies of the human insulin gene are made.

A bacterium with a plasmid.

The plasmid is taken out of the bacterium and cut open using a restriction enzyme.

The insulin gene is inserted into the bacterium's plasmid.

The plasmid is put back into a bacterium.

The bacterium divides to make more bacteria, all containing the human insulin gene.

△ Fig. 5.24 The process of genetic modification (genetic engineering).

Inserting a genetically modified plasmid into bacteria and culturing the now genetically modified bacteria in a fermenter makes it possible to produce large quantities of human insulin. Many other chemicals are produced in this way, such as growth hormone and blood-clotting factor.

EXTENSION

If new genes are introduced into the the cells of an animal or plant embryo in an early stage of development, then, as the cells divide, the new cells will also contain copies of the new gene.

The new gene can only be passed on to offspring if it is contained in the DNA of the gamete cells.

 SCIENCE IN CONTEXT   **INSULIN**

Insulin is a hormone that is essential for the control of blood glucose concentration (see page 170). Some people are unable to produce insulin – they have Type 1 diabetes. They need to inject insulin regularly to keep their blood glucose concentration under control.

Before the development of transgenic bacteria containing the human insulin gene, insulin was extracted from the pancreases of domesticated animals grown especially for this purpose. It was not only a slow and expensive process but the insulin produced by other species is slightly different from human insulin, so it could cause other problems in people.

Since the gene expressed in transgenic bacteria is the human gene, insulin that is identical to human insulin is produced by them. Also, growing the bacteria in a fermenter produces much larger amounts of insulin, more rapidly and more cheaply than the old process.

QUESTIONS

1. Explain the meaning of the following terms in genetic engineering:
   a) *restriction enzyme*
   b) *ligase enzyme*
   c) *vector*.

2. Draw a flow diagram to explain how a transgenic bacterium is created.

3. Explain why transgenic bacteria can produce human insulin and not a bacterial form of insulin.

GENETICALLY MODIFIED PLANTS

There are several ways in which to genetically modify plants. We might create brighter flowers, introduce different scents, create more flavoursome food crop, or crops that grow faster or give higher yields, or crops resistant to disease, pests, drought or frost. In fact selective breeding, which humans have been practising for centuries, is a form of genetic modification and has allowed us to produce varieties of plant better adapted to certain situations. However, the new techniques do the job faster and also allow us to combine characteristics from different species, which can not be done by selective breeding.

The procedure for producing transgenic plants is similar to producing transgenic bacteria.

- The required gene is obtained, for example by extracting it from the DNA of the source organism using a restriction enzyme.
- The gene is inserted into a bacterial plasmid using a ligase enzyme.
- The plasmid is then inserted into a particular type of bacterium, in this case one that not only causes infection in plant cells but also works by inserting plasmids into the plant cells.
- The bacterium multiplies and is used to 'infect' plant cells – this means the recombinant DNA of the plasmid is carried into the nucleus of each of the plant cells and inserts the gene into the plant's DNA.
- The plant cells are tested to make sure they contain the recombinant DNA.
- Those plant cells with the inserted gene are grown to produce whole new plants.
- As the plants grow, every time a transgenic cell divides, the new cells produced will also contain the inserted gene and so can show the desired characteristic.

◁ Fig. 5.25 These cotton plants have e been genetically modified to make them resistant to pests.

The production of new varieties of crops has obvious benefits. For example, new pest-resistant varieties should allow us to grow larger amounts of food more cheaply because farmers will not need to use pesticide. This should cost less, save time and may also provide a health advantage to farmers in exposing them to lower levels of pesticides. These benefits must be weighed against the fact that seeds of transgenic crop plants are more expensive than the seeds of traditional crop plants.

There are concerns, however, that genetically modifying crops may lead to problems. Through pollination and fertilisation, the recombinant gene could possibly spread from the modified crop species to another closely related wild one. This plant might have an advantage over other plants, and this could unbalance the food web.

There are also concerns that by introducing new genes into food crops we may accidentally introduce something harmful to human health. However, all the research so far suggests this is unlikely.

QUESTIONS

1. Explain how a plant may be genetically modified so that all its cells express the recombinant DNA.

2. Give one example of genetic modification of plants.

3. Describe one advantage and one disadvantage of the genetic modification of plants.

End of topic checklist

Genetic modification (genetic engineering) is the transfer of a gene from one organism into another so that the gene can be expressed.

A **ligase enzyme** is an enzyme that joins pieces of DNA together, and is used in genetic engineering.

Recombinant DNA is DNA that contains a gene inserted during genetic engineering.

A **restriction enzyme** is an enzyme that cuts DNA at a specific site.

A **transgenic organism** is an organism that contains DNA from a different species.

A **vector** is a structure, such as a virus or bacterial plasmid, for carrying genetic material into a cell in genetic engineering.

The facts and ideas that you should know and understand by studying this topic:

○ Restriction enzymes cut DNA at specific sites and can be used in genetic modification to cut out particular genes or to cut open plasmids.

○ Ligase enzymes can be used to join an inserted gene to other DNA to form recombinant DNA.

○ Viruses and bacterial plasmids can be used as vectors for carrying recombinant DNA into cells.

○ Large amounts of human insulin can be produced by genetically modified bacteria cultured in fermenters.

○ Food production can be improved by using genetically modified crop plants.

○ Transgenic organisms contain some genetic material from another species.

End of topic questions

1. When a virus infects a cell, it takes control of the cell to make lots more viruses.

    a) Explain why this makes viruses useful vectors for genetic engineering. **(1 mark)**

    b) Describe what precautions must be taken before using a virus for creating transgenic cells. **(1 mark)**

2. Haemophilia is an inherited disease in humans where a person does not make a blood-clotting protein that other people make. This means that even a small cut can result in prolonged bleeding, with a risk of death. The problems caused by the disease can be controlled by injecting the protein regularly.

    The protein used to be extracted from donated blood from many people, but that carried the risk of infections from pathogens in the blood. Now, much of the protein is produced by transgenic sheep that have been genetically modified to contain the gene for the human protein. The protein is produced in the sheep's milk, and is extracted and purified for human use.

    a) What is meant by *transgenic sheep*? **(1 mark)**

    b) The gene for the human protein is inserted into a sheep embryo. When the embryo divides, all its cells contain the gene. What would have been used to get the gene into the embryo? **(1 mark)**

    c) Describe the role of restriction and ligase enzymes in the process of genetic modification. **(3 marks)**

    d) Give two advantages of producing the protein like this for use in humans. **(2 marks)**

3. There have been newspaper headlines about transgenic plants like this: 'Frankenstein foods could kill us all'.

    a) Suggest why the headlines are written like this. **(1 mark)**

    b) Evaluate this headline in terms of the potential advantages and disadvantages of transgenic plants. **(4 marks)**

4. As the human population increases, we will need to grow more food. One way of doing this is to produce high-yielding crop varieties that are resistant to common diseases.

   a) One way of producing new varieties is by selective breeding. Explain how selective breeding could produce a disease-resistant, high-yielding crop variety.

   (2 marks)

   b) Another way of producing new varieties is by genetic modification. Explain how GM crops could produce a disease-resistant, high-yielding crop variety.

   (2 marks)

   c) State one disadvantage of using selective breeding compared with genetic engineering for the production of disease-resistant, high-yielding crop varieties.

   (1 mark)

   d) Suggest one disadvantage related to the environment of using genetic engineering rather than selective breeding for producing disease-resistant, high-yielding crop varieties. Explain your answer.

   (2 marks)

Cloning

INTRODUCTION

Clones are individual organisms that are genetically identical. Imagine having several clones of you. What advantages or disadvantages would there be? How would your friends and family react?

Remember, though, that just because clones have identical genes it does not mean they are exactly alike. Environmental factors create variation too. Identical twins are genetic clones, because they were formed when a fertilised egg cell divided in half during the early stages of cell division, creating two separate individuals by mitosis. We do not consider identical twins as completely identical, but as separate individuals, with their own characters developed by their environment as well as their genes. So would lots of clones of you be a good idea?

△ Fig. 5.26 Identical twins are clones in the genetic sense, but they are also individuals.

KNOWLEDGE CHECK

✓ Mitosis is cell division that produces two genetically identical cells.
✓ Identical plants can be produced from cuttings, a form of asexual reproduction.
✓ Many of the characteristics of an organism are controlled by the genes in its cell nuclei.

LEARNING OBJECTIVES

✓ Describe the process of micropropagation (tissue culture) in which explants are grown *in vitro*.
✓ Understand how micropropogation can be used to produce commercial quantities of genetically identical plants with desirable characteristics.
✓ Describe the stages in the production of cloned mammals by nuclear transfer.
✓ Understand how cloned transgenic animals can be used to produce human proteins.

CLONING OF PLANTS

Plant clones can be produced by taking cuttings (see page 201). This is a form of asexual reproduction, by which genetically identical individuals are produced through mitotic cell division. This can be a useful technique for plant growers who want more plants with identical characteristics, such as roses with an unusual colour. It is also useful for some crop plants, such as bananas, which do not grow well from seed.

A modern version of taking cuttings is **tissue culture**. This is used to grow large numbers of plants quickly, as only a tiny part of the original is needed to grow a new plant. This method is also called **micropropagation**. The procedure is carried out in laboratory conditions, often described as *in vitro*, which literally means 'in glass', referring to the glassware used in laboratory procedures.

Special procedures have to be followed.

- Many small pieces are cut from, for example, the growing tips of shoots of the chosen plant. These pieces are called **explants**.
- The pieces are sterilised by washing them in mild bleach to kill any microorganisms.
- In sterile conditions the explants are transferred onto a jelly-like growth medium that contains nutrients as well as plant hormones to encourage cell division.
- When there are many cells, they can be separated into smaller pieces again under sterile conditions and grown again.
- Some explants are placed onto a growth medium with different concentrations of plant hormones to encourage the cells to differentiate and produce roots, shoots and leaves.

◁ Fig. 5.27 New plants growing from explants after micropropagation.

When the tiny plants are large enough to be handled they can be transferred to other nutrient media and eventually to a normal growth medium like compost.

Micropropagation is increasingly being used to produce large numbers of genetically identical plants for commercial purposes. It is a fairly expensive technique because it has to be done in the laboratory, but for valuable plants it is worthwhile. For example, many of the orchids grown for the horticultural trade are produced this way, because orchids are more difficult to grow from seed and micropropagation guarantees large numbers of plants with the desired flower colour and form.

Developing investigative skills

It is possible to produce plants by micropropagation in a school laboratory.

Devise and plan investigation

❶ Write a plan to describe how you would produce clones of a houseplant using micropropagation, and what precautions you would need to take.

Demonstrate and describe techniques

❷ Explain why it is important to use sterile apparatus and to sterilise the explants.

Analyse and interpret data

❸ **a)** Will the form of the plants, and colour of flowers (if they have colour), that you produce using this method be identical, quite like, or very different from those of the original plant?

b) Explain your answer.

Micropropagation has another benefit for some crop plants. Bananas are normally grown from larger cuttings of existing plants. But the spread of viral diseases has threatened the growing of bananas in many areas. The cells taken from banana plants for micropropagation are less likely to contain viruses than the larger cuttings. As bananas do not grow well from seed, this is a good way to produce new, virus-free banana plants.

Δ Fig. 5.28 An orchid nursery in Thailand where thousands of plants are being reared using micropropagation.

QUESTIONS

1. Explain the meaning of the term *micropropagation*.

2. Draw a flowchart to show how new plants can be produced by micropropagation.

3. Explain why plants produced by micropropagation are genetically identical.

CLONING OF MAMMALS

Cloning can also now be used with mammals. For example, cells from a developing embryo (such as from a sheep) can be split apart before they become specialised. They are then developed into identical embryos, and each one is transplanted into the uterus of another female, called a host mother, to develop until ready for birth.

Another way of cloning animals involves making ordinary body cells grow into new animals. Dolly the sheep was the first mammal clone grown from the cell of a fully grown organism: in Dolly's case it was a cell from the udder of a sheep.

Dolly the sheep was produced in the following way.

- A mature diploid cell was removed from the udder of an adult female sheep, and the nucleus was taken out.
- An unfertilised egg cell was taken from a female sheep of a different breed.
- The haploid nucleus of the unfertilised egg cell was removed (the cell was **enucleated**) and thrown away.
- The diploid nucleus from the udder cell was inserted into the cytoplasm of the enucleated egg cell.
- The cell was treated with an electric shock so that it started to divide by mitosis.
- The dividing cell formed an embryo which was implanted into the uterus of another female sheep.
- The lamb that was born was called Dolly. She had the characteristics of the sheep from which the udder cell had been taken.

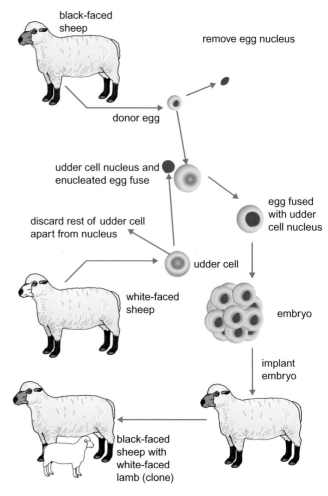

△ Fig. 5.29 Cloning Dolly the sheep.

REMEMBER

When describing an example of cloning mammals, make it clear that the characteristics of the new individual produced come from the genes in the original nucleus. So the cloned individual will be genetically identical to the animal from which that nucleus came.

USING CLONED TRANSGENIC ANIMALS TO PRODUCE HUMAN PROTEINS

The method used to produce Dolly the sheep has the potential to help produce transgenic animals. If a gene from another organism is inserted into the DNA in the nucleus before it is placed in the enucleated egg cell, then, as a result of mitosis, all the cells of the developing embryo will have nuclei containing the recombinant DNA. Once you have produced one transgenic animal then it can be cloned, producing genetically identical transgenic animals.

◁ Fig. 5.30 A transgenic goat. Its milk contains a chemical that is normally produced in humans.

One way that transgenic animals can be used is for the production of human proteins for use as medicinal drugs. There are already goats that produce a human blood anti-clotting chemical called antithrombin (AT) in their milk because they contain the human gene for making it. It is easy to separate AT from the milk and purify it for treating people who are genetically unable to produce it themselves. (People without AT are much more likely to form unwanted blood clots.) Cells from mice have been genetically engineered to produce human antibodies and then cultured in fermenters to produce large quantities of antibodies for use in the treatment of cancer, heart disease and other illnesses.

While there are many potential advantages in the use of transgenic animals, there are also concerns about the use of animals in this way. There are concerns for the welfare of the animals involved, and concerns that such modifications may have unforeseen and unwelcome effects. For example, Dolly the sheep developed diseases associated with much older sheep, and was put down at the age of 6. This may have just been bad luck, but some scientists think it may have been the result of cloning.

It may prove too expensive to use cloned transgenic animals in this way to produce drugs and organs. Many people also think it is unethical (not right) to create and use animals like this.

QUESTIONS

1. Explain how Dolly the sheep was produced.

2. Which sheep was Dolly a clone of: sheep A from which the egg cell was taken, sheep B from which the body cell was taken, sheep C which carried the embryo in its uterus and gave birth to Dolly? Explain your answer.

3. Explain why transgenic animals can be useful to humans.

End of topic checklist

Enucleate means to remove the nucleus from a cell.

Explants are tiny pieces of plant used in micropropagation.

In vitro describes a laboratory process taking place in a test tube or dish.

Micropropagation is the culture of explants in the laboratory to produce many clones of a single plant.

Tissue culture is another name for micropropagation.

The facts and ideas that you should know and understand by studying this topic:

○ New plants can be made from explants grown *in vitro* in the right conditions, using the technique called micropropagation (tissue culture).

○ Micropropagation can produce commercial quantities of genetically identical plants with desirable characteristics.

○ The production of cloned mammals, such as Dolly the sheep, involves introducing a diploid nucleus from a body cell of one animal into the enucleated egg cell of another animal.

○ Cloned transgenic animals can be used to produce human proteins.

End of topic questions

1. A plant grower wants to grow new plants. In each of the following cases, explain which technique would be best to use, and why.

   a) He has a disease-resistant plant with yellow flowers but wants to produce disease-resistant plants with flowers of other colours. **(2 marks)**

   b) He has a plant with unusual blue flowers and wants to produce many other plants like this because they will sell well. **(2 marks)**

2. Transgenic animals can be used to produce human proteins in their milk, for example antithrombin which can be used to help prevent blood clots in human patients.

   a) Suggest an explanation for why it is helpful to humans for the goats to produce antithrombin in their milk, and not in some other part of their bodies. **(2 marks)**

   b) Suggest an explanation for why human patients respond better to human antithrombin than natural goat antithrombin. **(2 marks)**

3. Bananas and other plantains are important sources of food in the Great Lakes region of eastern Africa. These plants naturally reproduce by asexual reproduction, and growers produce new plants by taking cuttings from older plants. Over the past 10 years, crop yield has decreased as a result of infection by soil pests and bacterial disease. Use of micropropagation is helping to reverse this trend, increasing yield from about 20 tonnes to around 45 tonnes per hectare.

   a) Explain why new banana plants produced by traditional methods are likely to be infected with pests and diseases. **(1 mark)**

   b) Suggest why the yield of bananas has fallen over the past decade or so. **(1 mark)**

   c) Describe how many new banana plants could be produced by micropropagation. **(4 marks)**

   d) Explain as fully as you can why micropropagation is helping to improve banana crop yield. **(2 marks)**

   e) Bacterial wilt is a major disease of banana plants. Scientists are trying to genetically modify bananas to protect them against this disease by transferring a gene for resistance to the bacteria from another plant. When a successful transgenic plant is produced, micropropagation is then used to produce many transgenic plants. Explain as fully as you can why micropropagation is useful for this. **(3 marks)**

Exam-style questions
Sample student answers

Question 1

The diagram shows the structure of an industrial fermenter used to culture the fungus *Penicillium*. Under certain conditions, *Penicillium* produces the antibiotic penicillin.

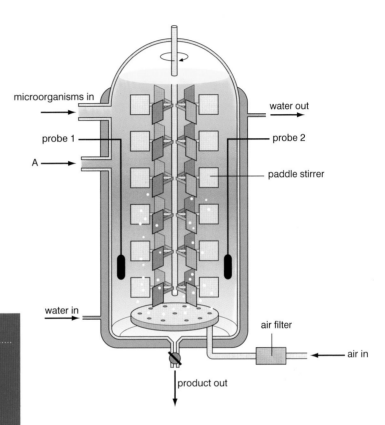

a) i) Suggest what may be pumped into the fermenter at point A. **(1)**

   nutrients ✔ ①

   ii) Suggest two factors that probes 1 and 2 may be designed to measure. **(2)**

   temperature ✔ ①

   pH ✔ ①

Exam-style questions continued

b) In the fermenter, explain fully the use of:

   i) the paddle stirrer   **(2)**

> *To mix the contents in the*
> *fermenter* ✔ ①

   ii) water circulating around the fermenter.   **(3)**

> *To cool the fermenter down* ✗

c) Give one way in which aseptic conditions in the fermenter are maintained.   **(1)**

> *The air filter* ✔ ①

d) What type of respiration is *Penicillium* mould carrying out in the fermenter? Explain your answer.   **(2)**

> *Aerobic* ✔ ①
> *Air is pumped into the fermenter* ✔ ①

(Total 11 marks)

$\dfrac{7}{11}$

Question 2

a) Yeast, a fungus, is used to make bread.

   i) Give the word equation for anaerobic respiration in yeast.   **(3)**

   ii) Explain why yeast is used to make bread.   **(1)**

b) Anaerobic respiration in yeast is affected by several factors.

   i) Describe an experiment to investigate the production of carbon dioxide by yeast at different temperatures.   **(5)**

   ii) Give two other factors that affect the production of carbon dioxide by yeast.   **(2)**

c) Apart from bread, name another food that is made using yeast.   **(1)**

(Total 12 marks)

---

b) i) Correct but the answer could have been more specific, and the candidate has only picked up one marking point. The paddle stirrer ensures that the microorganism (*Penicillium*), nutrients, oxygen and heat are distributed evenly throughout the fermenter.

ii) This is not incorrect, but does lack detail. An explanation is required. The respiration of the *Penicillium* releases heat. It is important that the temperature is kept constant, so that the enzymes of the *Penicillium* work to their optimum. Circulating water is used to regulate the temperature when it becomes raised.

c) The air filter removes microorganisms from air entering the fermenter and helps to maintain aseptic conditions for the growth of the *Penicillium*.

Other suitable answers include the sterilisation of the fermenter using steam, the addition of a pure culture of *Penicillium* to the fermenter (i.e. no other microorganisms are present), and the use of sterile nutrient solutions.

d) Correct.

Question 3

In the 1960s, the insecticide DDT was sprayed onto the water in Long Island Sound, New York, to kill mosquitoes. Insect control specialists were careful to use concentrations of the insecticide that were not lethal to other wildlife.

In a study carried out in 1967, scientists showed that some of the organisms living in the water had very high levels of DDT in their bodies. Some of their results are shown below:

Source	Concentration of DDT, parts per million	Notes on organism
water	0.00005	
plankton	0.04	
silverside minnow	0.23	small fish – feeds on plankton
sheepshead minnow	0.94	small fish – feeds on plankton
needlefish	2.07	predator of smaller fish
tern	3.91	bird – feeds on large fish
cormorant	26.40	bird – feeds on large fish

a) i) Draw a food web to show the interaction between the species in Long Island Sound. (6)

   ii) Explain why the concentrations are different in different organisms. (3)

b) Scientists now know that DDT reduces the reproduction of birds by affecting the production of their eggs.

   Predict the effect that DDT would have had on this food chain. (5)

(Total 14 marks)

Question 4

Production of fish in fish farms has increased considerably in the past 30 years.

a) Explain the importance of fish as a food. (2)

b) To ensure the production of maximum yields of healthy fish, certain factors must be controlled.

List three of these factors and explain the methods used to control them. (3)

c) A study was carried out in Norway on the farming of fish in cages in the sea. The nitrogen-containing wastes discharged into the sea were measured over a 22 year period.

Year	Nitrogen-containing waste discharged to the sea	
	Kilograms of waste per tonne of fish	**Total in tonnes**
1985	77.5	2 683
1990	83.4	12 484
1992	82.3	10 754
1995	49.9	13 780
1998	49.3	20 237
2001	39.8	20 206
2004	41.9	26 253
2007	50.2	40 877

i) What was the mass of fish produced in the following years? Show your working.

- 1985 (2)

- 2007 (2)

ii) Describe the effect of the release of nitrogen compounds such as nitrates into natural waters. (5)

(Total 14 marks)

Question 5

A student investigated the production of yoghurt in the laboratory. His results are shown in the graph below.

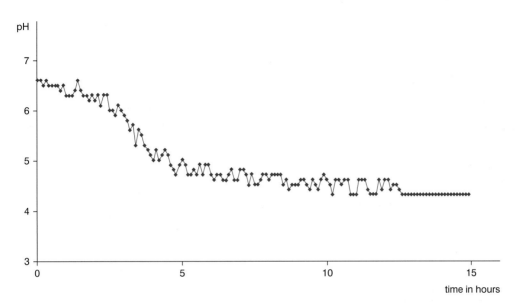

a) Name a bacterium involved in the production of yoghurt. **(1)**

b) i) Describe the change in pH over the 15-hour period. **(2)**

   ii) Explain the change in pH during yoghurt production. **(1)**

c) Name one condition required for the optimum growth of yoghurt-producing bacteria. **(1)**

(Total 5 marks)

Question 6

The passage below describes stages in the genetic modification of a bacterium to produce 'human' insulin.

Use suitable words to complete the sentences in the passage. **(5)**

Many copies of the human insulin gene are made.

Bacterial .... are cut open using a .... enzyme.

The pieces of human and bacterial DNA are then mixed together, and an enzyme called a .... added to join them.

The plasmids are tested to check which ones contain .... DNA. The plasmids act as .... when they are inserted into bacteria. These bacteria can now produce human insulin.

(Total 5 marks)

Question 7

Read the passage below. Use the information in the passage, along with your own knowledge, to answer the questions that follow.

Line:

1 Scientists have found bacteria and other organisms that will kill insect pests. A bacterium found in the soil called *Bacillus thuringiensis* (*Bt*) produces a protein that kills the caterpillars of butterflies and moths. The caterpillars of many moths and some butterflies cause serious losses to the yields of many crops.

5 When eaten, the bacterium forms spores in the gut of a caterpillar. The spores produce proteins called Insecticidal Crystal Proteins (ICP). These are insoluble in the gut of most insects, and just pass through, but in butterflies and moths, the high pH of their gut makes the proteins soluble. The toxic protein binds to the gut wall, creating holes in it, and the caterpillar is killed.

10 Spores of the bacterium can be applied to plants to kill these pests. *Bt* has been used as a spray for around 30 years by home gardeners and organic growers. But spraying crops with the bacterium means that it can be taken in by butterflies and moths that are *harmless* to crops. These insects can be important pollinators of plants. In addition, *Bt* will not kill caterpillars that feed on plant roots.

15 Now genetically modified crop plants are being created by injecting *Bt* genes into plant cells using a device called a gene gun. Scientists at several biotechnology companies have inserted genes from *Bt* into several crop plants, including maize and cotton. With genetically engineered *Bt* plants, only those insects that attack the crop should be exposed to *Bt* toxins. There should be no risk to other types of
20 insect.

Many people, however, are concerned about the possible effect of *Bt* plants on other insects. Caterpillars of the monarch butterfly feed on leaves of the milkweed plant, which often grows in and around fields of maize in the USA. A study in 1999 suggested that pollen from *Bt* maize, blown onto milkweed plants, killed the
25 caterpillars.

A more recent study, however, suggests that the pollen is only toxic in very high doses. Scientists found that leaves coated with pollen at densities at less than 1000 pollen grains per cm^2 had no effect on monarch butterfly caterpillars. Above this density, the caterpillars were smaller. They did find, however, that pollen from
30 one of the earliest forms of *Bt* maize (*Bt* 176) *was* toxic at very low levels.

a) Where is the bacterium *Bacillus thuringiensis* found in nature? (line 2) **(1)**

b) Why are people concerned about the application of *Bt* insecticide to the leaves of crop and garden plant leaves? (lines 10–15) **(2)**

c) What technique is used to insert the *Bt* genes into maize plants? (line 16) **(1)**

d) Suggest what form of pollination takes place in maize. Explain your reasoning. **(2)**

e) Using the information in the passage, evaluate the use of genetically modified maize in controlling insect pests. **(5)**

f) Suggest one further study on the pollen of genetically modified crops that could be investigated. **(1)**

(Total 12 marks)

The International GCSE Examination

INTRODUCTION

The International GCSE examination tests how good your understanding of scientific ideas is, how well you can apply your understanding to new situations and how well you can analyse and interpret information you have been given. The assessments are opportunities to show how well you can do these.

To be successful in exams you need to:

✓ have a good knowledge and understanding of science
✓ be able to apply this knowledge and understanding to familiar and new situations
✓ be able to interpret and evaluate evidence that you have just been given.

You need to be able to do these things under exam conditions.

OVERVIEW

The International GCSE course is designed to provide a basis for progression to further study in International Advanced Subsidiary and Advanced Level Biology, GCE Advanced Subsidiary and Advanced Level Biology, and the International Baccalaureate. The relationship of the assessment to the qualifications available is shown below:

Biology Paper 1 2 hours + **Biology Paper 2** → **International GCSE in Biology**

1 hour 15 minutes

+

Chemistry Paper 1 2 hours + Chemistry Paper 2 → International GCSE in Chemistry

1 hour 15 minutes

+

Physics Paper 1 2 hours + Physics Paper 2 → International GCSE in Physics

1 hour 15 minutes

↓

International GCSE in Science (Double Award)

Paper 1 is marked out of 110 and contributes 61.1% of the total International GCSE marks.

Paper 2 is marked out of 70 and contributes 38.9% of the International GCSE marks.

On the Edexcel Double Award Science course Paper 1 accounts for 33.3% of the overall marks.

There is no separate assessment of investigative skills – the assessment is included in the two written papers.

There will be a range of compulsory, short-answer structured questions, along with some multiple choice questions, and questions requiring longer answers, in both papers. You will be required to perform calculations, draw graphs and describe, explain and interpret biological ideas and information. In some of the questions the content may be unfamiliar to you; these questions are designed to assess data-handling skills and the ability to apply biological principles and ideas in unfamiliar situations.

ASSESSMENT OBJECTIVES AND WEIGHTINGS

The assessment objectives and weightings are as follows:

✓ AO1: Knowledge and understanding (38–42%)

✓ AO2: Application of knowledge and understanding, analysis and evaluation (38–42%)

✓ AO3: Experimental skills, analysis and evaluation of data and methods (19–21%).

The types of questions in your assessment fit the three assessment objectives shown in the table.

Assessment objective	Your answer should show that you can...
AO1: recall the science	Recall, select and communicate your knowledge and understanding of biology.
AO2: apply your skills and knowledge	Apply skills, including evaluation and analysis, knowledge and understanding of biological contexts.
AO3: use experimental skills	Use the skills of planning, observation, analysis and evaluation in practical situations.

EXAMINATION TIPS

To help you get the best results in exams, there are a few simple steps to follow.

Check your understanding of the question

✓ **Read the introduction to each question carefully before moving on to the questions themselves**.

✓ Look in detail at any **diagrams, graphs** or **tables**.

✓ Underline or circle the **key words** in the question.

✓ **Make sure you answer the question that is being asked** rather than the one you wish had been asked!

✓ Make sure that you understand the meaning of the '**command words**' in the questions.

Remember that any information you are given is there to help you to answer the question.

COMMON COMMAND WORDS

✓ '**Give**', '**state**', '**name**' are used when recall of knowledge is required, for example you could be asked to give a definition or make a list of examples.

✓ '**State what is meant by**' is used when the meaning of a term is expected but there are different ways for how these can be described.

✓ '**Identify**' is usually used when you have to select some key information from a text or diagram in the question.

✓ '**Describe**' is used when you have to give the main feature(s) of, for example, a biological process or structure. You do not need to include a justification or reason.

✓ '**Explain**' is used when you have to give reasons, e.g. for some experimental results or a biological fact or observation. You will often be asked to **justify** or 'explain your answer', i.e. give reasons for it.

✓ '**Suggest**' is used when you have to come up with an idea to explain the information you're given – there may be more than one possible answer, no definitive answer from the information given, or it may be that you will not have learnt the answer but have to use the knowledge you do have to come up with a sensible one.

✓ '**Calculate**' means that you have to work out an answer in figures, showing relevant working.

✓ '**Determine**' means you must use data from the question, or must show how the answer can be reached quantitatively. To gain maximum marks, there must be a quantitative element to the answer.

✓ '**Estimate**' means find an approximate value, number or quantity from a diagram/given data or through a calculation.

✓ '**Plot**', '**draw a graph**' are used when you have to use the data provided to produce graphs and charts. This includes drawing a line of best fit through the points you have plotted. A suitable scale and appropriately labelled axes must be included if these are not provided in the question.

✓ **'Sketch'** means produce a drawing by hand. For a graph, this would need a line and labelled axes with important features indicated. The axes are not scaled.

✓ **'Draw'** means produce a diagram either using a ruler or by hand.

✓ **'Add label'** is used when you have to add or label something given in the question, for example labelling a diagram or adding units to a table.

✓ **'Complete'** means complete a table/diagram given in the question.

✓ **'Comment on'** is used when you have to bring together a number of variables from data/information to form a judgement.

✓ **'Deduce'** means draw/reach conclusion(s) from the information provided.

✓ **'Discuss'** is used when you have to:

- identify the issue/situation/problem/argument that is being assessed within the question
- explore all aspects of an issue/situation/problem/argument
- investigate the issue/situation etc. by reasoning or argument.

✓ **'Evaluate'** means review information (e.g. data, methods) then bring it together to form a conclusion, drawing on evidence including strengths, weaknesses, alternative actions, relevant data or information. Come to a supported judgement of a subject's quality and relate it to its context.

✓ **'Give a reason/reasons'** is used when a statement has been made and you only have to give the reason(s) why.

✓ **'Justify'** means give evidence to support (either the statement given in the question or an earlier answer).

✓ **'Design'** means plan or invent a procedure from existing principles/ideas.

✓ **'Predict'** means give an expected result.

✓ **'Show that'** means verify the statement given in the question.

Check the number of marks for each question

✓ Look at the **number of marks** allocated to each question.

✓ Make sure you include at least as many points in your answer as there are marks.

✓ Look at the **space provided** to guide you as to the length of your answer. However, be aware that there may be more space than you need.

What to do if you need extra space to answer

✓ If you need more space to answer than provided, then either use the nearest available space, e.g. at the bottom of the page, or ask for extra paper.

✓ However, do NOT use any blank pages in the examination paper as these will not be scanned for the Examiner.

✓ You MUST state clearly WITHIN the marked-out answer space where you have continued your answer, e.g. 'continued at the bottom of page 12', otherwise your additional material may not be seen by the Examiner.

✓ You MUST make it clear IN YOUR ANSWER which question you are answering.

✓ If you use extra paper you MUST make sure you also include your name, candidate number and centre number.

REMEMBER

Beware of continually writing too much because it probably means you are not really answering the questions.

Use your time effectively

✓ Do not spend so long on some questions that you do not have time to finish the paper.

✓ You should spend approximately **one minute per mark**.

✓ If you are really stuck on a question, leave it, finish the rest of the paper and come back to it at the end.

✓ Even if you eventually have to guess at an answer, you stand a better chance of gaining some marks than if you leave it blank.

ANSWERING QUESTIONS

Multiple choice questions

✓ Select your answer by placing a cross (not a tick) in the box.

Short and long answer questions

✓ In short-answer questions, **do not write more than you are asked for**.

✓ You will not gain any marks, even if the first part of your answer is correct, if you have written down something incorrect later on or which contradicts what you have said earlier. This just shows that you have not really understood the question or are guessing.

✓ In some questions, particularly short-answer questions, answers of only one or two words may be sufficient, but in longer questions you should aim to use **good English** and **scientific language** to make your answer as clear as possible.

✓ Present the information in a logical sequence.

✓ Do not be afraid to also use **labelled diagrams** or **flow charts** if it helps you to show your answer more clearly.

Questions with calculations

✓ **In calculations always show your working**. Even if your final answer is incorrect you may still gain some marks if part of your attempt is correct. If you just write down the final answer and it is incorrect, you will get no marks at all.

✓ Write down your answers to as many **significant figures** as are used in the numbers in the question (and no more). If the question does not state how many significant figures then a good rule of thumb is to quote 3 significant figures.

✓ Do not round off too early in calculations with many steps – it is always better to give too many significant figures in your working than too few.

✓ You may also lose marks if you do not use the correct **units**. In some questions the units will be mentioned, e.g. calculate the mass in grams, or the units may also be given on the answer line. If numbers you are working with are very large, you may need to make a conversion, e.g. convert joules into kilojoules, or kilograms into tonnes.

Finishing your exam

✓ When you've finished your exam, **check through** your paper to make sure you have answered all the questions.

✓ Check that you have not missed any questions at the end of the paper or turned over two pages at once and missed questions.

✓ Cover over your answers and read through the questions again and check that your answers are as good as you can make them.

REMEMBER

In the two written papers, you will be asked questions on investigative work (Assessment objective AO3). It is important that you understand the methods used by scientists when carrying out investigative work.

More information on carrying out practical work and developing your investigative skills are given in the next section.

Developing experimental skills

INTRODUCTION

As part your International GCSE Biology course, you will develop practical skills and have to carry out investigative work in science.

This section provides guidance on carrying out an investigation.

Many investigations follow a common route:

1. Planning the investigation and assessing the risk.

2. Carrying out the practical work safely and skilfully.

3. Making and recording observations and measurements.

4. Analysing the data and drawing conclusions.

5. Evaluating the data and methods used.

1. PLANNING AND ASSESSING THE RISK

Learning objective: to devise and plan investigations, drawing on biological knowledge and understanding in selecting appropriate techniques.

Questions to ask

What do I already know about the area of biology I am investigating, and how can I use this knowledge and understanding to help me with my plan?

✓ Think about what you have already learned and any investigations you have already done that are relevant to this investigation.

✓ List the factors that might affect the process you are investigating.

What is the best method or technique to use?

✓ Think about whether you can use or adapt a method that you have already used.

✓ A method, and the measuring instruments, must be able to produce **valid** measurements. A measurement is valid if it measures what it is supposed to be measuring.

You will make a decision as to which technique to use based on:

✓ The accuracy and precision of the results required.

Investigators might require results that are as accurate and precise as possible but if you are doing a quick comparison, or a preliminary test to check a range over which results should be collected, a high level of accuracy and precision may not be required.

✓ The simplicity or difficulty of the techniques available, or the equipment required; is this expensive, for instance?

✓ The scale, e.g. using standard laboratory equipment or on a micro-scale, which may give results in a shorter time period.

✓ The time available to do the investigation.

✓ Health and safety considerations.

What am I going to measure?

✓ You need to decide what you are going to measure.

✓ You need to choose a range of measurements that will be enough to allow you to plot a graph of your results and so find out the pattern in your results.

✓ You might be asked to explain why you have chosen your range rather than a lower or higher range.

✓ The factor you are investigating is called the **independent variable**. A **dependent variable** depends on the value of the independent variable that you select.

How am I going to control the other variables?

✓ These are **control variables**. Some of these may be difficult to control. This may be especially difficult if you are carrying out an ecology investigation in the field, where factors such as varying weather conditions are impossible to control.

✓ You must decide how you are going to control any other variables in the investigation and so ensure that you are using a **fair test** and that any conclusions you draw are valid.

What equipment is suitable and will give me the accuracy and precision I need?

✓ The **accuracy** of a measurement is how close it is to its true value.

✓ **Precision** is related to the smallest scale division on the measuring instrument that you are using, e.g. when measuring the distance moved by the bubble in Method 3 (see page 383), a rule marked in millimetres will give greater precision that one divided into centimetres only.

✓ A set of precise measurements also refers to measurements that have very little spread about the mean value.

✓ You need to be sensible about selecting your devices and make a judgement about the degree of precision. Think about what is the least precise variable you are measuring and choose suitable measuring devices. There is no point having instruments that are much more precise than the precision you can measure the least precise variable to.

What are the potential hazards of the equipment, chemicals, organism and technique I will be using and how can I reduce the risks associated with these hazards?

✓ Investigators find out about the hazard associated with chemicals using CLEAPSS Student Safety Sheets or a similar resource. Information on biological hazards can be found in the CLEAPSS Laboratory Handbook.

✓ In the exam, be prepared to suggest safety precautions when presented with details of a biology investigation.

EXAMPLE 1

You have been asked about the design and planning of an investigation on carbon dioxide production by yeast under different conditions.

What do I already know?

Previously you have learnt about the role of yeast in bread making. In the equation for anaerobic respiration of yeast, you can see that in the absence of oxygen, yeast uses sugar for anaerobic respiration, producing carbon dioxide and ethanol.

As sugar is a reactant in the process, conditions that might affect the production of carbon dioxide on the process include the *concentration of sugar*, and the *type of sugar* may also be a factor.

The chemical reactions involved in this process are controlled by the yeast's enzymes, so you would expect this process to be affected by *temperature*.

Conditions that might be expected to affect the process are therefore:

✓ concentration of sugar

✓ type of sugar

✓ temperature.

You might be questioned on how to measure the effect these factors, or one of these factors, has on carbon dioxide production.

All of these factors are independent variables. The amount of carbon dioxide produced is the dependent variable because it is affected by the independent variables – concentration of sugar, type of sugar and temperature.

What is the best method or technique to use?

An investigator needs to set up an experiment so that they can measure the carbon dioxide produced by the respiring yeast.

Several methods are available to produce valid measurements.

Method 1

A simple way is to add some yeast to a sugar solution in a conical flask, as shown in the equipment below and count the bubbles produced over a period of time, for example one minute.

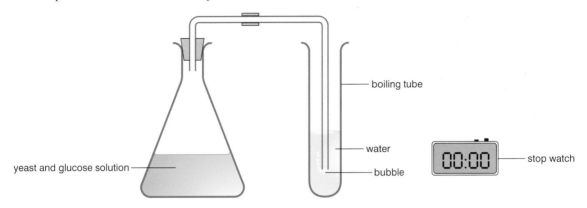

△ Fig. 6.1 Apparatus for method 1.

Method 2

An alternative method is to set up the yeast and sugar solution as before, but this time connect the glass tube to a gas syringe. This time, you can measure the volume of carbon dioxide produced over a period of time, e.g. one hour.

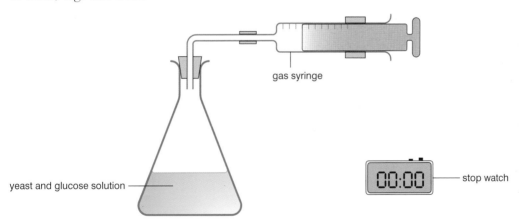

△ Fig. 6.2 Apparatus for method 2.

Method 3

Another method is based on a smaller scale. The yeast and sugar solution is placed in a syringe, which is connected to a pipette that contains a bubble of water. The movement of the bubble is measured over a period of time, e.g. one minute. It is less suitable for the temperature investigation, however, as it would be inadvisable to immerse the syringes in water baths, but it could be set up appropriately with care.

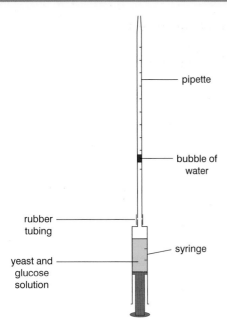

Δ Fig. 6.3 Apparatus for method 3.

Choosing the method:

✓ Accuracy and precision: methods 2 and 3 would be preferred when accurate and precise results are needed. Method 1 cannot be used to actually measure volume of carbon dioxide, but would be best for a preliminary test or a quick comparison.

✓ Micro-scale: method 3 might be most suitable here.

✓ Time available: method 3 might be most suitable, as smaller volumes are involved and the investigation could be carried out more quickly.

✓ Health and safety: as yeast and glucose are not currently classified as hazardous, the safety considerations are for handling glassware, and are the same for all three methods.

What am I going to measure?

You need to make a measure of the amount of carbon dioxide produced using different concentrations of sugar, different types of sugar, or at different temperatures.

The **independent variables** are the factors under investigation: the concentration of sugar, the type of sugar and the temperature.

The **dependent variables** are the measurements made: the number of bubbles per minute, the volume of carbon dioxide per minute, or the distance moved by the bubble.

Different concentrations of sugar (Example 1a)

A sensible range of sugar concentrations to use might be based on concentrations that yeast might encounter in nature. The investigator might choose to use concentrations ranging from 0 mol dm^{-3} to 0.25 mol dm^{-3} (at intervals of 0 mol dm^{-3}, 0.05 mol dm^{-3}, 0.10 mol dm^{-3},

0.15 mol dm^{-3}, 0.20 mol dm^{-3} and 0.25 mol dm^{-3}), as a preliminary test would show that these concentrations would give suitable results in the time period allocated. Higher concentrations might not be chosen as these might have an osmotic effect; water could be drawn from the yeast cells by osmosis, and so they would not function normally.

The investigation of six different concentrations would be sufficient to the investigator to plot a graph of the results and so find out any pattern in the results.

Different types of sugar (Example 1b)

Here, the investigator might decide to use sugars commonly found in nature that yeast might use for respiration. These could include fructose, glucose, lactose, maltose and sucrose.

Different temperatures (Example 1c)

It would be sensible to choose a range of temperatures that yeast might encounter in nature or in a kitchen. Respiration in yeast is a series of enzyme controlled reactions. In many cases, enzymes work best around 40 °C, and cease to function above around 60 °C. It might be decided, therefore, to measure carbon dioxide production at six different temperatures (10 °C, 20 °C, 30 °C, 40 °C, 50 °C and 60 °C). Again, this would be sufficient to allow the investigator to plot a graph of results and so find any pattern in the results.

How am I going to control the other variables?

The investigator must ensure that any differences in carbon dioxide production must be the result of, in the different investigations, sugar concentration, type of sugar and temperature, and not the result of some other factor. In other words, it must be a fair test and produce valid measurements.

So, the investigator must decide what other factors could affect the experiment and try to keep these constant. These are the control variables.

Factor under investigation/ independent variable	Factors to be kept constant				
	Yeast concentration	Sugar concentration	Type of sugar	Temperature	Duration of investigation
Sugar concentration	yes	no – vary	yes	yes	yes
Type of sugar	yes	yes	no – vary	yes	yes
Temperature	yes	yes	yes	no – vary	yes

△ Table 6.1: Variables in Example 1.

What equipment is suitable and will give me the accuracy and precision I need?

You now know what you will need to measure and so can decide on your measuring devices.

Measurement	Quantity required	Device
Mass of yeast	1.00 g	Balance measuring up to two decimal places
Volume of sugar solution	100 cm^3	100 cm^3 volumetric flask
Temperature	10–60 °C	Thermometer, with 1 °C precision
Time	One minute intervals	Stop watch (1 s precision)

△ Table 6.2 Suitable equipment for experiment.

You will need to be sensible about selecting your devices and make a judgement about the degree of accuracy. The accuracies of equipment need to be comparable. It would be not appropriate to measure the yeast that was put in every conical flask accurately without measuring the volume of sugar solution poured onto each flask to a similar level of accuracy.

What are the potential hazards of the equipment and how can I reduce the risks?

In Example 1b, the chemical hazards are as follows:

Fructose solution: NOT CURRENTLY CLASSIFIED AS HAZARDOUS

Glucose solution: NOT CURRENTLY CLASSIFIED AS HAZARDOUS

Lactose solution: NOT CURRENTLY CLASSIFIED AS HAZARDOUS

Maltose solution: NOT CURRENTLY CLASSIFIED AS HAZARDOUS

Sucrose solution: NOT CURRENTLY CLASSIFIED AS HAZARDOUS

Yeast, dried: NOT CURRENTLY CLASSIFIED AS HAZARDOUS.

These indicate that there are no specific hazards the investigator needs to be aware of. However, when handling *any* chemicals, it would be sensible to wear eye protection.

In terms of the equipment and technique, the major hazards are:

✓ handling hot liquids (at 50 °C and 60 °C)

✓ when connecting fragile glass tubing together

✓ possible contact between water and electrical sockets.

2. CARRYING OUT THE PRACTICAL WORK SAFELY AND SKILFULLY

Learning objective: To demonstrate and describe appropriate experimental and investigative methods, including safe and skilful practical techniques.

Questions to ask:

How shall I use the equipment and chemicals safely to minimise the risks – what are my safety precautions?

✓ When writing a Risk Assessment, investigators need to be careful to check that they have matched the hazard with the concentration of a chemical used. Many acids, for instance, are corrosive in higher concentrations, but are likely to be irritants or of low hazard in the concentration used when working in biology experiments.

✓ Do not forget to consider the hazards associated with all the chemicals, biological materials and the products made, even if these are very low.

✓ In the exam, you may be asked to justify the precautions taken when carrying out an investigation.

How much detail should I give in my description?

✓ You need to give enough detail so that someone else who has not done the experiment would be able to carry it out to reproduce your results.

How should I use the equipment to give me the precision I need?

✓ You should know how to read the scales on the measuring equipment you are using.

✓ You need to show that you are aware of the precision needed.

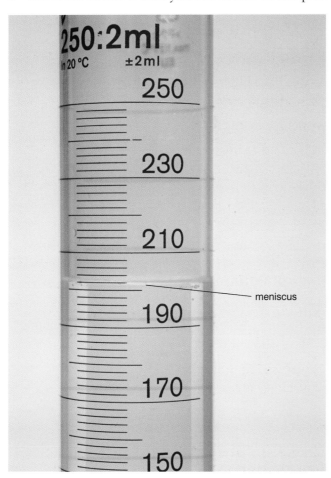

△ Fig. 6.4 The volume of liquid in a burette or measuring cylinder must be read to the bottom of the meniscus. The volume in this measuring cylinder is 202 cm³ (ml), not 204 cm³.

EXAMPLE 2

This is an extract from a student's notebook. It describes how the student carried out an experiment to investigate the production of carbon dioxide by yeast at different temperatures.

What are my safety precautions?

a) *Chemicals.*

Using the CLEAPSS Student Safety Sheets, I have looked up

the hazards associated with the chemical I am using:

Glucose (solutions from 0.05 - 0.25 mol dm^{-3}): LOW HAZARD

Although it is only a low hazard, it is still best to wear eye

protection when using the solutions, especially as some of

the liquids will be hot. It is also important to handle all

chemicals carefully, and wipe up any spills of liquid.

COMMENT

The student has used a data source to look up the chemical hazards.

b) *Organisms.*

I found my information on yeast from the Fisher Scientific

website:

Dried yeast may cause eye, skin, and respiratory tract

irritation. It is expected to be a low hazard for usual

handling. If the dust is inhaled, you should 'remove from

exposure and move to fresh air immediately'.

I will handle the powdered yeast carefully when making up

my suspension, trying to avoid making any dust.

COMMENT

The student has used a data source to look up the biological hazards.

c) *Equipment*

I must be careful when using the water bath not to get water near the electrical sockets.

I need to handle the glassware (conical flask, gas syringe and glass tubing) carefully. In particular, I need to protect my hands with a towel (or glove) when linking together the glass delivery tubes from the rubber bung in the conical flask to the gas syringe with rubber tubing. The tubing needs to be lubricated with water and I need to hold my hands close together to limit the movement of glass if a break occurs.

COMMENT

The student has suggested some sensible precautions.

How much detail should I give in my description?

The student's method is given below:

1. *I would transfer 100 cm³ of 0.25 mol dm⁻³ glucose solution to each of six conical flasks.*

2. *I would then transfer the conical flask to a water bath at 20 °C. I would leave it for a few minutes to reach the temperature, which I would check with a thermometer.*

3. *I would then add 1.00 g of yeast, swirl the mixture to mix in the yeast, then start the stop watch.*

4. *I would then place the tubing on the bung to connect it to the gas syringe, then place the bung in the conical flask.*

5. *I would then record the volume of carbon dioxide in the gas syringe every minute.*

6. *I would calculate the average volume of carbon dioxide produced per minute.*

7. *I would repeat the experiment three times and calculate the average rate of carbon dioxide produced.*

8. *I would then carry out the investigation at 10 °C, 30 °C, 40 °C, 50 °C and 60 °C.*

9. *I would then draw a graph of the average rate of carbon dioxide production against temperature.*

COMMENT

The method is detailed and well-written. The student has appreciated that it is important for the sugar solution to reach the temperature being investigated before the yeast is added, and to only connect the gas syringe after the yeast has started respiring (otherwise the syringe will initially measure the simple expansion of air when heated).

3. MAKING AND RECORDING OBSERVATIONS AND MEASUREMENTS

Learning objective: to make observations and measurements with appropriate precision, record these methodically, and present them in a suitable form.

Questions to ask:

How many different measurements or observations do I need to take?

✓ Sufficient readings have been taken to ensure that the data are consistent.

✓ It is usual to repeat an experiment to get more than one measurement. If an investigator takes just one measurement, this may not be typical of what would normally happen when the experiment was carried out.

✓ When repeat readings are consistent they are said to be **repeatable**.

Do I need to repeat any measurements or observations that are anomalous?

✓ An **anomalous result** or **outlier** is a result that is not consistent with other results.

✓ You want to be sure a single result is accurate (as in the example below). So you will need to repeat the experiment until you get close agreement in the results you obtain.

✓ If an investigator has made repeat measurements, they would normally use these to calculate the arithmetical mean (or just mean or average) of these data to give a more accurate result. You calculate the mean by adding together all the measurements, and dividing by the number of measurements. Be careful though; anomalous results should not be included when taking averages.

✓ Anomalous results might be the consequence of an error made in measurement. But sometimes outliers are genuine results. If you think an outlier has been introduced by careless practical work, you should omit it when calculating the mean. But you should examine possible reasons carefully before just leaving it out.

✓ You are taking a number of readings in order to see a changing pattern. For example, measuring the volume of gas produced in a reaction every 10 seconds for 2 minutes (so 12 different readings). It is likely that you will plot your results onto a graph and then draw a **line of best fit**.

✓ You can often pick an anomalous reading out from a results table (or a graph if all the data points have been plotted, as well as the mean, to show the range of data). It may be a good idea to repeat this part of the practical again, but it is not necessary if the results show good consistency.

✓ If you are confident that you can draw a line of best fit through most of the points, it is not necessary to repeat any measurements that are obviously inaccurate. If, however, the pattern is not clear enough to draw a graph then readings will need to be repeated.

How should I record my measurements or observations – is a table the best way? What headings and units should I use?

✓ A table is often the best way to record results.

✓ Headings should be clear.

✓ If a table contains numerical data, do not forget to include units; data are meaningless without them.

✓ The units should be the same as those that are on the measuring equipment you are using.

✓ Sometimes you are recording observations that are not quantities. Putting observations in a table with headings is a good way of presenting this information.

EXAMPLE 3

How many different measurements or observations do I need to take?

A student cut a number of cylinders from a potato and weighed them. These were placed in sucrose solutions of different concentrations.

After one hour, the cylinders were removed, blotted dry and reweighed. The student calculated the percentage change in mass for each cylinder. The results are shown below.

Concentration of sucrose ($mol \; dm^{-3}$)	Percentage change in mass of potato cylinders				Average percentage change in mass	Texture of potato cylinders
	Experiment 1	Experiment 2	Experiment 3	Experiment 4		
0.0	+31.4	+33.7	+31.2	+32.5	+32.2	firm
0.2	+20.9	+33.4	+22.8	+21.3	+21.7	firm
0.4	−2.7	−1.8	−1.9	−2.4	−2.2	slightly soft
0.6	−13.9	−12.8	−13.7	−13.6	−13.5	soft
0.8	−20.2	−19.7	−19.3	−20.4	−19.9	floppy
1.0	−19.9	−20.3	−21.1	−20.3	−20.4	floppy

△ Table 6.3 Results for Example 3.

In this table of results:

✓ the description of each measurement is clear

✓ the units are given

✓ the data are recorded to the same number of decimal places, and decimal points are aligned

✓ calculations of means are recorded to the appropriate number of significant figures.

The student has recorded four measurements for each concentration investigated.

With the exception of the cylinder in Experiment 2 in a concentration of 0.2 $mol \; dm^{-3}$ sucrose (highlighted in the table), the repeats show good consistency.

Do I need to repeat any measurements or observations that are anomalous?

The result from the cylinder in Experiment 2 in a concentration of 0.2 $mol \; dm^{-3}$ sucrose is not consistent with the other results for this concentration. It an anomalous result and is highlighted in the table. The student has not included this result in the calculation of the mean for this concentration.

EXAMPLE 4

How should I record my measurements or observations?

Here are some results of food tests carried out by an investigator:

Food substance tested	Colour change on heating with Benedict's solution
10% glucose solution	blue → green → yellow → orange → red
Biscuit	blue → greenish blue
Grape	blue → green → yellow → orange → red
Honey	blue → green → yellow → orange → red-brown
Potato	remained blue

△ Table 6.4 Results of some food tests.

Note the clear table headings. The right-hand column does not simply say 'colour' but refers to colour *changes*.

COMMENT

Do not forget that some investigations might benefit from including both numerical data and observations, e.g. in the osmosis experiment in Example 3, the student also found it useful to include information on the firmness of the potato cylinders at the end of the experiment.

---

4. ANALYSING THE DATA AND DRAWING CONCLUSIONS

Learning objectives: to analyse and interpret data to draw conclusions from experimental activities which are consistent with the evidence, using biological knowledge and understanding, and to communicate these findings using appropriate specialist vocabulary, relevant calculations and graphs.

Questions to ask:

What is the best way to show the pattern in my results? Should I use a bar chart, line graph or scatter graph?

✓ Graphs are usually the best way of demonstrating trends in data.

✓ A bar chart or bar graph is used when one of the variables is a **categoric variable**, for example when one of the variables is the type of leaf, or species of organism.

✓ A line graph is used when both variables are **continuous**, e.g. time and temperature, time and volume.

✓ Scatter graphs can be shown to show the intensity of a relationship, or degree of *correlation*, between two variables.

✓ Sometimes a line of best fit is added to a scatter graph, but usually the points are left without a line.

When drawing bar charts or line graphs:

✓ Choose scales that take up most of the graph paper.

✓ Make sure the axes are linear and allow points to be plotted accurately, for example, it is much easier if one big square = 5 or 10 units, and not 3 units.

✓ Label the axes with the variables (with the independent variable on the x-axis).

✓ Make sure the axes have units.

✓ If more than one set of data is plotted use a key to distinguish the different data sets.

If I use a line graph, should I join the points with a line or a smooth curve?

✓ When you draw a line, do not just join the dots!

✓ Remember there may be some points that do not fall on the curve – these may be incorrect or anomalous results.

✓ A graph will often make it obvious which results are anomalous and so it would not be necessary to repeat the experiment.

✓ If following the biological rhythms of an organism over a period of time, you should join the data points, point-to-point.

Do I have to calculate anything from my results?

✓ It will be usual to calculate an **arithmetic mean (average)** from the data. More rarely, you might have to work out the **mode** (most common value) or **median** (middle value).

✓ Sometimes it is helpful to make other calculations, before plotting a graph (see Example 8). Other types of calculation might include:

  • the energy content of food *per gram* when burning a sample of food

  • the volume of water taken up by a plant from the distance moved by a bubble in a potometer (volume = $\pi r^2 \times$ distance)

  • the density of plants, for example as plants per m², or frequency of plants or animals, from quadrat data

  • conversions from one unit to another of a different magnitude, e.g. between millimetres (mm) and micrometres (µm)

  • rearrangement of algebraic equations to solve them.

✓ Investigators also look for numerical trends in data, for example, the doubling of a reaction rate every 10 °C, or the doubling of numbers of microorganisms every 20 minutes.

✓ Sometimes you will have to make some calculations before you can draw any conclusions.

Can I draw a conclusion from my analysis of the results, and what biological knowledge and understanding can be used to explain the conclusion?

✓ You need to use your biological knowledge and understanding to explain your conclusion.

✓ It is important to be able to add some explanation which refers to relevant scientific ideas in order to justify your conclusion.

EXAMPLE 5

What is the best way to show the pattern in my results?

A student did an experiment to compare the loss of water from leaves of three different species of tree – hazel, lime and oak.

The student measured the mass of 10 leaves of similar size and hung the leaves on a line. After three hours, the student removed the leaves and measured the masses of the leaves again.

Species	Average loss of water (g/hour)
Apple	0.30
Hazel	0.05
Oak	0.01

△ Table 6.5 Results of Example 5.

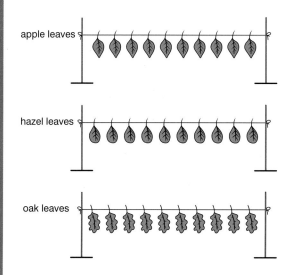

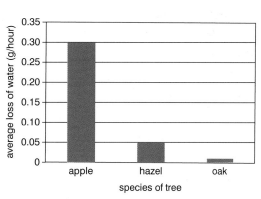

△ Fig. 6.5 Bar chart showing water loss from different leaves.

A bar chart or bar graph is used to display the data in this instance, as the type of leaf is a categoric variable.

EXAMPLE 6

What is the best way to show the pattern in my results?

A student investigated the effect of different light intensities on photosynthesis in pondweed.

The student measured the oxygen collected over a number of days.

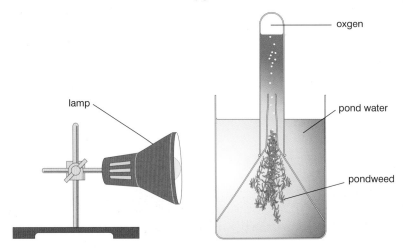

△ Fig. 6.6 Apparatus for Example 6.

A line graph is needed. as both the volume of gas and time are continuous variables.

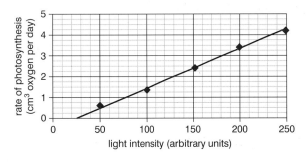

◁ Fig. 6.7 Experimental results of Example 6.

If I use a line graph, should I join the points with a line or a smooth curve?

In this case it is clear that the line of best fit should be a straight line.

EXAMPLE 7

What is the best way to show the pattern in my results?

A student collected a set of data from an investigation on the effect of the application of different amounts of fertiliser on plots of a plant on the school grounds. The student used a scatter graph to display the data.

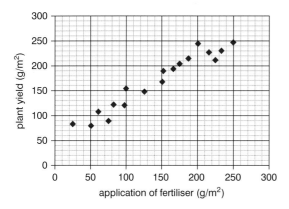

△ Fig. 6.8 Experimental results for Example 7.

EXAMPLE 8

If I use a line graph, should I join the points with a line or a smooth curve?

In an investigation on the activity of an enzyme at different temperatures, a student obtained a set of results that shows different phases.

Photographic film is made from a sheet of plastic coated with light-sensitive silver particles bonded by the protein gelatine. When the gelatine is broken down by a protease, the silver particles fall off, and the film becomes clear.

Equal-sized strips of photographic film were placed in a series of test tubes. An identical volume of protease was added to the tubes, each of which was incubated at a different temperature.

The amount of time taken for each strip of film to become clear was measured and recorded.

Temperature (°C)	Average time taken for breakdown of gelatine (s)
4	3450
13	667
25	175
30	130
40	133
50	7140

△ Table 6.6 Table of results for Example 8.

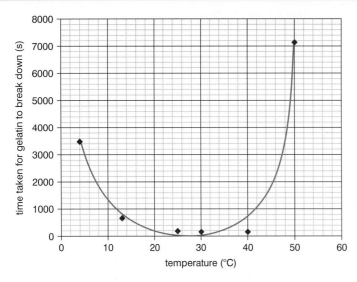

△ Fig. 6.9 Results for Example 8 with line of best fit plotted.

Do I have to calculate anything from my results?

The trend of how enzyme activity is affected by temperature is not well illustrated by the graph. It is shown better if the *rate of reaction* is calculated and plotted against temperature. The rate is the inverse of the time taken to break down the gelatine, i.e. 1 ÷ the time taken.

Temperature (°C)	Average time taken for breakdown of gelatine (s)	Average rate of breakdown of gelatine (=1/time)(s⁻¹)
4	3450	0.00029
13	667	0.00150
25	175	0.00571
30	130	0.00769
40	133	0.00752
50	7140	0.00014

△ Table 6.7 Table of results with rate column added.

The graph on page 399 shows one way of plotting the rate of breakdown of gelatine (rate of reaction) using the values as shown in the table above. Sometimes it can be easier to convert very small numbers (or very large numbers) to **standard form** before plotting. In this case for example you might convert all the numbers to: 2.9×10^{-4}, 15.0×10^{-4}, and so on.

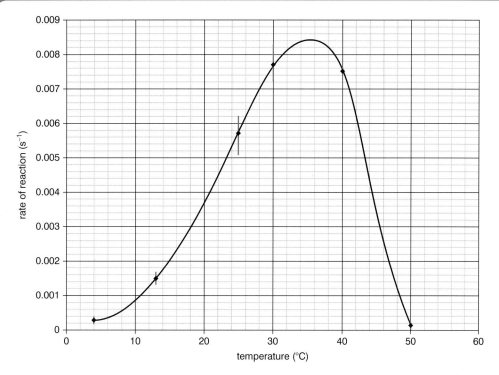

△ Fig. 6.10 Rate of reaction for Example 8.

Note that on this graph, the student has also drawn range bars to show how the data points are arranged around the mean.

Rates can also be calculated by looking at the gradient (steepness) of a line graph, or part of a graph.

Can I draw a conclusion from my analysis of the results?

The student wrote:

Enzymes work best at a particular temperature. My graph suggests that the optimum temperature for protease is around 35 °C. At lower temperatures, enzymes work slowly because the molecules have less energy and move around more slowly, so there are fewer successful collisions between enzyme and substrate (gelatine) molecules.

Enzymes work by a lock and key mechanism, with the substrate fitting into the enzyme. At temperatures that are too high, the structure of an enzyme will be changed so that it will not work. This change is irreversible and the enzyme is said to be denatured.

This is a good, concise conclusion and links the data to the mechanism of enzyme action.

5. EVALUATING THE DATA AND METHODS USED

Learning objective: to evaluate data and methods.

Questions to ask:

Do any of my results stand out as being inaccurate?

✓ You need to look for any anomalous results or outliers that do not fit the pattern.

✓ You can often pick this out from a results table (or a graph if all the data points have been plotted, as well as the mean, to show the range of data). In Example 3, the results from the cylinder in Experiment 2 in a concentration of 0.2 mol dm^{-3} sucrose are not consistent with the other results for this concentration. The anomalous result or outlier is highlighted in the table. The student has not included this result in the calculation of the mean for this concentration. It may be a good idea to repeat this part of the practical again, but it is not necessary if the other results show good consistency.

What reasons can I give for any inaccurate results?

✓ When answering questions like this it is important to be specific: answers such as 'experimental error' will not score any marks.

✓ It is often possible to look at the practical technique and suggest explanations for anomalous results.

✓ When you carry out the experiment you will have a better idea of which possible sources of error are more likely.

✓ Try to give a specific source of error and avoid statements such as 'the measurements must have been wrong'.

Your conclusion will be based on your findings, but must take into consideration any uncertainty in these introduced by any possible sources of error. You should discuss where these have come from in your evaluation.

Error is a difference between a measurement you make, and its true value.

The two types of errors are:

✓ random error

✓ systematic error.

With **random error**, measurements vary in an unpredictable way. This can occur when the instrument you are using to measure lacks sufficient precision to indicate differences in readings. It can also occur when it is difficult to make a measurement. If two investigators measure the height of a plant, for instance, they might choose different points on the soil, and the tip of the growing point to make their measurements.

With **systematic error**, readings vary in a controlled way. They're either consistently too high or too low. One reason could be down to the way you are making a reading, e.g. taking a burette reading at the wrong point on the meniscus, or not being directly in front of an instrument when reading from it.

What an investigator *should not* discuss in an evaluation are problems introduced by using faulty equipment, or by using the equipment inappropriately. These errors can be, or could have been, eliminated, by:

✓ checking equipment

✓ practising techniques before the investigation, and taking care and having patience when carrying out the practical.

Overall, was the method or technique I used accurate enough?

✓ If your results were good enough to provide a confident answer to the problem you were investigating the method probably was good enough.

✓ If you realise your results are not precise when you compare your conclusion with the actual answer it may be you have a systematic error (an error that has been made in obtaining all the results). A systematic error would indicate an overall problem with the experimental method.

✓ If your results do not show a convincing pattern then it is fair to assume that your method or technique was not precise enough and there may have been a random error (i.e. measurements vary in an unpredictable way).

If I were to do the investigation again, what would I change or improve upon?

✓ Having identified possible errors it is important to say how these could be overcome. Again you should try and be absolutely precise.

✓ When suggesting improvements, do not just say 'do it more accurately next time' or 'measure the volumes more accurately next time'.

✓ For example, if you were measuring small volumes, you could improve the method by using a burette to measure the volumes rather than a measuring cylinder.

✓ Other errors arise from accuracy of measurement. An investigation can also often be improved by extending the range, e.g. temperature, time, pH, etc., over which it is carried out.

EXAMPLE 9

What reasons can I give for any inaccurate results?

In this example, the pH of yoghurt was monitored during its production using a data logger. The data logger gave very precise, consistent readings. Before the experiment started, the data logger had to be set, or *calibrated*, using solutions of known pH (called buffers).

In Example 3, it is possible that the potato cylinder had not been blotted dry properly before its mass was measured.

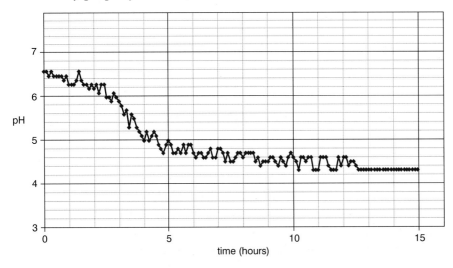

△ Fig. 6.11 Experimental results for Example 9.

EXAMPLE 10

In an investigation on the energy content of food, an investigator used the apparatus below to find the energy content of a piece of pasta.

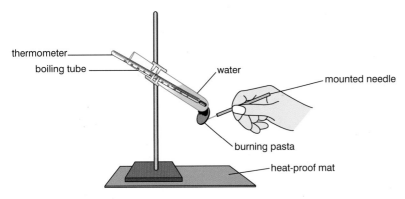

△ Fig. 6.12 Apparatus for Example 10.

Was the method or technique I used accurate enough?

The value obtained by the investigator was 1272 joules of energy per gram of pasta. This value was much lower than the value of 13 440 joules per gram of pasta printed on the packet.

The main problem with this method is that it relies on the transfer of all the chemical energy in the food to heat energy, which is used to warm the water. But using this equipment, the transfer is nowhere near 100 percent efficient. Reasons for this are:

✓ conversion of the chemical energy to other forms of energy

✓ incomplete combustion of the food

✓ poor transfer of heat from the burning pasta to the water.

Measurements of the temperature rise may also be inaccurate because the heat energy is not evenly distributed through the water.

If I were to do the investigation again, what would I change or improve upon?

It is possible to improve the accuracy of energy values of food in the school laboratory by modifying the method above or using different equipment. The transfer of heat to the container holding the water (in this case, a boiling tube) can be improved, and the container could also be insulated. Many investigators use a device called a bomb calorimeter to overcome these problems. The food is burned inside the bomb calorimeter.

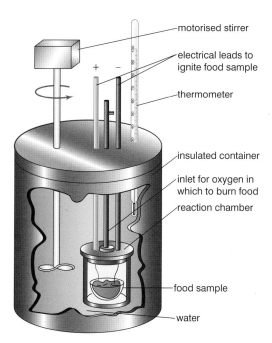

△ Fig. 6.13 Improved apparatus for Example 10.

Mathematical skills

The table below shows the mathematical skills that you will need to use during your International GCSE course. You will also need to be able to use these skills during your examinations.

		B
1	**Arithmetic and numerical computation**	
A	Recognise and use numbers in decimal form	✓
B	Recognise and use numbers in standard form	✓
C	Use ratios, fractions, percentages, powers and roots	✓
D	Make estimates of the results of simple calculations, without using a calculator	✓
2	**Handling data**	
A	Use an appropriate number of significant figures	✓
B	Understand and find the arithmetic mean (average)	✓
C	Construct and interpret bar charts	✓
D	Construct and interpret frequency tables, diagrams and histograms	✓
E	Understand the principles of sampling as applied to scientific data	✓
F	Understand simple probability	✓
G	Understand the terms mode and median	✓
H	Use a scatter diagram to identify a pattern or trend between two variables	✓
I	Make order of magnitude calculations	✓
3	**Algebra**	
A	Change the subject of an equation	✓
B	Substitute numerical values into algebraic equations using appropriate units for physical quantities	✓
C	Solve simple algebraic equations	✓

4	**Graphs**	
A	Translate information between graphical and numerical form	✓
B	Plot two variables (discrete and continuous) from experimental or other data	✓
C	Determine the slope and intercept of a linear graph	✓
5	**Geometry and trigonometry**	
C	Calculate areas of triangles and rectangles, surface areas and volumes of cubes	✓

Glossary

abiotic factors Non-living or physical factors, such as temperature or rainfall, that affect the population sizes and distributions of organisms.

absorption The movement of digested food molecules from the gut into the blood stream.

abundance The number of organisms of a particular species in an area, related to the population size.

acid rain Rain that contains higher than normal concentrations of dissolved acidic gases, such as sulfur dioxide and nitrogen oxides, which causes the rain to have a lower pH than normal.

active site A 3D-shaped space in the molecule of an enzyme that matches the shape of another molecule that reacts with it (the substrate).

active transport The movement of molecules across a cell membrane using energy from respiration. This movement is often against a concentration gradient.

adaptation The features or characteristics of an organism that makes it suited to its particular environment.

adenine(A) A base found in DNA and RNA; it pairs with thymine (T).

ADH (antidiuretic hormone) The hormone involved in regulating water content in the blood by changing the permeability of the collecting duct of nephrons.

adrenaline A hormone, produced by the adrenal glands, that stimulates heart rate as preparation for 'fight or flight'.

adrenal glands Glands that produce the hormone adrenaline.

aerobic respiration Respiration (the breakdown of glucose to release energy) using oxygen.

agitation Mixing up, such as in a fermenter to keep the microorganisms in contact with the ideal amounts of nutrients and oxygen.

alimentary canal The tubular part of the digestive system, from mouth to anus, that food passes along.

allele One form of a gene, producing one form of the characteristic that the gene produces.

alveoli (single: alveolus) The air sacs in lungs where gases diffuse between the air in the lungs and the blood.

amino acid The basic unit of a protein.

amniotic fluid Fluid surrounding developing fetus in the uterus, which protects the fetus from mechanical damage.

amylase A digestive enzyme that breaks down starch to maltose.

anaerobic respiration Respiration (the breakdown of glucose to release energy) without oxygen. In animal cells it produces lactic acid; in plant and fungi cells it produces ethanol and carbon dioxide.

anther The male part of a flower that produces pollen.

antibiotic A chemical that kills bacteria.

antibiotic resistance Resistance to the effect of an antibiotic by bacteria which are normally killed by that antibiotic.

antibodies Chemicals produced by lymphocytes to defend against infection by pathogens. Antibodies are developed to be specific to different pathogens.

anticodon A section of three bases on a tRNA molecule that joins with the codon on the mRNA during protein synthesis.

artery A blood vessel that carries blood away from the heart.

aseptic precautions Steps that involve killing all unwanted microorganisms to prevent their growth, such as in a fermenter.

asexual reproduction The production of new individuals without fertilisation, from division of body cells in the parent.

ATP (adenosine triphosphate) A substance made when sugar is broken down in respiration. It is ATP that directly provides the energy that cells need.

atrium (plural: atria) One of two chambers of the heart that receive blood from veins and pump it into the ventricles.

auxin A plant hormone that stimulates growth in the growing tip.

balanced diet Intake of food that supplies all the protein, lipid, carbohydrate, vitamins, minerals, water and dietary fibre that the body needs in the right amounts.

base One of four molecules (adenine [A], thymine [T], cytosine [C] and guanine [G]) that join in pairs (A with T, C with G) that link the two strands within DNA.

biconcave The shape of a red blood cell in which the middle is pressed inwards, making the cell thinner in the middle than at the edges.

bile A highly alkaline liquid, produced by the liver and stored in the gall bladder, which emulsifies lipids.

bioaccumulation When a toxic substance, such as a pesticide, is stored in living tissue.

biodiversity A measure of the variety of organisms in an area.

biological catalyst Catalysts in living cells that control metabolic reactions i.e. enzymes.

biological control Using organisms, such as a predator or parasite, to control the numbers of pests.

biomass The mass of living material, such as the mass of a living organism.

biotic factors Factors that affect the distribution of organisms, caused by other organisms, such as competition for food.

bladder Where urine is held in the body before it is released into the environment.

Bowman's capsule The cup-shaped structure at the start of a nephron where ultrafiltration occurs.

brain Organ that together with the spinal cord makes up the central nervous system.

breed Animals of the same species but with distinct characteristics of their own.

bronchi (single: bronchus) The two divisions of the trachea as it joins to the lungs.

bronchioles The tiny tubes in the lungs that carry air to the alveoli.

cancer The uncontrolled division of cells.

capillaries Tiny blood vessels in the tissues. Every cell in the body is in very close proximity to a capillary, enabling gaseous exchange.

carbohydrate Molecules such as starch, glycogen or sugars, often used to provide energy.

carbon cycle The movement of different forms of the element carbon between living organisms and the environment, represented as a diagram.

carcinogenic Something that causes cancer (the uncontrolled division of cells).

carnivore A meat-eater, such as a secondary consumer or a tertiary consumer.

carpel The female structure in flowers which contains one or more ovaries and their stigmas and styles.

catalyst A substance that increases the rate of a chemical reaction, such as an enzyme.

cell The 'building blocks' of which tissues are composed; some microorganisms are just one cell.

cell differentiation When cells develop into new types of specialised cell.

cell membrane The structure surrounding cells that controls what enters and leaves the cell.

cell sap The liquid inside plant cell vacuoles containing water and dissolved substances.

cell wall A wall around cells, giving them support and shape, made of cellulose in plants, and chitin in fungi.

cellulose The material that makes up a plant cell wall.

central nervous system The part of the nervous system that coordinates and controls responses, consisting of brain and spinal cord.

chemical digestion The breakdown of large molecules into smaller ones using enzymes.

chitin A fibrous carbohydrate that makes up the cell walls of fungi.

chlorophyll The green chemical in chloroplasts that captures light energy for photosynthesis.

chloroplast An organelle found only in plant (and some protoctist) cells, where photosynthesis takes place in the cell.

chromosome A long DNA molecule that is found in a cell nucleus.

cilia Tiny hairs lining the trachea, bronchi and bronchioles which help to remove dirt and microorganisms.

coordination Detecting and responding appropriately to a stimulus.

clone A genetically identical copy.

codominance When both alleles for a gene are expressed (have an effect) in the phenotype.

codon A sequence of 3 bases on DNA or mRNA that codes for a specific amino acid.

collecting duct Tube that connects a nephron to a ureter in the kidneys.

colon Main part of the large intestine.

combustion Burning, such as the burning of fossil fuels.

community All the populations of organisms that live in an area or ecosystem.

compensation point The point at which the rate of photosynthesis and respiration in a plant are equal, so that there is no net production or uptake of carbon dioxide or oxygen.

competition The battle for available resources between organisms that live in the same place.

concentration gradient The difference in the concentration of a substance between two areas. Diffusion is usually down the concentration gradient: molecules move from the area of high concentration to the area of low concentration.

cone cells Light sensitive cells in the retina that respond to different wavelengths of light; responsible for colour vision.

conservation The processes of protecting an environment or habitat from change, so that the natural state is maintained and the organisms that live there can flourish.

consumer (primary/secondary) An animal in a food web. Primary consumers eat plants; secondary consumers eat primary consumers.

convoluted tubules Two sections of the nephron closely associated with a capillary. Selective reabsorption of glucose takes place in the proximal convoluted tubule.

core temperature The internal temperature of the body, about 37 °C, necessary for life processes to go on effectively.

cornea The thick clear part of the eye that bends light rays that enter the eye, focusing them on the retina.

coronary heart disease The narrowing of coronary arteries and formation of blood clots as a result of smoking; it can lead to a heart attack.

cotyledon A food store surrounding the zygote in a plant.

crop rotation Growing a different crop each season in a particular field, alternating leguminous (pea/bean) and non-leguminous crops, to keep the soil fertile.

cutting A part taken from a plant (such as piece of stem, root or leaf) and treated so that it grows into a new plant, a form of artificial asexual reproduction of plants.

cytoplasm The jelly-like liquid inside the cell which contains the organelles and where many chemical reactions take place.

cytosine (C) A base found in DNA and RNA; it pairs with guanine (G).

daughter cell A cell produced by division of a parent cell.

decomposer An organism that causes decay of dead material, such as many fungi and bacteria.

decomposition The decay of dead plants and animals and their waste materials.

deforestation The destruction of large areas of forest or woodland.

denatured When the shape of an enzyme's active site is permanently changed and the enzyme can no longer function.

denitrifying bacteria Bacteria that convert nitrates in the soil to nitrogen gas which is released to the atmosphere.

dialysis A technique used to filter the blood artificially if the kidneys do not function adequately.

diaphragm The muscular sheet at the bottom of the lungs that controls breathing.

diffusion The net movement of molecules down a concentration gradient; it is a passive process (it does not use energy).

digestion The breakdown of food; physical or mechanical digestion breaks food down into smaller pieces (such as by chewing and the muscular action of the stomach) ready for chemical digestion by enzymes.

diploid A cell that contains two sets of chromosomes.

distal convoluted tubule Part of each nephron in the kidney, connected to the proximal convoluted tubule by the loop of Henle.

distribution How organisms are spread out in an area, where they are found.

DNA (deoxyribonucleic acid) The chemical that forms chromosomes and carries the genetic code.

dominant allele The allele, which if present, is expressed in the phenotype (such as spotted coat in leopards).

dormant In a suspended state; seeds stay dormant until conditions are right for germination.

double helix The shape of the DNA molecule, rather like a twisted ladder.

duodenum The first part of the small intestine.

ecology The study of living organisms and their environment.

ecosystem All the organisms (biotic factors) and physical (abiotic) factors in a fairly self-contained area, such as a lake or desert.

effector The part of the body that brings about a response to a stimulus.

egestion The removal of undigested material from the body (faeces). (compare with excretion)

egg cell The female gamete in animals and plants.

electrical impulse The form in which information is sent along nerves.

embryo A developing young, where cell division and differentiation are taking place rapidly. In plants it develops in the seed. In humans, the embryo is the stage between zygote and fetus.

emulsify To break up the large droplets of a lipid in an aqueous solution into smaller droplets.

endocrine gland A collection of cells that secrete hormones into the blood.

enhanced greenhouse effect An increase in the greenhouse effect, most likely caused by the release of additional greenhouse gases from human activity.

enucleate To remove the nucleus from a cell.

environment Organisms and the factors that affect them.

enzyme A protein that acts as a catalyst, speeding up reactions.

epidermis (or epidermal tissue in plants) The layer of cells on the outer surface of a body or organ, such as a leaf.

eukaryote An organism whose cells contain a nucleus and other organelles such as mitochondria and chloroplasts. Plants, animals, fungi and protoctists are all eukaryotic organisms.

eutrophication The addition of nutrients to water, which may lead to water pollution.

evaporation When particles in a liquid (such as water) gain enough energy to move fast enough and become a gas (as in water vapour).

evapotranspiration The transfer of water from the ground to the atmosphere by the combined effect of both evaporation from the soil and transpiration from plants.

evolution The change in the characteristics of a species over time; this occurs through the process of natural selection.

excretion Removal of waste (often toxic) substances that have been produced from chemical reactions inside the body, such as carbon dioxide and urea in animals.

exhalation Breathing out.

explants Tiny pieces of plant used in micropropagation.

extinction When an entire species ceases to exist.

extracellular Outside a cell, for example, extracellular secretion or extracellular digestion.

faeces The undigested material that remains after food is digested and absorbed in humans and other animals.

family pedigree A diagram that shows the inheritance of different forms of a characteristic through the generations within a family.

fatty acid One of the basic units of a lipid, along with glycerol.

fermenter A large industrial vessel that provides the optimum conditions for the growth of microorganisms.

fertilisation Fusion (joining) of male and female sex cells.

fertiliser Added to soil to provide mineral ions to improve plant growth, either natural fertiliser (e.g. compost or manure) or artificial fertiliser.

fetus The name given to the developing baby in the uterus.

fibre (dietary fibre) A plant material that is difficult to digest and keeps the food in the alimentary canal soft and bulky, aiding peristalsis.

food chain/web Diagram showing the flow of energy between organisms; a food chain is a simple 'line' (plant, primary consumer, secondary consumer); food webs are a combination of food chains.

fossil fuels Fuels formed from organic material, such as peat, coal, oil and natural gas.

fruit The usually soft, fleshy structure surrounding a seed of a plant, formed from the ovary after fertilisation.

FSH (follicle-stimulating hormone) A hormone produced by the pituitary gland that helps control the menstrual cycle.

gamete A sex cell.

gas exchange In the lungs, diffusion of oxygen from alveoli into blood and of carbon dioxide from blood into alveoli.

gene A section of DNA that codes for a specific protein to produce a particular characteristic.

genetic code The code formed by the order of the bases in DNA that instructs cells how characteristics should be produced; it does this by telling cells how to make particular proteins.

genetic diagram A diagram that displays how a characteristic may be inherited by offspring from the parents' alleles.

genetic engineering / genetic modification The transfer of a gene from one organism into another so the gene can be expressed.

genome All of an organism's DNA or genes.

genotype The description of an organism's alleles for a particular characteristic; its genetic makeup.

geotropism A growth response in plants in response to gravity.

germination The start of plant growth from a seed, which only occurs when there is the right amount of water, oxygen and an appropriate temperature.

gestation The time when a fetus develops in the uterus, about 40 weeks in humans.

global warming Warming of the Earth's surface and atmosphere, possibly as a result of an enhanced greenhouse effect.

glomerular filtrate The liquid in the Bowman's capsule produced by ultrafiltration.

glomerulus A small knot of capillaries associated with a Bowman's capsule.

glucose A simple sugar.

glycerol One of the basic units of a lipid, along with fatty acids.

glycogen A carbohydrate stored in animals and fungi. It is made from excess glucose.

grazing When plants are eaten by animals.

greenhouse effect The warming effect caused by greenhouse gases in the atmosphere that prevent some of the heat energy radiated by the Earth's surface escaping into space.

greenhouse gases Atmospheric gases, such as carbon dioxide and water vapour, that trap some of the heat radiated from the Earth's surface and prevent it escaping into space.

growth The permanent increase in body size and dry mass of an organism, usually from an increase in cell number or cell size (or both).

guanine (G) A base found in DNA and RNA; it pairs with cytosine (C).

habitat A small part of an ecosystem where a species lives.

haemoglobin The red chemical in red blood cells that combines reversibly with oxygen.

haploid A cell that contains only one set of chromosomes, such as gametes.

heart rate The number of heart beats in a given time, for example, beats per minute.

herbivore A plant-eater or primary consumer.

heterozygous When the two alleles for a gene are different in the genotype.

homeostasis The maintenance of a constant internal environment in the body (such as constant temperature, water balance, etc.).

homozygous When the two alleles for a gene are the same in the genotype.

hormonal system A chemical response system in humans where hormones produced by the endocrine glands are carried in the blood to target organs where they affect the cells.

hormone A chemical that is released from an endocrine gland and travels in the blood to another site in the body (its target organ), where it has its effect; hormones are described as chemical messengers.

humidity A measure of the concentration of water molecules in the air.

hypha (plural: hyphae) A single thread of fungal mycelium.

hypothalamus The part of the brain that monitors the water content of the blood.

ileum The last part of the small intestine.

immune system The system that protects the body against infection and includes white blood cells.

ingestion The taking of food into the alimentary canal.

inhalation Breathing in.

inherited Passed on from parent to offspring via the genes.

insulin A hormone produced by the pancreas that causes muscle and liver cells to take glucose from the blood.

interspecific predation Predation of one species by another, for example predator and prey.

intercostal muscles The muscles between the ribs that can control breathing.

intraspecific predation Predation of individuals in a species by other individuals of the same species.

in vitro A laboratory process taking place in a test tube or dish.

ionising radiation Radiation, such as gamma rays, x rays and ultraviolet radiation, that can damage cells and produce mutations in genes.

kidneys Organs involved in excretion; these are where urea, excess water and excess ions are removed from the blood.

Lactobacillus The type of bacteria that are used to turn milk into yoghurt.

leaching The loss of dissolved mineral nutrients in soil water as it soaks deep into the ground beyond the reach of plant roots.

LH (luteinising hormone) A hormone produced by the pituitary gland that helps control the menstrual cycle.

ligase enzyme An enzyme that joins pieces of DNA together, used in genetic engineering.

limiting factor A factor (such as light, temperature) that controls the rate of photosynthesis; it is the condition that is least favourable.

line transect A line along which quadrats are placed to sample organisms.

lipase A digestive enzyme that breaks down lipids to fatty acids and glycerol.

lipid A fat or oil, such as cholesterol, often made of the basic units of three fatty acids and one glycerol.

loop of Henle A large loop of the nephron in the kidney between the proximal and distal convoluted tubules.

lungs Organs involved in breathing; they contain many alveoli where gas exchange occurs.

lymphocytes A type of white blood cell that makes antibodies to attack a pathogen.

maltase A digestive enzyme that breaks down maltose to glucose.

maltose A small carbohydrate molecule.

meiosis The form of cell division that produces four genetically different haploid cells from a diploid parent cell, producing gametes.

memory cell A type of blood cell produced by lymphocytes that 'remembers' pathogens that have previously been destroyed by antibodies, and can quickly make the same antibodies in large quantities so the same pathogens cannot re-infect.

menstrual cycle The continuous sequence of events in a woman's reproductive organs; each cycle of ovulation (ripening and release of an egg) and menstruation (shedding of the unwanted uterus lining) takes about 28 days and is controlled by a number of hormones.

mesophyll Tissue in plants that packs the spaces between other tissues.

metabolic reactions The chemical reactions that cause the life processes of organisms to occur.

microbes / microorganisms Tiny, usually single-celled, organisms (such as bacteria); some cause disease.

micropropagation The culture of explants in a laboratory to produce many clones of a single plant.

microvilli Tiny finger-like extensions of the cell membrane of the surface cells of villi.

mineral ions (minerals) Nutrients that plants and animals need in small amounts, such as nitrates that are needed for making amino acids, or iron that is needed to make new red blood cells.

mitochondria (single: mitochondrion) Cell organelle in which respiration takes place.

mitosis The form of cell division that produces two identical diploid daughter cells from a diploid body cell, used for growth and repair in the body and in cloning and asexual reproduction.

molecule A particle, composed of several atoms joined together, that makes up substances like water, carbon dioxide and food.

monohybrid cross The inheritance of a characteristic produced by one gene.

mRNA (messenger RNA) Molecules of RNA that are copies of the DNA code for a particular protein, that travel to the ribosomes for protein synthesis.

multicellular Organisms made of more than one cell, often many millions.

mutagen A chemical that produces mutations in genes.

mutation A random change in a gene, sometimes producing a new allele.

mycelium A mass of hyphae that form the body of a fungus.

natural selection The process by which evolution happens, as first described by Charles Darwin; it is the influence of the environment on survival and/or reproduction, so that organisms with some characteristics are more successful at producing offspring than others.

nephron A tiny kidney tubule where ultrafiltration and reabsorption take place to produce urine.

nerves Bundles of cells that connect receptors to the central nervous system and the central nervous system to effectors.

nervous system A response system in humans that uses electrical impulses between receptor cells, nerve cells and effector cells to produce a response to a stimulus.

net movement When particles are moving in different directions but there are more moving in one direction than another.

neurone A nerve cell, specially adapted for carrying nerve impulses.

neurotransmitters Chemicals that pass from one neurone to another across a synapse.

nitrifying bacteria Soil bacteria that convert ammonium ions to nitrite ions, and nitrite ions to nitrate ions, important in the nitrogen cycle.

nitrogen cycle The movement of the element nitrogen in different forms between living organisms and the environment, shown as a diagram.

nitrogen-fixing bacteria Bacteria found in the soil and in root nodules of some plants that can convert atmospheric nitrogen into a form of nitrogen that plants can use.

nucleic acid DNA and RNA are types of nucleic acid.

nucleus The organelle in plant and animal cells that contains the genetic material.

nutrition The process by which an organism gets the chemicals and energy it needs (nutrients) from its surroundings.

oesophagus (gullet) Tube that takes food from the mouth to the stomach.

oestrogen A hormone produced by the ovaries that helps to control the menstrual cycle and produces secondary sexual characteristics in girls.

omnivore An animal that eats both plants and animals.

organ A group of tissues that work together to carry out a particular function, such as the stomach or the heart.

organelle A structure within a cell that carries out a particular function, such as nucleus, vacuole.

organism A living thing, ranging in size from a single-celled microorganism to an elephant.

osmoregulation The regulation of the concentration of water in the blood.

osmosis The movement of water molecules through a partially permeable membrane, from a weak solution to a stronger solution.

ovary (in plants and humans) A structure that contains immature egg cells.

ovulation In a woman's menstrual cycle, when an egg is released from the ovary, usually about halfway through the cycle.

ovule The female structure in flowers that contains one egg cell.

oxygenation The adding of oxygen to water, usually by bubbling air through it.

oxygen debt The result of anaerobic respiration; extra oxygen is needed after exercise to break down the lactic acid formed during anaerobic respiration.

palisade cell Cells in the upper part of a leaf that contain the most chloroplasts and carry out most of the photosynthesis.

pancreas Organ that secretes digestive enzymes and the hormone insulin.

parasite An entity that lives on or in another organism. Viruses are parasites, as well as some animals, plants, fungi and protoctists.

partially permeable Describing a membrane (such as the cell membrane) that has tiny pores (holes) which allow some molecules (such as water) but not larger molecules to pass through.

passive The opposite of *active*: something that happens without the need for additional energy.

pathogen An organism that causes disease in another living organism. Examples are found in fungi, bacteria, protoctists and viruses.

peristalsis The rhythmic muscular contractions of the walls of the alimentary canal that move food along.

pest An organism that causes damage to crop plants or farm animals.

pesticide A chemical used to kill pests.

phagocyte A type of white blood cell that engulfs and destroys pathogens.

phenotype The physical characteristics of an organism; it is affected by both an organism's genes and its environment.

phloem Tubes formed from many living cells that carry dissolved substances, such as sucrose and amino acids, for example from the leaves to other parts of a plant.

photosynthesis The chemical process in which plants use light, water and carbon dioxide to create glucose and oxygen.

phototropism A growth response in plants in response to light.

physical digestion The breakdown of large food pieces, such as by chewing in the mouth.

pituitary gland Gland that releases the hormone ADH. NB – This needs to be marked as Separate Biology Only.

placenta The structure formed by a fetus that attaches to the uterus wall and exchanges essential substances between the mother's blood and the blood of the fetus.

plasma The liquid, watery part of blood which carries dissolved food molecules, urea, hormones, carbon dioxide and other substances around the body and also helps to distribute heat energy.

plasmid A small circle of genetic material found in some bacteria in addition to the circular chromosome.

platelets Fragments of much larger cells that cause blood clots to form at sites of damage in blood vessels.

pleural membranes The two membranes surrounding the lungs.

pollen grain A male structure in plants that contains the male gamete.

pollen tube A tube that develops from a pollen grain down through the style, carrying the male gamete to the female gamete.

pollination The process in which pollen from the anther of one flower is transferred to the stigma of another flower, before fertilisation can take place.

pollution Damage to the environment, people and other organisms, often as a result of adding chemicals to the air, water or land.

polygenic Characteristics that are controlled by the alleles of more than one gene.

population All the organisms of one species living in the same habitat.

predation When prey animals are killed and eaten by predators.

predator An animal that kills and eats prey animals.

prey An animal that is killed and eaten by a predator.

primary consumer An animal that eats plants (also called a herbivore).

producer An organism that produces its own food, such as plants using light energy in photosynthesis to produce glucose.

progesterone A hormone produced in the ovaries that helps to control the menstrual cycle.

prokaryote An organism whose cells do not contain a nucleus or organelles such as mitochondria or chloroplasts. Bacteria are prokaryotic organisms.

protease A digestive enzyme that breaks down proteins to amino acids.

proteins Large molecules, such as found in muscle tissue, made of many amino acids.

protein synthesis The process in a cell by which new proteins are made, occurs in the ribosomes.

proximal convoluted tubule Part of each nephron in the kidney, between the Bowman's capsule and the loop of Henle.

Punnett square A form of genetic diagram.

pyramid of biomass A diagram that shows the biomass in different trophic levels of a food chain, often a pyramid shape.

pyramid of energy A diagram that shows the energy content of different trophic levels of a food chain, always a pyramid shape.

pyramid of number A diagram that shows the number of individual organisms in different trophic levels of a food chain, often a pyramid shape.

quadrat A square frame used for sampling the abundance and distribution of organisms.

receptor A structure that detects a stimulus.

recessive allele The allele that is only expressed if no corresponding dominant alleles are present.

recombinant DNA DNA that contains a gene inserted during genetic engineering.

rectum Last part of the large intestine.

reflex The simplest response to a stimulus, such as blinking.

reflex arc The pathway along which nerve impulses travel during a simple reflex (such as withdrawing your hand from a hot object).

reproduction (asexual/sexual) The process of creating new members of a species; asexual reproduction does not involve sex cells; sexual reproduction is the combination of sex cells.

respiration The chemical process in which glucose is broken down inside the mitochondria in cells, releasing energy in the form of ATP and producing carbon dioxide and water.

respiratory system The body system that includes the lungs, diaphragm and other organs involved in breathing.

restriction enzyme An enzyme that cuts DNA at a specific site, used in genetic engineering.

retina The part of the human eye, at the back of the eye, containing light-sensitive cells.

ribosome Very small cell organelles in the cytoplasm, where protein synthesis occurs.

RNA (ribonucleic acid) The chemical that mRNA, tRNA and ribosomes are made from; it is similar to DNA but is usually single stranded and contains the base uracil (U) instead of thymine (T); carries the genetic code in some viruses.

rod cells Cells in the retina that are sensitive to light intensity but cannot distinguish different colours; responsible for night vision.

root hair cells Cells in the epidermis of roots that have a long extension of cytoplasm to increase the surface area to volume ratio, where uptake of substances from soil water occurs.

runner A stem that grows along the ground, making it easy to put down new roots and so reproduce asexually.

saprotrophic nutrition The digestion of dead food material outside the body, as in fungi.

secondary consumer An animal that eats primary consumers.

secondary sexual characteristics Physical characteristics that develop at puberty, such as facial hair and deeper voices in boys, or the menstrual cycle in girls.

secretion Releasing chemicals that have been made inside a cell into the fluid outside the cell.

seed A hard-shelled structure formed from an ovule that contains the plant embryo and food stores.

selective breeding Choosing individuals for breeding based on their characteristics.

selective reabsorption The reabsorption of more of some substances (such as glucose from filtrate) than others.

sense organs Organs containing receptor cells adapted for the detecting of a particular type of stimulus.

sensitivity The detection of changes (stimuli) in the surroundings by a living organism, and its responses to those changes.

sewage Human waste, faeces and urine, which needs to be disposed of.

sex chromosomes Chromosomes that control the sex of the individual. In humans these are the XX chromosomes in women and XY in men.

sexual reproduction The production of new individuals by the fusion of male and female gametes.

simple sugar A basic sugar unit (such as glucose) that can join together with other sugar units to make large carbohydrates such as starch and glycogen.

soil erosion The washing away of soil as a result of rainfall when there is little vegetation to hold on to the soil.

specialisation The process by which cells have developed (often in terms of structure) to perform a particular function more efficiently; also called cell differentiation.

specific Describing the action of an enzyme or antibody that is targeted to a particular other substance to produce a particular reaction or result.

sperm The male gamete in humans.

spinal cord Part of the central nervous system found inside the spinal column of the skeleton.

spinal reflex A reflex processed by the spinal cord rather than the brain, such as when you pull your hand away after touching a hot object.

spongy mesophyll The layer of cells in the lower part of the leaf where there are many air spaces, so allowing the movement of gases (carbon dioxide and oxygen) within the leaf.

starch A form in which carbohydrates are stored for energy in plants.

stamen The male structure in flowers that contains the anther.

stem cell Cells that are able to differentiate to form specialised cells.

stigma The female structure in flowers where pollen grains attach in pollination.

stimulus A change in conditions that triggers a response in an organism.

stomata (single: stoma) Tiny holes in the surface of a leaf (mostly the lower epidermis) which allow gases to diffuse in and out.

style A structure that supports the stigma in a flower.

substrate A molecule that reacts to a particular enzyme.

sucrose A form in which carbohydrates are stored for energy in plants.

synapse A small gap between two neurones across which neurotransmitters travel.

system A group of organs that work together to carry out a particular function, such as the mouth, stomach and intestines in the digestive system.

target organ An organ of the body containing cells that respond to a particular hormone.

tertiary consumer An animal that eats secondary consumers.

testes (single: testis) The site of sperm production in men, testosterone is also produced here.

testosterone A hormone produced by the testes that produces secondary sexual characteristics in boys.

thorax The centre part of the body, protected by the ribs, which contains the lungs and heart.

thymine (T) A base found in DNA; it pairs with adenine (A).

tissue A group of similar cells that have a similar function, such as muscle tissue.

tissue culture Another name for micropropagation.

trachea The tube leading from the mouth to the bronchi, sometimes called the windpipe.

transcription The first part of protein synthesis in which a mRNA copy is made of a DNA gene; it happens in the nucleus.

transgenic organism An organism that contains DNA from a different species.

translation The second part of protein synthesis in which amino acids are joined together in the order coded for by mRNA; it happens on the ribosomes in the cytoplasm.

translocation Movement of dissolved sugars and other molecules through a plant, through the phloem.

transpiration Evaporation of water vapour through the stomata of a plant.

tRNA (transfer RNA) Small RNA molecules that transport amino acid molecules to the ribosomes for protein synthesis.

trophic level A feeding level in a food chain or food web, such as producer, primary consumer.

tropism A growth response to a particular stimulus in plants.

ultrafiltration Filtration using pressure, as happens between the glomerulus and Bowman's capsule.

uracil (U) A base found in RNA; it pairs with adenine (A).

urea A substance produced in the liver from the breakdown of amino acids not needed in the body.

ureters Tubes that connect the kidneys to the bladder.

urethra A tube that connects the bladder to the outside of the body.

urinary system The body system that includes the kidneys, ureters, bladder and urethra.

urine A liquid waste produced by kidneys, containing water, urea and ions.

uterus The womb, where a human fetus develops before birth.

vaccination Giving a vaccine (to a person or animal) to stimulate the immune system and protect against infection.

vacuole A large structure containing cell sap, which is found in the middle of many plant cells.

valves Flaps in the heart, and in veins, that prevent the flow of blood in the wrong direction.

variation Differences between individuals.

variety Plants of the same species but with different characteristics of their own.

vascular bundle A plant vein made up of xylem and phloem vessels.

vasoconstriction Narrowing of blood vessels.

vasodilation Widening of blood vessels.

vector A structure, such as a virus or bacterial plasmid, for carrying genetic material into the nucleus of a cell in genetic engineering.

vein A blood vessel that carries blood to the heart. Or, another word for a plant vascular bundle.

ventilation Another word for breathing.

ventricles The two chambers of the heart that receive blood from the atria and pump it out through arteries.

villus (plural villi) A finger-like projection of the small intestine wall where absorption of digested food molecules occurs.

vitamins Nutrients needed by the body in tiny amounts to remain healthy, for example vitamins A, C and D.

xylem Tubes, formed from dead cells in the vascular bundles of a plant, which carry water and dissolved substances from the roots to the leaves and other parts of the plant.

yeast A type of fungus, used for example in bread making.

yield The amount of food produced from a crop or a farm animal.

zygote A fertilised egg, formed from the fusion of a male gamete and female gamete.

Answers

SECTION 1 THE NATURE AND VARIETY OF LIVING ORGANISMS

Characteristics of living organisms

Page 11

1. **a)** Any suitable answers for human, such as:

    movement – walking; respiration – combination of oxygen with glucose to release energy, carbon dioxide and water; sensitivity – vision; homoeostasis – control of core body temperature; growth – increase in height; reproduction – having a baby; excretion – producing urine; nutrition – eating food.

    b) Any suitable answers for a specific animal such as:

    movement: crawling; respiration: combination of oxygen with glucose to release energy, carbon dioxide and water; sensitivity: smell; homoeostasis: control of body water content; growth: increase in length; reproduction: producing young; excretion: losing carbon dioxide through respiratory surface; nutrition: eating food.

    c) Any suitable answers for a plant, such as:

    movement: growing towards light; respiration: combination of oxygen with glucose to release energy, carbon dioxide and water; sensitivity: detecting direction of light; homoeostasis: controlling loss of water through stomata by opening and closing them; growth: increase in height; reproduction: producing seeds; excretion: diffusion of waste products out of leaf for photosynthesis (oxygen) and respiration(carbon dioxide); nutrition: taking in nutrients from soil and making glucose by photosynthesis.

2. Movement – to reach best place to get food or other conditions favourable for growth

    respiration – to release energy from food that can be used for all life processes

    sensitivity – to detect changes in the environment

    homoeostasis – to prevent damage being done to cells as a result of changes in the body

    growth – to increase in size until large/mature enough for reproduction

    reproduction - to pass genes on to next generation

    excretion – to remove harmful substances from body

    nutrition – to take in substances needed by the body for growth and reproduction.

Page 12

1. Reproduction

2. Movement – viruses cannot move themselves. Respiration – viruses do not respire. Sensitivity – viruses do not sense or respond to changes in the environment. Homeostasis – viruses do not need to control internal conditions. Growth – viruses do not build new materials within their structure to produce a permanent increase in size. Excretion – viruses do not carry out metabolism, so they do not have waste products to excrete. Nutrition – viruses do not take in food substances for use in respiration or to build new materials.

3. Any suitable argument with justification, such as non-living materials do not reproduce, so viruses must be living.

4. Any suitable argument with justification, such as viruses cannot reproduce independently of other organisms and show just one of the eight characteristics of living organisms, so they cannot really be classed as living.

Variety of living organisms

Page 16 (top)

1. Multicellular means made of many cells, such as human, plant, etc.

2. The xylem vessels in the stems and leaves are strong, so help to support the plant. Water in the cells, when the cells are turgid, also support the plant as it pushes against the cell walls.

3. Chloroplast.

Page 16 (bottom) – 17

1. **a)** Nucleus, cytoplasm, cell membrane.

    b) Cell wall, large central vacuole, possibly chloroplasts.

2. Animals have nervous systems and organs for movement that plants do not have.

3. Animals need to move around to find food, or to find suitable habitats to live in. Plants need to take water and nutrients from the ground, so they need roots in the ground.

Page 18

1. **a)** Cell walls and central vacuole, cannot move around.

    b) No chloroplasts, may store carbohydrate as glycogen.

2. Saprotrophic nutrition is the digestion of dead material by external digestion, where enzymes are secreted on to the food and the digested food absorbed into the body. Animals eat the food (sometimes from living/just killed organisms) and digest the food inside their alimentary canal.

3. Toadstools and mushrooms are the reproductive structures of fungi. The main body of the fungus is the mycelium, the mass of tiny thread-like hyphae which is often below the surface of the ground.

Page 19

1. Some contain a chloroplast and can photosynthesise as some plant cells do, others do not have chloroplasts and feed on other organisms, so are more like single animal cells.

2. Nucleus, cytoplasm, cell membrane, single-celled.

3. No, because mosquitoes only carry the *Plasmodium* which is the actual cause of the disease.

Page 20

1. Any three from:

   plant cells have cellulose cell walls, bacterial cells may have a cell wall made of other chemicals

   plant cells have a nucleus containing chromosomes, bacterial cell has free chromosome in cytoplasm

   bacterial cell may contain plasmid, plant cell does not

   plant cells have large vacuole/chloroplasts, bacterial cells do not.

2. Bacteria have no nucleus, they have a cell wall and single chromosome lying free in cytoplasm; protoctists have a nucleus containing the chromosomes, only some have a cell wall, they may contain chloroplasts and are much larger than bacterial cells.

3. Bacterial chromosomes are: single; circular; lying free in cytoplasm / not in a nucleus.

Page 22

1. Outer protein coat surrounding genetic material.

2. They do not have most of the characteristics of a living cell, and behave like particles until they have infected a cell.

3. Viruses are much smaller than bacteria, ~100 nm (100×10^{-9} m) compared with 2 μm (2×10^{-6} m).

SECTION 2 STRUCTURE AND FUNCTIONS IN LIVING ORGANISMS

Level of organisation

Page 31

1. Any two from: nucleus, vacuole, chloroplast.

2. a) Any two suitable such as: muscle, nervous, epithelium (lining), blood, bone.

   b) Each from a) with suitable adaptations, such as:

   muscle tissue can contract to move bones or other parts of body

   nervous tissue can carry electrical impulses

   epithelium tissue has a smooth surface and may have adaptations such as cilia to move substances across surface

   blood contains cells that carry oxygen, plasma that carries dissolved food substances etc.

   bone tissue is hardened to make strong bones.

3. a) Any two suitable such as: xylem, phloem, epidermal, mesophyll.

   b) Each from a) with suitable adaptations, such as:

   xylem: long tubes through which water moves from roots through stems to other parts of the plant

   phloem: living cells that transport food around plant

   epidermal tissues: in continuous layers across surfaces of plants to prevent pathogens entering

   mesophyll: packs other tissues to hold them all in place.

Page 33

1. Any two suitable such as:

   heart: pumps blood round body

   kidneys: remove waste substances from body

   liver: controls many processes in body

   brain: coordinates thought and response to stimuli.

2. Any two suitable such as:

   leaf contains cells with chloroplasts for photosynthesis

   stem contains strengthening tissue in xylem etc to support other parts of plant

   roots have root hair cells where water and dissolved nutrients are absorbed into the plant.

3. A body system is a group of organs that work together to carry out the life processes in an organism.

Cell structure

Page 37

1. a) Drawing should be drawn with thin, clear pencil lines, no crossing out, to show the outline of the cell in the photograph and the central shape.

   b) Diagram should be labelled to show nucleus, cytoplasm and cell membrane.

Page 38

1. a) Chloroplast.

   b) (Large) vacuole.

   c) Cell wall.

Page 40

1. Long length so can carry impulses from one place to another; nerve endings to connect to other nerve cells or organs.

2. To reduce the risk of rejection.

Biological molecules

Page 47

1. **a)** Fatty acids and glycerol.

   b) Simple sugars.

   c) Amino acids.

2. Protein is formed from amino acids, carbohydrates from simple sugars; carbohydrates often made from one kind of simple sugar, proteins from many different kinds of amino acids.

3. **a) i)** The blue solution would change colour and an orange-red precipitate would form, because glucose is a reducing sugar.

   ii) The solution would not change colour as there is no starch present.

   b) i) There would be no change in colour because sucrose and the starch in wheat flour are not reducing sugars.

   ii) The brown solution would turn blue-black because of the starch in flour.

Page 51

1. A chemical that is found in living organisms that speeds up the rate of reactions.

2. Proteins.

3. Without enzymes, the metabolic reactions of a cell would happen too slowly for life processes to continue.

4. As temperature increases, the rate of the reaction will increase, up to a maximum point (the optimum) after which it decreases rapidly as the enzyme is denatured.

Movement of substances into and out of cells

Page 60

1. Any answer that means the same as the following:

   net movement – the overall direction of movement

   diffusion – the overall movement of particles from an area of higher concentration to an area of lower concentration (in a gas or a solution or across a partially permeable membrane).

2. Passive, because no energy is provided by the cell for it to happen.

3. **a)** Glucose can but starch cannot.

   b) Only particles that are small enough to pass through the holes in the membrane can diffuse. Larger molecules cannot diffuse through the membrane.

Page 62

1. Any answer that means the same as the following: the net movement of water molecules from a region of their higher concentration to a region of their lower concentration.

2. **a)** It is a passive movement of molecules as the result of a concentration gradient, moving from where they have a higher concentration to where they have a lower concentration.

   b) Osmosis only considers the movement of water molecules, diffusion considers solute molecules. Osmosis always involves movement across a partially permeable membrane, diffusion may or may not involve movement across a membrane.

3. Diagram should show water molecules leaving the red blood cell as a result of osmosis and entering the solution.

Page 64

1. Concentration gradient, distance, temperature and surface area.

2. Any answer that means the same as the following: if there is a greater difference in concentration between two areas then there will be a greater net movement from the area of higher concentration to the area of lower concentration, as there will be more particles moving away from the area of higher concentration than towards it. So rate of diffusion is faster.

3. As size increases, area increases as the square of the length, but volume increases by the cube of the length. This means that volume increases faster than area, so the ratio of surface area to volume must get smaller.

Page 65

1. Active transport is the movement of a substance across a cell membrane against its concentration gradient, using energy.

2. Uptake of nitrate ions by root cells in plants, uptake of glucose from kidney tubules into the blood in animals.

Nutrition

Page 70

1.

$$6CO_2 + 6H_2O \xrightarrow[\text{light energy}]{\text{chlorophyll}} C_6H_{12}O_6 + 6O_2$$

2. CO_2 from air, H_2O from soil water, $C_6H_{12}O_6$ used in cells for respiration or converted to other chemicals for use in cells, O_2 released into air if not needed in respiration.

3. Any four from: as glucose for respiration; converted to sucrose for storage in fruits; converted to starch for storage; converted to cellulose to form cell walls; converted to oils for storage; converted to proteins for growth.

1. Thin broad leaves; chlorophyll in cells; veins containing xylem tissue that transports water and mineral ions to the leaves and phloem tissue that takes products of photosynthesis to other parts of the plant; transparent epidermal cells; palisade cells tightly packed in a single layer near top of leaf; stomata to allow gases into and out of leaf; spongy mesophyll layer with air spaces and a large internal surface.

2. A large surface area helps to maximise the rate of diffusion, in this case diffusion of carbon dioxide into cells for photosynthesis and oxygen out of cells so that it can be released into the air.

3. Allows as much light as possible through to the chloroplasts in the palisade cells below.

Page 77 (top)

1. Light intensity, carbon dioxide concentration, temperature.

2. As light intensity increases, so rate of photosynthesis increases because more energy is supplied to drive the process.

   As carbon dioxide concentration increases, so rate of photosynthesis increases because there is more reactant for the process.

   As temperature increases, the rate of photosynthesis increases because the particles in the reaction, including enzymes, are moving faster and bump into each other more. There is a maximum temperature above which the rate of photosynthesis decreases because the enzymes that control the process start to become denatured.

3. The only part of the leaf that can photosynthesise is the green part where there is chlorophyll. So only in the green parts can glucose be produced. Starch is formed from glucose, so it is only formed in the green parts where photosynthesis has taken place.

Page 77 (bottom)

1. Plants make their own foods, and need to convert the carbohydrates made by photosynthesis into other substances, such as proteins, which contain additional elements.

2. **a)** Limited growth, lack of green colour in leaves.

   b) Lack of green colour in leaves.

3. **a)** Nitrogen is an essential element for making substances other than carbohydrates, such as amino acids and proteins. Without proteins, the plant cannot make new cells so the plant will not grow well. It is also needed for making chlorophyll, so without nitrogen the plant will not be able to make as much of the green pigment.

   b) Magnesium is needed to make chlorophyll, which is the green substance in plants. Without enough magnesium the plant will lose the green colour and become yellow.
   Any magnesium in the plant is transported to the new leaves, so that photosynthesis can continue there for making food for growth.

Page 78

1. **a)** The leaves of the plant with limited nitrogen are paler green / more yellow than those of the plant with a lot of nitrogen.

   b) The plant with plenty of nitrogen is larger, bushier and has more leaves than the plant with limited nitrogen.

2. The leaf cells of the plant with limited nitrogen will contain less chlorophyll because nitrogen is needed to make this substance.

3. The plant with plenty of nitrogen is not only able to make more chlorophyll and therefore photosynthesise more and produce more carbohydrate, it also has sufficient nitrogen to convert some of that glucose into proteins. So it can make more new cells more rapidly than the plant with limited nitrogen, and so grow taller and bushier and produce more leaves.

4. When the crop plants grow, they take in nitrogen as nitrate ions from the soil and use them to make substances such as proteins and chlorophyll in the plant tissues. When the plant is harvested, the nitrogen compounds in the tissues are taken as well. This leaves fewer nitrate ions in the soil for the next crop. With a smaller amount of nitrate ions in the soil, the new crop will not grow as well as the previous crop. Additional nitrate ions in the form of nitrogen-containing fertilisers, make sure the new crop has sufficient nitrogen for rapid and healthy growth.

Page 80

1. More intense sunlight, so likely to make sufficient vitamin D naturally in skin and less dependent on vitamin D in diet.

2. Higher latitude so less intense sunlight means less vitamin D produced in skin, and poor diet would mean increased risk of too little vitamin D in diet.

3. Fish liver oil is formed from liver which is a good source of vitamin D.

4. If diets include more sources of vitamin D, such as fish, eggs, cheese and milk, then the risk of vitamin D deficiency is reduced.

5. Lack of light on the skin means little chance of natural vitamin D production, and a vegetarian diet reduces the amount of vitamin D available in the diet. Vitamin D supplements in the diet can help to avoid the deficiency, but care must be taken not to take too much vitamin D over a long period as high levels in the body can be toxic. So supplements during the winter maybe more advisable than supplements all the time.

1. Carbohydrates, proteins and lipids.

2. **a)** USDA: Vegetables and grains. UK: Fruit and vegetables, and bread, rice, potatoes, pasta.

   b) Vitamins, minerals, fibre and carbohydrates.

   c) Foods and drinks high in fat and/or sugar.

   d) To avoid taking in too much energy, which can lead to obesity.

3. Any answers along the lines of: different people need different amounts of energy every day, for example active people need more than people who are seated for much of the day; men have a larger average body mass than women so will need more energy to support that extra tissue; some groups of people need more of a particular group of nutrients than others, for example, pregnant women need additional iron.

4. Food that contains more energy than the body uses is converted into body fat. High levels of body fat cause obesity, which is associated with many health problems.

1. Sketch should show the following labels correctly attached to organs shown on the diagram:

   mouth, where food is broken down by physical digestion (chewing) and amylase enzyme starts digestion of starch in food

   oesophagus moves food from mouth to stomach by peristalsis

   stomach, where churning mixes food with protease enzymes and acid to start digestion of protein molecules

   duodenum, where alkaline bile neutralises the acid and enzymes from pancreas help complete digestion of proteins, lipids and carbohydrates

   ileum, where digested food molecules are absorbed into the body

   colon, where water is absorbed from undigested food

   rectum, where faeces is held until it is egested through the anus

   liver, where bile is made

   gall bladder, where bile is stored until needed

   pancreas, where proteases, lipases and amylase are made, which pass to the small intestine.

2. Egestion is the removal of undigested food from the alimentary canal – food that has never crossed the intestine wall into the body. Excretion is the removal of waste substances that have been produced inside the body.

3. Waves of contraction of the circular and longitudinal muscles push the food bolus further along the alimentary canal.

1. Chemical digestion uses chemicals (enzymes) to help break down large food molecules into smaller ones. Physical digestion is the breaking of large pieces of food into smaller ones, for example by chewing.

2. The digestive enzymes break down food molecules that are too large to cross the wall of the small intestine into smaller ones which can diffuse across cell membranes and so enter the body. If we did not have enzymes, we would not be able to absorb many nutrients from our food.

3. **a)** Amylase.

   b) Maltose.

4. **a)** The acid increases the acidity in the stomach, providing the right conditions for enzymes that digest food in the stomach.

   b) Bile neutralises the acidity of food from the stomach, providing the right conditions for enzymes that digest food in the small intestine. It also emulsifies lipids, providing a larger surface area for lipase enzymes to work on.

1. Long length, villi that cover the intestine wall increase the surface area; microvilli on surface of villi cells increase surface area for absorption even further; extensive blood supply removes absorbed food molecules quickly so maintaining a high concentration gradient for diffusion, lacteals in the villi carry absorbed lipid molecules away to the rest of the body.

2. Some foods may be needed quickly by the body; if digested food is not absorbed quickly as food passes along the intestine it may be lost in faeces.

Respiration

1. Inside cells.

2. **a), b) and c)** Glucose (from digested food from alimentary canal via the blood) + oxygen (from air via lungs) $\rightarrow$ + carbon dioxide (excreted through lungs) + water (used in cells or excreted through kidneys) (+ ATP (provides energy for cell processes))

   d) Most water used in cells because the camel is much better than humans at using the water from respiration.

3. Aerobic respiration uses oxygen from the air.

1. Aerobic respiration uses oxygen / anaerobic does not. Aerobic respiration releases more ATP/energy than anaerobic. Aerobic produces carbon dioxide and

water, but anaerobic produces either lactic acid (animals) or ethanol and carbon dioxide (plants/fungi).

2. During vigorous exercise, they may not be able to get enough oxygen from the blood for all the energy they need for contracting. So the additional energy comes from anaerobic respiration.

3. Similarities: use glucose as substrate; produce energy; do not need oxygen. Differences: animals produce lactic acid; plants produce ethanol and carbon dioxide.

4. Inside germinating seeds where the cells may be too deep to get oxygen by diffusion, or where the seed is surrounded by water which prevents oxygen from the air reaching the seed.

Gas exchange

Page 107

1. All the time, because they continually need energy for making new substances and other life processes.

2. When the light intensity is high enough / during daylight hours, because photosynthesis needs energy from sunlight.

3. It is the point for a plant when the rate of photosynthesis / oxygen production / carbon dioxide uptake is the same as the rate of respiration / carbon dioxide production / oxygen uptake.

4. From the graph, between the hours of about 9:30 and 15:30 when the rate of photosynthesis exceeds the rate of respiration.

Page 109

1. Carbon dioxide is soluble and acidic, so when more gas is being produced, such as during respiration, the solution becomes more acidic. When carbon dioxide is removed from the solution, such as during photosynthesis, the solution becomes less acidic.

2. Sketch should show: thin leaf to maximise area for gas exchange and minimise distance that gases have to diffuse between air and photosynthesising (palisade) cells. Spongy mesophyll cells and air spaces connected to air via stomata, to maximise the internal surface area. Stomata that control gases moving into and out of leaf.

3. The gas molecules are small and so can diffuse across cell membranes, into and out of cytoplasm and into and out of chloroplasts. The rate of photosynthesis is, in part, controlled by the rate of diffusion of gases between the chloroplasts and air.

Page 111

1. Trachea – carries air from mouth down to lungs; bronchi – the two large divisions of the trachea as it reaches the lungs, supported with rings of cartilage to prevent collapse during breathing; bronchioles – the fine tubes in the lungs that carry air to alveoli; alveoli have a large surface area and are very thin for efficient diffusion of gases; pleural membranes – that surround the lungs are involved in ventilation; ribs and intercostal muscles – protect the lungs but also help expand the volume of the thorax during forced or deep breathing; diaphragm – muscular sheet below lungs which controls relaxed breathing.

2. Gas exchange is the movement of gases into and out of the cells of the body, or between the lungs and the blood. Ventilation (breathing) is the movement of air into and out of the lungs.

Page 115

1. a) Inhalation: The diaphragm contracts and flattens, pulling downwards; the intercostal muscles contract lifting the ribs out and up; these actions increase the volume of the thorax, causing the volume of the lungs to increase. This decreases the pressure inside the lungs, and so causes air to enter the lungs from outside.

   b) Exhalation: The diaphragm muscle relaxes, and moves upwards; the intercostal muscles relax, so the ribs fall back and down; both actions reduce the volume of the thorax, so reducing the volume of the lungs. This increases the pressure inside the lungs compared with the air outside, so this pushes air out of the lungs.

2. a) Just two, the wall of the alveolus and the wall of the blood capillary.

   b) To ensure that oxygen rapidly diffuses from the area of high concentration, in the alveoli, to the area of low concentration, in the blood.

   (Do not forget that there is also a concentration gradient for carbon dioxide.)

3. Sketch similar to Fig. 2.77 on page 114, with annotations showing: thin alveolar wall and wall of capillary allows rapid diffusion; high concentration gradients for gases between blood and air in alveolus due to continuous blood flow through capillary and ventilation of alveolus (lungs); large area of contact between capillary and alveolus, maximising area over which diffusion can occur.

Page 119 (top)

1. So it is easier to compare the values, because there will have been different numbers of mothers/babies in each group. This makes it easier to see any pattern in the results.

2. The chart shows that even some non-smoking mothers had low birth weight babies. However, the proportion goes up as the number of cigarettes per day goes up, and heavy smokers have more than twice the proportion of low birth weight babies as the non-smokers.

3. The conclusion is incorrect, because some babies born to non-smoking women have low birth weight, so there must be other causes. However, the chart does show that increasing levels of smoking increases the risk of having a low birth weight baby in this sample of women.

   A better conclusion would say something like: In this sample, women who smoked a little during pregnancy had an increased risk of having a low birth weight baby, and those that smoked a lot had an even greater risk.

4. This study shows the results for Canadian women only, and there may be other factors in this population which could produce this result. By comparing many studies from different countries, with women who have different lifestyles, it can balance out the effect of other factors. So any remaining relationship between birth weight and smoking becomes more definite and reliable.

5. Carbon monoxide could have this effect, by reducing the amount of oxygen that gets to the fetus's cells. Since oxygen is needed for respiration, if the amount of oxygen is reduced, the rate of respiration will be reduced, which will reduce the rate at which energy is released that can be used for building new cells (growth).

Page 119 (bottom)

1. Bronchitis – respiratory system; emphysema – respiratory system; cancer – any part of the body but mainly respiratory system; stroke/heart attack – circulatory system.

2. Carcinogenic means cancer-causing / a chemical that can cause cells to become cancerous so they grow and divide without stopping.

3. Emphysema is the breaking down of some of the surface of the alveoli. This leaves a smaller area for gas exchange /absorption of oxygen. So the person may not get enough oxygen for activity. Additional oxygen in the gas they breathe can help to get more oxygen into their blood and so to their cells.

4. Small molecules from the smoke, such as nicotine and carbon monoxide, can diffuse into the blood and be carried around to all parts of the body.

Transport

Page 127

1. Over the distance of several cells, diffusion and osmosis work too slowly to supply substances the cell needs to carry out all the life processes as quickly as needed.

2. Xylem tubes carry water and dissolved substances from the roots to other parts of the plant including the leaves.

3. Phloem cells carry dissolved food materials, such as sucrose and amino acids, from the leaves where they are formed to other parts of the plant that use them for life processes or where they will be stored, or from storage regions to growing regions.

Page 128

1. Osmosis.

2. Diagram should included annotations like the following, at the appropriate point: soil water has higher concentration of water molecules than cytoplasm of cells in the root; water molecules enter root hair cells by osmosis; water molecules pass from cell to neighbouring cell by osmosis until they reach the xylem.

Page 131

1. Diagram should include annotations like the following, at the appropriate point: water molecules evaporate from surfaces of cells into air spaces; water molecules from air spaces move into and out through stomata into the air – diffusion (net movement) from inside leaf to outside; osmosis causes water molecules to move from xylem into neighbouring leaf cells, and then from cell to cell until they reach a photosynthesising cell; transpiration is the evaporation of water from a leaf.

2. Closing stomata reduces diffusion of water molecules out of the leaf. At night, carbon dioxide is not needed for photosynthesis, so keeping stomata open would lose water unnecessarily. (Although some oxygen is needed for respiration, only very small amounts are needed, and this can diffuse in through even almost completely closed stomata.)

3. a) When temperature is higher, water evaporates more easily, and particles move faster, so water molecules will diffuse out of the leaf more quickly.

   b) When air humidity is low, there is a low concentration of water molecules in the air. So far more water particles will move out of the stomata into the air than are moving from the air into the leaf. This means the rate of transpiration will be faster.

Page 133

1. Haemoglobin binds with oxygen when it is at high concentration such as in the lungs, and releases oxygen when it is at low concentration, such as in respiring tissues where the oxygen has reacted with glucose to produce carbon dioxide and water.

2. Breathing might become more rapid because with each breath the haemoglobin is combining with less oxygen than the body is used to. So it will deliver less oxygen to cells and the body response will be to increase breathing rate and depth.

3. More red blood cells in a given volume of blood means there will be more haemoglobin in that

volume. More haemoglobin can combine with more oxygen, so each cm^3 of blood will carry more oxygen and deliver more oxygen to the body cells.

4. Training at high altitude for several weeks will cause the red blood cell count to increase. This will increase the oxygen-carrying capacity of the blood. When the athlete then competes at low altitude, their blood will be delivering more oxygen to their muscle cells than if they had trained at low altitude. So their muscles will be able to work harder aerobically than after low-altitude training.

Page 135

1.

Blood component	Function
Plasma	Carries dissolved substances, such as carbon dioxide, glucose, urea and hormones; also transfers heat energy from warmer to cooler parts of the body
Red blood cells	Carry oxygen
White blood cells	Protect against infection
Platelets	Cause blood clots to form when a blood vessel is damaged

2. The biconcave disc shape increases surface area to volume ratio, so rate of diffusion of oxygen into and out of cell is maximised. Haemoglobin inside cell binds with oxygen when oxygen concentration is high and releases oxygen when oxygen concentration is low. Cell has no nucleus, so there is as much room as possible for haemoglobin. Cell has flexible shape so can squeeze through the smallest capillaries and reach all tissues.

3. Phagocytes engulf pathogens inside the body and destroy them. Lymphocytes produce antibodies that attach to the pathogens, either attracting phagocytes or causing the pathogens to break open and die. This all helps to prevent pathogens causing damage when they infect us.

4. Damage to a blood vessel can create an easy route of infection into the body. So forming a blood clot where there is damage, as quickly as possible, helps to reduce the risk of infection.

Page 138

1. a) Renal arteries.

    b) Aorta.

    c) Hepatic vein.

2. Arteries are large vessels with thick, elastic muscular walls; capillaries are tiny blood vessels with very thin walls that are often only one cell thick; veins are large vessels with thin walls, a large lumen and valves to prevent backflow of blood.

3. The walls stretch as blood enters them, and slowly recoil as the blood flows through, balancing out the pressure so that the change in pressure is reduced.

Page 142

1. Vena cava, right atrium, right ventricle, pulmonary artery, capillaries in the lungs, pulmonary vein, left atrium, left ventricle, aorta.

2. Valves between the ventricles and atria, and at the base of the blood vessels leaving the heart can close to prevent blood moving in the wrong direction.

3. Heart rate increases with increasing level of exercise so that the blood is carried round the body faster and can carry oxygen from the lungs more quickly to the muscle cells, glucose from the alimentary canal (or released from liver cells) to the muscle cells, and carry carbon dioxide from the muscle cells more quickly to the lungs.

Excretion

Page 149

1. Leaf, because this is where the gaseous waste products of photosynthesis and respiration are excreted from the plant into the environment.

2. Lungs – carbon dioxide from respiration excreted from body; skin – water and ions excreted from body; kidneys – urea and other waste products in excess such as ions and water.

3.

Structure	Function
Kidneys	Produce urine by filtering waste substances from the blood
Ureters	Carry urine from kidneys to bladder
Bladder	Stores urine until it is released to the environment
Urethra	Short tube linking bladder to environment

Page 152

1. Blood in glomerulus, nephron (Bowman's capsule, proximal convoluted tubule, loop of Henle, distal convoluted tubule, collecting duct), ureter, bladder, urethra.

2. a) Bowman's capsule (in conjunction with glomerulus of blood capillaries).

    b) Proximal convoluted tubule for most substances, also collecting duct for additional water.

3. a) Glucose.

    b) Active transport.

Page 153

1. Osmoregulation is the control of the concentration of water in the blood.

2. **a)** ADH.

   b) Hypothalamus in brain. (It is later stored and released by the pituitary gland.)

   c) Cells of the collecting ducts in kidneys.

3. ADH makes collecting duct walls more permeable to water, so more water is reabsorbed from the filtrate, making more concentrated urine.

Coordination and response

Page 160

1. A tropism is a growth response of a plant to a stimulus.

2. **a)** Shoots grow towards light.

   b) Roots grow in the direction of the force of gravity.

3. Auxin is produced in the tip of the growing shoot and diffuses down the shoot. Auxin on the bright/light side of the shoot moves across the shoot to the darker side as it diffuses down the shoot. Cells on the dark side of the shoot elongate more than the cells on the light side of the shoot, so the shoot starts to bend as it grows so that the tip is pointing towards the light.

Page 165

1. **a)** The cornea is transparent so light passes through it easily into the eye. The cornea also refracts light to help focus it onto the retina.

   b) The pupil is a hole surrounded by the iris, that controls the amount of light that passes through the iris to the back of the eye.

   c) The retina contains the light-sensitive cells that respond to light.

2. As light intensity increases, the pupil gets smaller, reducing the amount of light that can enter the eye. This change happens because the radial muscles in the iris relax and circular muscles contract. As light intensity decreases, the pupil gets larger, increasing the amount of light that can enter the eye. This change happens because the radial muscles in the iris contract and circular muscles relax.

3. Light entering the eye from a near object needs to be refracted more than light from a distant object in order to focus it on the retina. The ciliary muscles contract, which reduces the tension on the ligaments that are attached to the lens. This allows the lens to become thicker and more rounded, which means it refracts light more.

Page 167

1. A reflex response is a simple response of receptor>nerve>spinal cord>nerve>effector, that often does not include the brain. This makes it possible to respond to a stimulus very quickly. Reflex responses are usually important in survival, for example to protect you from holding on to something dangerous, or blinking to protect the eye if something comes toward it.

2. Heat causes the receptor cells in the skin to respond. The receptor cells cause nerve cells (sensory neurones) to send electrical impulses to the spinal cord. The nerve cells pass the electrical impulses to nerve cells in the spinal cord (relay neurones), which pass the electrical impulses to nerve cells in the nerves leading to the arm muscles (motor neurones). These nerve cells pass the electrical impulses to the effector cells i.e. the muscles, which cause the muscle cells to contract and move the hand away from the heat.

Page 169

1. Homeostasis is the maintenance of conditions inside the body within limits that allow cells to work efficiently.

2. Control of core body temperature, and control of concentration of water in the blood.

3. Skin blood vessels dilate when the core body temperature is too high. This allows heat energy carried by the blood to reach the skin surface more easily and so be transferred to the environment more rapidly. Skin blood vessels constrict when the core body temperature falls too low. This reduces blood flow to near the skin's surface, so heat energy cannot be transferred as easily to the skin surface and so cannot be transferred to the environment as quickly. This keep more heat energy within the body.

Page 171 (top)

1. Homeostasis is the regulation of conditions inside the body so that they remain fairly constant. So the concentration of blood glucose is kept within a narrow range.

2. Insulin is released from the pancreas when blood glucose concentration is high. It causes muscle and liver cells to take in glucose and convert it to glycogen for storage, so that blood glucose concentration falls.

3. The more insulin they inject the more glucose will be taken up by cells, so they have to make sure that blood glucose concentration will not fall too low, because that is dangerous.

4. Exercise needs energy from respiration, and that is supplied by the breakdown of glucose in muscle cells. This will mean that blood glucose concentration falls because the glucose will move down its concentration gradient into the muscle cells. With a lower blood glucose concentration, less insulin is needed in the blood.

Page 171 (bottom)

1. a) A chemical messenger in the body, that produces a change in the activity of some cells.

    b) A gland that secretes hormone.

    c) An organ that contains cells which are affected by hormones.

2.

Hormone	Where produced	Effects
Adrenaline	Adrenal glands	Prepares body for action, e.g. increases heart rate, increases breathing rate, dilates pupils, causes glucose to be released into blood
Insulin	Pancreas	Causes liver and muscle cells to remove glucose from blood
Testosterone	Testes	Produces secondary sexual characteristics in boys and needed for sperm production
Progesterone	Ovaries	Involved in control of menstrual cycle
Oestrogen	Ovaries	Involved in control of menstrual cycle, produces secondary sexual characteristics in girls

SECTION 3 REPRODUCTION AND INHERITANCE

Reproduction

Page 192

1.

Sexual reproduction	Asexual reproduction
• Fusion of male gamete with female gamete	• New individuals produced from division of body cell of parent
• Offspring genetically different from parents and from each other	• Offspring genetically identical to parent
• (Usually) two parents	• Only one parent
• Slower because male and female need to find each other and mate	• Faster because no search for mate

2. Any suitable example that refers to changing environment where genetic variability in offspring increases chance of survival of offspring that are genetically different to parent.

3. Answer should include:
   summer not long so asexual reproduction increases numbers more rapidly;
   summer short period, so variability of environment not as big a problem, so genetic variability would not be an advantage;
   if parent is feeding well on food plant, the offspring from that parent are equally likely to survive and grow well on that food plant, so variability would be a disadvantage.

Page 195

1. It produces the pollen.

2. In pollen grains.

3. Stigma where pollen grains attach. Style which supports the stigma. Ovary which surrounds and protects the ovule, inside which is the female gamete.

Page 197

1. Wind-pollinated flowers are usually small, no colourful petals, anthers and stigmas hang outside flower, make a lot of lightweight pollen.

2. The lightweight pollen can be carried far in wind, anthers hang outside the flower so pollen more likely to be caught by wind.

3. Insect-pollinated flowers are often large, brightly coloured, produce nectar and sometimes scent, make small amounts of larger pollen grains.

4. Scent, nectar (as food), large petals with bright colours all help to attract insects to the flower. As the insects feed, they pick up pollen which is then carried to other flowers.

Page 199

1. Pollination is the transfer of pollen from one flower to another. Fertilisation is the fusion of a male gamete with a female gamete to form a zygote.

2. Pollination (pollen grain lands on stigma); pollen tube grows down through style to ovule; pollen tube delivers male gamete to egg cell; nucleus of male gamete and nucleus of female gamete fuse to form zygote.

3. Female gamete develops into embryo plant; ovule wall forms hard outer shell of seed; ovary forms fruit.

Page 202 (top)

1. The food reserves provide the food needed for respiration, so that cell growth can take place, until photosynthesis supplies food.

2. a) Essential for germination.

b) Increased temperature (up to the point when enzymes are denatured) increases rate of enzyme action and so growth.

3. Most seeds germinate below the surface of the ground, where there is no light. If they needed light, they would not germinate.

4. Take piece of existing plant (such as piece of stem, root or shoot) > dip into rooting hormone mixture > place in compost > keep moist > after a few weeks roots will develop.

Page 202 (bottom)

1. Sketch as Fig. 3.18 on page 202.

2, 3. Labels and annotations as follows:
testes, where sperm (male gametes) produced;
sperm duct, carries sperm to urethra;
prostate gland and seminal vesicles, produce liquid in which sperm swim;
penis, when erect delivers sperm into vagina of female;
urethra, tube that carries sperm from sperm ducts to outside the body.

Page 203

1. Sketch as Fig. 3.19 on page 203.

2, 3. Labels and annotations as follows:
ovaries – where egg cells form;
oviducts – carry the eggs to the uterus and where fertilisation by sperm takes place;
uterus – where embryo implants into lining and fetus develops;
cervix – base of uterus where sperm are deposited during sexual intercourse;
vagina – where penis is inserted during sexual intercourse, and along which baby passes as it is born.

Page 204–205

1. Characteristics of the body that develop at puberty and prepare the body for sexual reproduction.

2. The shedding of the thickened lining of the uterus and the start of the development of another egg in the ovary.

3. Circle numbered 1 to 28 around the circle;
   a) ovulation at about day 14;
   b) menstruation about days 28/0–4;
   c) increase in oestrogen about days 8–12, decrease about days 12–14;
   d) increase in progesterone about days 14–18, decrease about days 23–28.

Page 205

1. a), b) FSH: secreted from pituitary gland, target cell egg in ovary.
oestrogen: secreted from developing egg in ovary, target cells include cells lining uterus wall and cells in pituitary gland
LH: secreted by pituitary gland, target cells in ovary, causing ovulation.
progesterone: secreted by ovary cells, target cells include cells lining uterus wall and cells in pituitary gland (inhibiting FSH and LH secretion).

2. In the blood.

3. a), b) Sketch should include curves for oestrogen and progesterone as shown in Fig. 3.20 on page 204, plus curves for FSH and LH as shown in Fig. 3.21 on page 205.
Annotations should indicate:
rise in FSH at start due to low levels of progesterone and oestrogen, fall when increasing progesterone inhibits production;
rise in oestrogen as a result of developing egg, fall in oestrogen after egg ovulated;
rise in LH due to high oestrogen, fall after progesterone increases and inhibits production;
rise in progesterone after ovulation, fall at end of cycle if egg not fertilised.

4. The actions happen over a long time, and target cells may be in more than one part of the body. Hormones can control these kinds of responses much more easily than the nervous system.

5. Although each hormone has a different effect, the coordination of all four hormones acts to keep the cycle within limits and produces a repeating pattern.

Page 208

1. **a)** The cell produced by the fusion of a male gamete and female gamete.

   b) The dividing ball of cells formed from the zygote, which implants in the uterus wall lining.

   c) Developing baby in the uterus (womb) from the point where the placenta has developed.

2. **a)** Protects the baby from bumps from the outside world and from wide variation in temperature.

   b) Provides nutrients from mother's blood and carries waste to mother's blood to be excreted ; provides a barrier preventing the mother's and baby's blood mixing, so baby is not harmed by the mother's higher blood pressure, or by pathogens or many harmful chemicals.

3. For the exchange of materials. Substances such as glucose (food molecules), oxygen and other nutrients diffuse from the mother's blood into the fetus's blood, and waste products such as carbon dioxide and urea diffuse from the fetus's blood into the mother's blood. The rate of diffusion is faster over a large surface area and when the distance that needs to be crossed is as short as possible. The placenta provides both of these so that diffusion is as rapid as possible.

Inheritance

1. Gene, chromosome, nucleus, cell.

2. a) The shape of the DNA molecule, a twisted ladder shape with two strands joined by pairs of bases.

   b) The 'rungs' of the DNA double helix, which are formed from an AT pair or a GC pair.

3. Different forms of the same gene that code for different variations of the same characteristic, such as different eye colours.

4. RNA is (usually) single stranded and contains the base uracil (U) instead of thymine (T).

5. a) Transcription involves: a DNA gene 'unzips' forming two single strands; one DNA strand acts as a 'template' for the formation of a corresponding strand of mRNA; the bases on the mRNA and DNA form pairs, C with G, A (mRNA) with T (DNA), and U (mRNA) with A (DNA). (The mRNA then leaves the nucleus.)

   b) Translation involves: mRNA attaches to a ribosome; every 3 bases on the mRNA is a codon, and matches with a corresponding anticodon on tRNA; tRNA molecules bring amino acids to the ribosome which are joined together to make up proteins; the mRNA base sequence determines the protein amino acid sequence.

Page 221

1. a) The characteristic is fully expressed in the phenotype even when the organism has only one allele of that form for that gene (heterozygous).

   b) The characteristic is only expressed in the phenotype when both alleles for that gene are of this form and is not expressed in a heterozygote.

   c) Having two identical copies of that allele for a particular gene.

   d) Having different alleles for a particular gene.

2. a) 2

   b) 1

   c) 2

Page 223

1. The inheritance of a characteristic produced by one gene.

2. Genotype (the alleles in the chromosomes) BB, phenotype (what the organism looks like) brown; genotype Bb, phenotype brown (because the brown allele is dominant); genotype bb, phenotype black (because the organism does not have the brown allele).

3. a) Answer may be presented as a full layout diagram or a Punnett square, showing the adult genotypes and phenotypes (male Bb brown and female Bb brown), the possible gametes produced (male B and b, female B and b), genotypes and phenotypes of possible offspring (BB brown, Bb brown, Bb brown, bb black).

   b) This cross produces a theoretical probability of one black rabbit for every three brown rabbits, a ratio of 1:3, probability of 1 in 4 or 0.25 or 25%.

Page 227 (top)

1. So that, when the plants were bred together, the results in the offspring were not confused by a mix of alleles in one or both of the parents, as both parents would be homozygous.

2. Random variation is possible in the results. So the larger the sample, the more likely that any random variation will be averaged out.

3. He removed the stamens from every flower, so they could not self-pollinate. He also covered each flower after he had hand-pollinated it, so that other pollen could not get to the stigma.

4. Any characteristic may be used, with alleles appropriately designated with capital letter for dominant and lower-case letter for recessive allele. Parents used should show one with phenotype of dominant allele, homozygous, e.g. BB, and one parent with phenotype of recessive allele, i.e. bb. First cross will produce all individuals with phenotype of dominant allele but heterozygous in genotype, i.e. Bb. Crossing of these individuals will produce characteristic 1 BB : 2 Bb : 1 bb in genotype and 3 dominant characteristic to 1 recessive characteristic in next generation.

5. If Mendel had not been as thorough about his method, then his results would not have been as clear and reliable. So he would not have been able to have drawn clear and repeatable conclusions about the way characteristics are inherited in pea plants.

Page 227 (bottom) – 228

1. They are polygenic characters, controlled by more than one gene.

2. When both alleles are expressed in the phenotype, and there is no dominance of one allele over the other.

3. a) I^A and I^B.

   b) Only I^o.

c)

			Father AB	
			Gametes	
			I^A	I^B
Mother O	Gametes	I^o	I^AI^o blood group A	I^BI^o blood group B
		I^o	I^AI^o blood group A	I^BI^o blood group B

d) There is a 1 : 1 ratio, 0.5 or 50% or 1 in 2 probability of blood group A and blood group B.

Page 229

1. 3
2. 2
3. 2
4. 'Freckles' are dominant because I has no freckles, but her parents C and D do. I must be homozygous recessive and C and D must both be heterozygous. If having no freckles was dominant then at least one of C and D would have had to have had no freckles.

Page 230

1. XX
2. XY
3. 0.5 or 50% (or 1 in 2), because there is an equal chance that an X sperm or a Y sperm will fertilise the X egg cell.

Page 233

1.

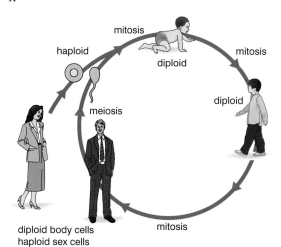

diploid body cells
haploid sex cells

2. Diploid body cell > chromosomes duplicate > chromosomes line up across middle of cell > copies of chromosomes separate and each copy goes to the opposite side of the cell > cell divides in two to produce two identical diploid body cells.

3. Diploid body cell > chromosomes duplicate > chromosome pairs line up across middle of cell > one chromosome from each pair moves to opposite sides of the cell > cell divides in two > chromosomes line up across middle of cells > chromosome copies separate and each copy goes to the opposite side of the cells > cells divide in two to form four haploid cells that are not genetically identical.

4. Meiosis produces non-identical cells, so there is variety in the gamete cells. When the gamete cells fuse, this will mean that the offspring will vary from each other.

Page 236

1. **a)** Genes, such as eye colour; environment, such as weight as a result of diet. (Other suitable examples are acceptable.)

   b) Any suitable example showing a combination of genes and environment, such as human height, which depends on genes and a healthy diet to achieve the potential of the genes.

2. A change in the DNA of a gene.

3. Ionising radiation, such as ultraviolet radiation, x-rays or gamma rays; chemical mutagens such as the chemicals in tobacco smoke.

Page 239

1. **a)** Evolution: change in the characteristics of a species over time.

   b) Natural selection: the influence of the environment on survival and/or reproduction, such that organisms with some characteristics are more successful at producing offspring than others.

2. **a)** If the individuals in a population are all the same, natural selection will favour (or disadvantage) them all equally.

   b) If individuals with a particular variation of a characteristic have a greater survival advantage, they are more likely to produce offspring that carry their genes, and so their genes will become more common in the next generation.

3. Diagrams should show the following: person infected with bacteria > bacteria grow in number inside patient > treatment of patient with antibiotics kills off least resistant bacteria but some resistant bacteria survive > some of these bacteria escape into the environment from the patient and infect another person > the same antibiotic cannot be used on that patient as the bacteria are resistant.

SECTION 4 ECOLOGY AND THE ENVIRONMENT

The organism in the environment

Page 258

1. An ecosystem is all the organisms and the environmental factors that interact within an area. Examples include a lake, desert, tropical rainforest, coral reef, or anything similarly large-scale that has reasonably definable boundaries.

2. A habitat is the space in which a species/population of one species lives, such as under a rotting log, in the open water of a pond or lake, or anything of similar scale.

3. Populations of different species that live in different habitats form the community of organisms that live in an ecosystem.

4. Biotic factors are those due to other living organisms, e.g. predation and competition. Abiotic factors are non-living, e.g. temperature and rainfall.

Page 262

1. A square of a specific size, used for defining a sample area.

2. The number of individuals within a given area.

3. Choice might favour some parts of the area over others, such as places where there are no stinging or thorny plants. This may change the results, increasing or decreasing the average for some species, and therefore will change the conclusions that are drawn.

4. It is a quicker method than counting using quadrats.

Page 264

1. How organisms are spread within an area.

2. A sampling line along which quadrats are placed for taking samples.

3. A line transect placed so that it runs from one habitat to another (from a pond edge to a field, or from shade to light under a large tree), can be sampled to show how the factor that is changing (i.e. water availability, or light intensity) affects the distribution of the organisms.

4. A measure of the variety of living things in an area.

5. a) For example, tropical rainforest.
   b) For example, Arctic ice cap.

Feeding relationships

Page 270

1. Any description that means the same as:
   a) Organism that produces its own food from simpler materials, e.g. plants making carbohydrates in photosynthesis.

   b) Animal that eats producers, also a herbivore.
   c) Animal that eats primary consumers (grouped with other meat-eaters as a carnivore).
   d) Organism that causes decay of dead material, such as some fungi and bacteria.

2. a) Primary consumers because they eat producers/plants.
   b) Any consumer level above primary consumer because they eat other animals.
   c) Any consumer level because they eat both plant and animal tissue.

3. They feed on dead/decaying tissue and waste materials from all trophic levels. Food chains only show the feeding relationships between living organisms.

Page 271

1. Insectivorous bird, toad snake, fox or owl.

2. The fox feeds as a secondary consumer on rabbits, rats, mice and seed-eating birds. It also feeds as a tertiary and quaternary consumer on insectivorous birds.

Page 276

1. a) A diagram showing the total numbers of organisms at each trophic level in a food chain or food web in an area.
   b) A diagram showing the total biomass of organisms at each trophic level in a food chain or food web in an area.
   c) A diagram showing the total energy in the organisms at each trophic level in a food chain or food web in an area.

2. Any suitable example that includes producers, primary consumers and secondary consumers from a reasonable food chain. Count the number of individuals feeding at each level within the same size area. Draw a pyramid of three layers, starting with producers at the bottom and ending with secondary consumers at the top, with the bar for each level drawn to scale.

3. A pyramid of biomass only shows the mass at a particular time in an area. If some trophic levels have a shorter life-span than others, they will be under-represented in the pyramid, which may cause an inverted shape.

Page 278

1. Humans eat plant and animal tissue.

2. Wheat grain > human; wheat grain > chicken > human.

3. For wheat grain > human, two steps with wheat at the bottom. For wheat grain > chicken > human, three steps with wheat grain at the bottom and human at the top. In both pyramids, the steps should get increasingly narrower from the bottom step up the pyramid.

4. As chickens transfer some energy to the environment as heat due to respiration, a smaller proportion of the energy in the wheat is available for humans to eat. This makes this energy transfer less efficient than humans eating the grain directly.

5. a) If we all become vegetarian, then more of the energy in the plants we grow is available to us than if we eat animals that eat the plants.

   b) Humans do not eat grass, so we would need to convert meadows (such as where sheep feed) to crop fields and not all the places where there are meadows will grow crops well. Also, fish are in a food chain that starts with microscopic plankton. Gathering sufficient photosynthetic plankton to eat could take more energy than harvesting the fish that eat them.

Page 279 (top)

1. Light energy from Sun (gain) > some reflected, some passes straight through, some wrong wavelength (losses) > light energy converted to chemical energy during photosynthesis > heat energy transferred to environment from photosynthetic reactions and from respiration (losses) > chemical energy in plant biomass.

2. Chemical energy in food (gain) > chemical energy in undigested food lost as faeces (loss) > chemical energy in absorbed food molecules converted to chemical energy in waste products such as urea lost in urine (loss) > heat energy from respiration transferred to environment (loss) > chemical energy in animal biomass.

3. Only a proportion of the energy taken in is used to make new biomass, the rest is lost to the environment either as chemical or heat energy. That means there is always less energy transferred to the next trophic level than was taken in, so the shape must always be a pyramid.

Page 279 (bottom)

1. Any element found in plant or animal tissue, such as carbon or nitrogen.

2. Both flow in one direction only, but energy is eventually lost to the environment in the form of heat energy whereas substances are retained or recycled back to the start of the chain.

Cycles within ecosystems

Page 285 – 286

1. a) Respiration releases carbon dioxide into the atmosphere from the breakdown of complex carbon compounds inside organisms.

   b) Photosynthesis is the fixing/conversion of carbon dioxide from the atmosphere into more complex carbon compounds in plant tissue.

   c) Decomposition is the decay/breakdown of dead plant and animal tissue by decomposers, releasing carbon dioxide into the atmosphere during respiration.

2. a) Carbon dioxide.

   b) Complex carbon compounds e.g. starch, cellulose, sucrose.

   c) Complex carbon compounds e.g. as found in oil.

3. Combustion of fossil fuels releases carbon dioxide into the atmosphere from complex carbon compounds where the carbon has been locked away for millions of years. This greatly increases the rate at which carbon dioxide is being returned to the atmosphere.

Page 288

1. a) Nitrifying bacteria increase the amount of nitrate ions in the soil by converting ammonium ions to nitrite ions and then to nitrate ions.

   b) Nitrogen-fixing bacteria convert atmospheric nitrogen gas directly into a form of nitrogen that plants can use.

   c) Denitrifying bacteria reduce the amount of nitrate ions in soil by converting them to nitrogen gas.

2. Nitrifying bacteria increase the fertility of soils because plants can only take in nitrogen in the form of nitrates dissolved in soil water. Without nitrogen the plants will not grow well, and become stunted.

3. Decomposers break down complex nitrogen compounds in dead plant and animal tissues and animal waste. This releases ammonium ions that nitrifying bacteria convert to nitrate ions that plants need. Without decomposers, the bacteria would have nothing to work on, and the concentration of nitrate ions in the soil would decrease.

Human influences on the environment

Page 298

1. a) Smoke/emissions from factories contains acidic gases, such as sulfur dioxide, which dissolve in water droplets in clouds that then fall as acid rain.

   b) The clouds containing the acidic water droplets can be blown over great distances away from the industrial areas by wind.

2. Damage direct to delicate tissues in lungs, to soft-skinned organisms such as fish and amphibians, and to single-celled organisms. Indirect damage by changing the acidity of the soil, affecting its fertility due to leaching of minerals, or making poisonous minerals more soluble. As a result of changes in food web other organisms may be affected due to interdependency.

3. Carbon monoxide combines with haemoglobin, replacing the oxygen that it normally carries in the blood. Tissues receive less oxygen for respiration, resulting in damage especially in rapidly growing or respiring tissues.

Page 301

1. The greenhouse effect is a natural process that warms the Earth's surface when greenhouse gases in the atmosphere prevent longer wavelength radiation escaping into space. The enhanced greenhouse effect is the additional warming caused by the addition of greenhouse gases to the atmosphere as a result of human activity.

2. a) Natural, respiration; human, combustion of fossil fuels.

    b) Natural, soil bacteria in the nitrogen cycle; human, addition of nitrogen-containing fertilisers to soil.

    c) Natural, digestion of food in animal guts and decay of waterlogged vegetation; human, increase in herd animals and artificial waterlogged vegetation in rice paddy fields.

3. Any from: increase in number and intensity of storms, more drought, more flooding, change to summer/winter temperatures and precipitation.

Page 304

1. The addition of nutrients to water.

2. Runoff of fertiliser into water as a result of heavy rainfall, leaching of soluble nutrients in fertiliser through soil into water systems.

3. Sewage added to water > adds nutrients to water = eutrophication > plant growth rate increases > increase in dead plant material > increase in microorganisms > respiration by microorganisms increases removing dissolved oxygen from water > less dissolved oxygen for other organisms which die = water pollution.

Page 305

1. They are decomposers.

2. Because the microorganisms are aerobic. If conditions became anaerobic, then other microorganisms would grow and the materials in the liquid would not be broken down in the same way. Some of these anaerobic microorganisms might be pathogenic which would make the water more dangerous for release into water systems.

3. The microorganisms digest the organic molecules in the sewage, causing them to break down into smaller inorganic molecules such as carbon dioxide and nitrate ions.

4. The microorganisms that produce methane are anaerobic.

5. Dried sludge still contains large amounts of organic nitrogen. When this is added to fields, it is broken down by bacteria to release inorganic nitrogen in a form that crop plants can absorb. The plants use these for making proteins and other nitrogen-containing organic molecules, so the sludge improves crop growth.

6. The chlorine kills any microorganisms left in the water, so that none are added to the water system.

7. The BOD of sewage is high because microorganisms can use the nutrients in it to grow and multiply rapidly. So respiration rate is high and the amount of oxygen taken from the water by the microorganisms is high. The BOD of treated water is low because the nutrients in sewage have been broken down to a form that does not encourage growth, or absorbed by the microorganisms in the treatment beds and used for their growth.

Page 306

1. a) Deforestation: the destruction / cutting down of large areas of forest and woodland.

    b) Soil erosion: the washing away of soil by heavy rainfall.

    c) Leaching: when soluble nutrients dissolve in soil water and soak away deep into the ground.

2. a) Trees no longer take up water from the ground and release it to the atmosphere through transpiration, however water can now evaporate more easily from exposed soil, which can affect the moisture in the air above the forest and so rainfall.

    b) Soil erosion because there are no tree roots to hold the soil, and increased leaching because there are few plant roots to absorb the nutrients, both remove soluble mineral nutrients from the soil, which will reduce the rate of plant growth.

    c) Burning or rotting of trees releases the carbon stored in the wood as carbon dioxide to the atmosphere at a much faster rate than normal, plus trees are no longer taking in carbon dioxide through photosynthesis, increasing atmosphere carbon dioxide.

SECTION 5 USE OF BIOLOGICAL RESOURCES

Food production

Page 323

1. Any four of the following: temperature, carbon dioxide, light, water, pests.

2. For each of the factors mentioned in **Q1**: optimum temperature allows enzymes to work at

fastest rate so maximising rate of growth;
carbon dioxide is usually a limiting factor in photosynthesis, so increasing carbon dioxide concentration in the air around the plants should increase the rate of photosynthesis and therefore growth;

light is needed for photosynthesis, so by keeping light levels high enough for photosynthesis throughout the day and night will increase the amount of growth of the plants;

plants need water for photosynthesis and for the transport of soluble materials, so having the right amount of water all the time will increase the rate of growth;

pests damage plants, limiting the rate of growth, so removing pests should increase the rate of growth.

3. Polytunnels help to protect a crop from the environment, because conditions inside them can be more controlled. So using polytunnels can make it possible to grow crops when it may be too cold at night or in cooler seasons, or when it gets dark early in the evening, etc.

Page 325

1. To add nutrients to the soil so that the plants grow bigger and produce more yield to harvest.

2. Planting different crops in the same field at different times, to use different nutrients and to add nitrogen back to the soil when legumes are planted.

3. Natural fertilisers include manure, guano and dried sludge from sewage treatment. Artificial fertilisers are made from chemicals in industrial processes.

Page 328

1. They damage the plants so that they do not grow as quickly, so they do not produce as much yield.

2. It will keep the predator/parasite in the same place as the pest and so maximise predation/parasitisation of the pest. It will also reduce the chance of new pests getting to the crop.

3.

	Advantages	Disadvantages
Pesticides	Removes large proportion of pests quickly	Pests may develop resistance
	Easy to apply	May damage other species
		May result in greater numbers of pests (if predators killed)

Biological control	Less likely for pest to develop resistance	Slower to act than chemicals
	Usually better targeted at pest so less damaging to other species (except for introduced species)	Introduced predators/ parasites may cause more problems to the environment and be difficult to control
	Very good within closed spaces, such as glasshouses	May not work well in open areas as they may move away

Page 329

1. Pesticide use increased greatly from about 1000 tonnes per year in 1980 to over 25 000 tonnes in 1995.

2. Yield generally increased from about 350 kg/ha in 1980 to a maximum of nearly 800 kg/ha in 1991. After this it fell to between 500 and 600 kg/ha.

3. Yield may have increased between 1980 to 1991 as pesticide use increased and controlled the whitefly. However, increasing pesticide use after that did not prevent a decrease in yield, suggesting that the whitefly were becoming resistant to the pesticide.

4. As the prey numbers increase, the predators have more food and so can produce more young.

5. Increasing pesticide use killed predators as well as pests. So, as pest numbers increased after the pesticide had lost effect, there were too few predators to control them.

6. As whitefly become resistant to the pesticide, they are unaffected by it. But their predators are still killed by its use. So the whitefly numbers can increase even further, which will increase the transmission of the virus that damages the cotton.

7. Using pesticides specific for whitefly will not kill the predators, so a combination of predators and chemicals can be used to keep the whitefly numbers under control so that damage to cotton is reduced and yield increases.

Page 331

1. Any four suitable foods, such as bread, cheese, yoghurt, single-cell protein.

2. Milk > pasteurised to kill microorganisms > inoculated with *Lactobacillus* > bacteria convert lactose to lactic acid > lactic acid causes milk proteins to coagulate and form yoghurt.

1. A large vessel in which microorganisms are grown in large numbers under controlled conditions.

2. a) Temperature, pH, oxygenation, nutrient concentration.

   b) Temperature will increase due to the reactions of respiration and other reactions of the microorganisms. If temperature rises too high, it may reduce rate of growth or kill the microorganisms.

   pH may change because of substances released by the microorganisms into the solution. This may reduce rate of growth.

   Oxygen concentration might fall as oxygen is used for respiration. Microorganisms are aerobic, so rate of growth will reduce if oxygen concentration falls.

   Nutrient concentration will fall as microorganisms use nutrients to make new cells. Rate of growth will fall if nutrients are not added to replace what is used.

3. a) Keeping things sterile.

   b) It prevents other microorganisms growing rapidly in the fermenter which could be harmful or compete with the added microorganisms.

Page 337

1. Keeping fish in contained areas, improving growing conditions to increase the rate of growth, and harvesting the fish.

2. Any three from the following: quality of water, food supply, predators, stocking limit, pests and diseases.

3. For each factor given in **Q2**:
   to prevent build-up of fish waste and uneaten food, and to provide as clean a water supply as possible;
   to provide food of the right sort at a high enough rate to maximise growth;
   to prevent predators killing fish;
   to prevent intraspecific predation in farmed fish that are carnivorous;
   to prevent pests and diseases from harming fish and so reducing the rate of growth or killing them.

Selective breeding

Page 345 (top)

1. a) Characteristics that can be bred for in selective breeding programmes are those that are controlled by genes.

   b) If the desired characteristic is not controlled by a gene, it cannot be selectively bred for.

2. Any three from:

   increasing size of part that we eat;

   decreasing size of parts we do not eat;

improving pest and disease resistance;

improved growth in adverse conditions.

3. For each of factors included in **Q2**:

   the part of the plant we eat is what is measured as the yield;

   decreasing other parts means the plant has more energy to grow the part we do eat;

   diseases and pests damage the plant so it does not grow as well as it could, so making plants resistant will help them to grow faster;

   adverse conditions (such as heat or drought) slow the rate of growth, so improving growth in these conditions will help to improve yield where these conditions happen.

4. Because people like different plants, and new varieties of plants are often considered more attractive than old varieties.

Page 345 (bottom)

1. In sexual reproduction, the offspring inherit half their alleles from one parent and half from the other. In selective breeding, only a few parent types are used for breeding. So this limits the range of variation in the alleles in the parents. So the offspring can only inherit from this limited range.

2. This means that when they cross-breed the seed produced is more likely to contain the characteristics that they have been bred for. Some of that seed planted in the following year will produce plants that still show the required characteristics.

3. The environment is continually changing, and may change a lot over the next century or so as a result of climate change. This means that characteristics that are useful now, may be less important than others in the future, such as the ability to withstand drought (if the climate gets hotter and drier in some places), or resistance to particular pests or diseases that are not common now but may become more common as climate changes. If the varieties we are growing do not have these characteristics, then crop yield will be reduced as growth of the crop plants is reduced.

4. Any suitable answer with appropriate justification, such as: in the short term this does not seem to make sense because we need to grow enough food for everyone to eat, but in the long term it does make sense because the environment changes and we cannot predict how it will change and what characteristics we will need for crops in the future.

Page 347

1. Choose male and female animals with characteristics that are nearest to the desired characteristic and breed them. From the offspring, select individuals with characteristics that are

nearest to the desired characteristic and breed them together. Repeat this for enough generations until you have individuals with the desired characteristic.

2. **a)** Muscle is the meat that we eat. So this is increased yield.

    b) Disease harms the animal so that it grows more slowly, or may even die. Resistant animals will grow better and need less treatment for infections with chemicals such as antibiotics.

    c) Docile animals are easier to handle and move around, which makes them easier to work with.

3. Different breeds of sheep have different characteristics which are often linked to their local environment, or whether they are farmed for meat or milk, so there can not be one breed that is suited to all environments and is equally good for meat and milk.

Genetic modification (genetic engineering)

Page 353

1. **a)** An enzyme that cuts DNA at a specific site.

    b) An enzyme that joins two pieces of DNA.

    c) Something which carries genetic material into another cell.

2. Required gene extracted from DNA of organism using a restriction enzyme > bacterial plasmid removed from bacterium and cut open with same restriction enzyme > required gene and plasmid mixed together with ligase enzyme > ligase enzyme joins inserted gene and plasmid together > plasmid inserted into another bacterium > inserted gene produces new characteristic (e.g. production of human insulin) in bacteria.

3. Because the human insulin gene has been inserted into the bacterial DNA, and is decoded by the bacterial cell exactly as it would be in a human cell, producing human insulin.

Page 355

1. Required gene inserted into bacterial plasmid > plasmid inserted into bacterium > bacterium 'infects' plant cell and inserts plasmid into cell > required gene inserted into plant DNA > plant cell cultured to divide to produce new plant > cells in new plant produced by mitotic division of original cell are genetically identical and contain a copy of the inserted gene.

2. Insertion of a gene for pest resistance, to kill caterpillars that try to eat the plant.

3. Any one of each of the following:

    Advantage: reduces need for pesticide (so less damage to environment), reduces time needed to look after crop (so reduces cost), farmer does not risk health from spraying pesticides

    Disadvantage: seed is more expensive, gene may transfer to wild plants and potentially affect the food web, there may be health concerns with the food produced from the crop.

Cloning

Page 361

1. Micropropagation: the cultivation of explants (small pieces of plant tissue) in the laboratory to create many clones of the original plant.

2. Explants taken from parent plant and sterilised > explants placed on nutrient medium containing chemicals for growth as well as hormones > hormones stimulate growth and differentiation to produce roots, stems and leaves > when large enough to handle, the tiny plants are moved on to different nutrient media > when large enough, moved into pots of compost.

3. They are all produced by the mitotic division of the cells from the original plant so they all contain the same genes.

Page 364

1. A diploid cell taken from an adult sheep > nucleus removed and placed in an enucleated egg cell from another sheep > cell stimulated to start dividing to produce embryo > embryo placed in the uterus of another sheep to develop ready for birth.

2. Sheep B, as the diploid nucleus contains the genes that code for the characteristics of Dolly.

3. They can produce chemicals, e.g. antibodies, to treat human diseases.

INDEX

Acknowledgements

Every effort has been made to trace copyright holders and to obtain their permission for the use of copyright material. The publishers will gladly receive any information enabling them to rectify any error or omission at the first opportunity. The publishers would like to thank the following for permission to reproduce photographs:

(t = top, b = bottom, c = centre, l = left, r = right)

Cover & p1 Roberto Sorin/Shutterstock, pp8-9 Stephen Coburn/Shutterstock, p10t Science Photo Library/Alamy, p10b formiktopus/Shutterstock, p14 Alexandru-Radu Borzea/Shutterstock, p15 Juan Camilo Bernal/Shutterstock, p16t Fong Kam Yee/Shutterstock, p16b Four Oaks/Shutterstock, p18 Martin Fowler/Shutterstock, p19 Péter Gudella/Shutterstock, pp28-29 Jubal Harshaw/Shutterstock, p30 Tinydevil/Shutterstock, p32 Eduard Härkönen/Shutterstock, p35 Carol and Mike Werner/Alamy, p36 Phototake Inc./Alamy, p37 Jubal Harshaw/Shutterstock, p38 Melba Photo Agency/Alamy, p40l Steve Gschmeissner/Science Photo Library, p40r Power and Syred/Science Photo Library, p43 Andrew Lambert Photography/Science Photo Library, p45 Andrew Lambert Photography/Science Photo Library, p46 Martin Shields/Alamy, p50 franssonkane/Shutterstock, p55 Dr. Stanley Flegler, Visuals Unlimited/Science Photo Library, p57 Picsfive/Shutterstock, p68 oksana.perkins/Shutterstock, p71 Triff/Shutterstock, p73 A.Krotov/Shutterstock, p75 sciencephotos/Alamy, p77t Nigel Cattlin/Science Photo Library, p77b Nigel Cattlin/Alamy, p79 HLPhoto/Shutterstock, p80 Jeff Rotman/Alamy, p81 Images of Africa Photobank/Alamy, p97 Phototake Inc./Alamy, p98 Nickolay Vinokurov/Shutterstock, p100 Jan Kratochvila/Shutterstock, p101t Maridav/Shutterstock, p101b Bogdan Wankowicz/Shutterstock, p102 olivier borgognon/Shutterstock, p105 Armin Hinterwirth/iStockphoto, p107 Matt Tilghman/Shutterstock, p109 Jubal Harshaw/Shutterstock, p113 Andrew Gentry/Shutterstock, p117 Westend61 GmbH/Alamy, p118 Olegusk/Shutterstock, p124 Sebastian Kaulitzki/Shutterstock, p125 Power And Syred/Science Photo Library, p126t Dr Keith Wheeler/Science Photo Library, p126b Zastol'skiy Victor Leonidovich/Shutterstock, p127 Nigel Cattlin/Alamy, p132 National Cancer Institute/Science Photo Library, p134 Dmitry Naumov/Shutterstock, p140 Levente Gyori/Shutterstock, p141 Racheal Grazias/Shutterstock, p146 Jubal Harshaw/Shutterstock, p148 S.Borisov/Shutterstock, p149 GaudiLab/Shutterstock, p153 Poco a poco, p157 Tudor Stanica/Shutterstock, p158 Anest/Shutterstock, p171 Dmitry Lobanov/Shutterstock, p175 Fotokostic/Shutterstock, pp188-189 Olena Timashova/iStockphoto, p190 Phototake Inc./Alamy, p192 WILDLIFE GmbH/Alamy, p194 Dr Jeremy Burgess/Science Photo Library, p195 glyn/Shutterstock, p196t piyato/Shutterstock, p196b Wildlife GmbH/Alamy, p197 Tim Gainey/Alamy, p198 Valentyn Volkov/Shutterstock, p201 epsilon_lyrae/Shutterstock, p206 Francis Leroy, Biocosmos/Science Photo Library, p207 Nic Cleave Photography/Alamy, p212 Galyna Andrushko/Shutterstock, p214 Andy Lim/Shutterstock, p218 CNRI/Science Photo Library, p219l Eric Isselée/Shutterstock, p219r Eric Isselée/Shutterstock, p230 C003/0730/Science Photo Library, p235t Sebastian Kaulitzki/Shutterstock, p235c ISM/Science Photo Library, p236 M. J. Mayo/Alamy, pp254-255 Dudarev Mikhail/Shutterstock, p256 wandee007/Shutterstock, p257 Khoroshunova Olga/Shutterstock, p258t AuntSpray/Shutterstock, p258b Cory Woodruff/Shutterstock, p259 Paul Glendell/Alamy, p262 neelsky/Shutterstock, p263l Paul Glendell/Alamy, p263r InnaVar/Shutterstock, p266 Sarah Pettegree/Shutterstock, p268 Colin Pickett/Alamy, p269 Anan Kaewkhammul/Shutterstock, p275 Linda Bucklin/Shutterstock, p281 Rick Wylie/Shutterstock, p283 Anest/Shutterstock, p284 Jane McIlroy/Shutterstock, p288 Frank Vincentz, p291 nadirco/Shutterstock, p292 Peter Gudella/Shutterstock, p293 Darrin Henry/Shutterstock, p294 Martyn F. Chillmaid/Science Photo Library, p295 Sue Kearsey, p297t Leslie Garland Picture Library/Alamy, p297b Dallas Events Inc/Shutterstock, p299 NASA/NSSDC, p302 Julio Etchart/Alamy, p303 xpixel/Shutterstock, p304 Andrey Kekyalyaynen/Shutterstock, p305 Tropical Rain Forest Information Center (TRFIC)/Basic Science and Remote Sensing Initiative (BSRSI)/Landsat 7 Project Science Office/Goddard Space Flight Center/NASA, p306 Earth Observations Laboratory, Johnson Space Center/NASA, pp320-321 Elenamiv/Shutterstock, p322 Peter Zachar/Shutterstock, p323 Videowokart/Shutterstock, p324t US Department of Agriculture/Science Photo Library, p324b Christopher Elwell/Shutterstock, p325 meunierd/Shutterstock, p326 Nigel Cattlin/Alamy, p327 Herb Pilcher, p328 Ryan M. Bolton/Shutterstock, p330b Lilyana Vynogradova/Shutterstock, p330c Nayashkova Olga/Shutterstock, p335 ilFede/Shutterstock, p337l Jaochainoi/Shutterstock, p337r Gustavo Miguel Fernandes/Shutterstock, p342 Vitaly Titov & Maria Sidelnikova/Shutterstock, p343 Nigel Cattlin/Alamy, p344 Photocrea/Shutterstock, p345 Francesco Tonelli/Alamy, p346 John Carnemolla/Shutterstock, p347 janecat/Shutterstock, p349r Mikael Damkier/Shutterstock, p349l Vangert/Shutterstock, p350 Eye of Science/Science Photo Library, p354 AGStockUSA/Alamy, p359 AISPIX/Shutterstock, p360 Prisma Bildagentur AG/Alamy, p361 Nigel Cattlin/Alamy, p363 Inga Spence/Alamy, p387 Ed Phillips/Shutterstock.